WEIGHING IMPONDERABLES AND OTHER QUANTITATIVE SCIENCE AROUND 1800

J.L. Heilbron

University of California Press
Historical Studies in the Physical and Biological Sciences
Supplement to Vol. 24, Part 1
1993

Historical Studies in the Physical and
Biological Sciences, Supplement to Vol. 24, Part 1

Berkeley Papers in History of Science, 13

Uppsala Studies in History of Science, 11

Bologna Studies in History of Science, 2

Order from
Office for History of Science and Technology
University of California
543 Stephens Hall
Berkeley, CA 94720

-or-

University of California Press
Journals Division
2120 Berkeley Way
Berkeley, CA 94720

Price $20.00

Copyright © 1993
The Regents of the University of California
ISBN: 0-918102-17-0
Library of Congress catalog card number: 91-62713

Designed and typeset by Diana Wear
Archival photography by Peter Houtzager
Office for History of Science and Technology

Contents

Introduction	1
I. The Standard Model	5
1. The scheme of imponderables	6
An epoch in physical science, 7; A world of as-if, 16	
2. Between calculus and chemistry	23
II. Some Means to the End	35
1. New angles on angles	35
As easy as π, 36; Completing the circle, 47	
2. Measuring imponderables	65
The first weightless weighings, 66; Caloric, 86	
III. Laplace's School	139
1. The spirit of Arcueil	141
The school tie, 141; The dead hand, 146	
2. A mathematical figleaf	150
Clairaut and Newton, 150; The Laplace transformation, 157	
3. Sound and fury	166
Lagrange and Newton, 166; Laplace's approach 171; Enter γ, 175; Again the figleaf, 178	
IV. Varieties of Mercantilist Mathematics	185
1. Cartographic control	185
A French business, 185; Enter the Army, 195; An export business, 198; The cross-channel link, 201; Imperial measures, 207	
2. The shape of science and the Earth	213
Newton's apple, 214; Elusive ellipticity, 221; Revolutionary geodesy, 231; No ellipsoid of revolution, 239	

V. The Measure of Enlightenment 243
  1. Plight of the people 243
      Cain's legacy, 243; Mathematics and the rights of man, 245; Republican time, 249
  2. Action of the Academy 257
      An anchor for revolutionary storms, 257; Fall and rise, 263
  3. Response of the people 268
      Resistance, 270; Spread, 273

Bibliography 279

Index 321

LIST OF ILLUSTRATIONS

2.1.1	Picard's 10-foot zenith sector, 1670	38
2.1.2	Bird's 8-foot mural quadrant, 1753	41
2.1.3	Ramsden's 3-foot theodolite, 1787	45
2.1.4ab	Hadley's octant, 1731	48
2.1.5ab	Mayer's geodetic instrument, 1752	51
2.1.6ab	Borda's first reflecting circle, 1752	52–53
2.1.7	Borda's second reflecting circle, ca. 1780	55
2.1.8	Lenoir's version of Borda's reflecting circle, ca. 1780	56
2.1.9ab	Borda-Lenoir geodetic instrument, ca. 1785	58–59
2.1.10	Biot and Arago's method of measuring the index of refraction of a gas, 1802	62
2.1.11ab	The geometry of Malus' goniometer, 1809	64
2.1.12	Malus' measurement of double refraction, ca. 1810	65
2.2.1ab	Coulomb's apparatus for measuring the force between magnetic poles, 1785	67–68
2.2.2	Geometry of Coulomb's method for finding the force between like magnetic poles, 1785	71
2.2.3ab	Robison's apparatus for measuring electrical force, ca. 1770	73–75
2.2.4abc	Coulomb's electrical balance, 1785	76

LIST OF ILLUSTRATIONS

2.2.5	The geometry of Coulomb's balance, 1785	77
2.2.6ab	Coulomb's apparatus for measuring the pull between like electrical charges, 1785	78–79
2.2.7	Henley's electrometer, 1772	83
2.2.8	Bennett's electroscope, 1786	84
2.2.9	Volta's demonstration of the influence of geometry on electrical capacity, 1778	85
2.2.10	Cavendish's procedure for fixing the boiling point of a thermometer, 1777	88
2.2.11	Nollet's procedure for fixing the boiling point, ca. 1740	89
2.2.12ab	Gay-Lussac's apparatus for determining the coefficient of thermal expansion of gases, 1802	93
2.2.13	Gay-Lussac's second method of obtaining the thermal dilation of gases, before 1815	94
2.2.14ab	Dulong and Petit's measurement of the expansion of mercury, 1818	99
2.2.15	Lavoisier and Laplace's ice calorimeter, ca. 1780	103
2.2.16	Delaroche and Bérard's apparatus for measuring the specific heat of gases, 1813	108
2.2.17	Delaroche and Bérard's apparatus for measuring the specific heats of gases and water, 1813	111
2.2.18	Clément and Desormes' method of obtaining specific heats of gases, 1813	113
2.2.19	Rumford's differential thermoscope, ca. 1802	122
2.2.20	John Leslie's differential thermometer, ca. 1801	125
2.2.21	Leslie's scheme to show the propagation of radiant heat in air, 1804	126
2.2.22	William Herschel's set-up for investigating the infra-red, 1800	129

LIST OF ILLUSTRATIONS

2.2.23	Herschel's graph of the relative intensity of the heat and the visible spectrum, 1800	130
2.2.24	Herschel's apparatus for studying the effects of screens on the heating power of solar rays, 1800	131
2.2.25	Leslie's apparatus for testing the heating effects of the different solar rays, 1804	133
2.2.26	Delaroche's plot of radiant-heat exchange, 1812	136
3.2.1	Newton's diagram of the path of light through a refractive medium, 1687	152
3.2.2	Geometry of Clairaut's theory of refraction, 1739	154
3.2.3	Clairaut's diagram of the forces implicated in capillarity, 1743	155
3.2.4	Geometry of Laplace's account of refraction, 1805	159
3.2.5	Geometry of Laplace's calculation of capillary force, 1806	160
3.2.6	Diagram to elucidate Laplace's theory of capillary action, 1806	162
3.2.7	Geometry of capillary rise	163
3.3.1	Diagram for Newton's theory of the propagation of sound, 1687	167
3.3.2	Newton's representation of simple harmonic motion, 1687	168
3.3.3	Diagram for Lagrange's theory of the propagation of sound, 1759	169
3.3.4	Diagram for Laplace's calculation of the force exerted by caloric around gas molecules, 1821	180
3.3.5	Diagram for finding the force exerted by an enclosed gas on a gaseous molecule, 1821	181
4.1.1	Ramsden's theodolite in action, 1787	204
4.2.1	Diagram for Newton's theory of the shape of the earth, 1687	215

LIST OF ILLUSTRATIONS

4.2.2	Further to Newton's theory of the earth's shape, 1687	216
4.2.3	Geometry of the Newtonian ellipsoid of revolution	219
4.2.4	Geometry of Maskelyne's measurement at Schehallien, 1775	232
5.1.1	Allegory of the reform of the calendar, ca. 1795	253
5.1.2	Clock face showing conversion of old and new hours, ca. 1793	255
5.2.1	Borda and Cassini's method for obtaining the length of a seconds pendulum, ca. 1792	265

PREFACE

The chapters in this book originated in lectures given at the first session of the International Summer School in History of Science, held in Bologna in 1988. In their original form they covered the range of material set out in my contributions to *The quantifying spirit in the 18th century*, edited by Tore Frängsmyr et al. (University of California Press, 1990). The original treatment included a brief section on population statistics, as a further example of connections among mathematics, the physical world, and the bureaucratic state. I had intended to develop the subject here and to extend it to cadasters, since both fit well with surveying and mensuration, which are discussed in detail, and both engaged Lavoisier and Laplace, who appear prominently in the book in other connections. Unfortunately neither time nor space permitted this desirable extension. A further study is underway.

I am indebted to my colleagues Tore Frängsmyr (Uppsala) and Giuliano Pancaldi (Bologna), co-organizers of the Summer School, for the opportunity to deliver the original lectures, and to the members of the School for their tolerance and comments. And I have the greatest pleasure in thanking Diana Wear and Peter Houtzager (Berkeley) who produced this book, for their inspired handling of a difficult text and persnickety author.

INTRODUCTION

The bible deals ambiguously with numbers. The first book of Chronicles says that Satan put David up to counting the people of Israel and Judah. The second book of Samuel says that God did it. The accounts agree, however, that David's precocious census was a bad mistake. Numbering might be divine or devilish, but not human. "I have done a very wicked thing," David said to God. "I pray thee remove thy servant's guilt, for I have been very foolish." God gave him a choice of punishments, fittingly expressed: three years of famine, three months of military defeat, or three days of pestilence. They negotiated pestilence, and 70,000 men died, but not the numerator David.[1]

Mankind's uneasy relations with numbers have left plain traces in European languages. The English word "calculating" means both "computing" and "scheming." The German word "vermessen" means to measure, to mismeasure, and to be presumptuous, arrogant, and insolent. An old meaning of the Italian word "matematizzare" was to put a hex, or cast a spell, on people. To many, the difference between a sorcerer and a mathematician was, and may be still, too small to measure.

These remarks are intended to suggest that the spread of quantitative methods and imagery throughout the science, near science, and practical activity of the late 18th and early 19th centuries, which forms the subject of this book, requires an explanation. A hundred years earlier, well-educated people got through school without learning the rule of three. Take Samuel Pepys, who did not

---

1. 1 Chron., 21:1–16; 2 Sam., 24:1–16. Cf. Exod., 30:12–15, where God makes a monetary charge on each man counted by Moses.

know the multiplication table at the age of thirty: he had finished his schooling literate in English and Latin, but innumerate, unable to do sums and unacquainted with vulgar fractions. When he became clerk to the Admiralty, he rose every day at five in the morning to learn his times tables.[2] The great virtuoso Thomas Hobbes, who hobnobbed with the likes of Descartes, did not come across geometry until he was forty, and then, as he tells us, only by accident.[3]

There have always been those who loved mathematics for itself. A double tradition of enthusiasm descended from antiquity, which may be followed into early modern times in the prefaces to printings of Euclid. One tradition emphasized the etymological origin of geometry in land measurement, and praised mathematics for its practical utility. The other observed God's tendency to create by number and measure, and made number the key to the first and final truths about the Universe. Here is how this second tradition spoke through one of its loudest followers, John Dee, who wrote in the 16th century. "O comfortable allurement, O ravishing perswasion, to deal with a Science, whose subject is so Auncient, so pure, so excellent, so surmounting all creatures....By *Numbers* propertie...we may...arise, clime, ascend, and mount up (with Speculative winges) in spirit, to behold in the Glas of Creation, the *Forme* of *Formes*, the *Exemplar Number* of all things Numerable....Who can remaine, therefore, unpersuaded, to love, allow, and honor the excellent science of Arithmatike?"[4] It was not mathematics in this second sense, as a route through the ridiculous to the sublime, but mathematics as a practical means of control of everyday transactions, that became endemic in the late 18th century.

Quantification in the physical sciences, as opposed to astronomy, mechanics, and geometrical optics, first succeeded after 1750. The earliest success came by analyzing the phenomena of electricity and magnetism in terms of distance forces, whose power Newton had demonstrated in his theory of planetary motions and hinted at in his queries about everything else. For Newton and his immediate

2. Cohen, *Calculating people* (1982), 26; Pepys, *Diary* (1970), 3, 131, 134–5, 137.
3. Shapin and Schaffer, *Leviathan* (1985), 318.
4. Dee, "Mathematical Preface" (1570), *ir, *iiijv; Heilbron, in Shumaker, ed., *Propaedeumata* (1978), 4, 7.

disciples, natural philosophy aimed at the truth of things; the scheme of distance forces would eventually help them to "arise, clime, ascend, and mount up" to the Form of Forms. For the applied mathematician of the high Enlightenment, however, the scheme of quantitative distance forces was merely an instrument of calculation.

The quantitative natural philosophy of the late 18th century was instrumentalist not only in its software, that is, in its theoretical models; but also in its hardware, in its reliance on measurements by instruments that had become much more reliable and precise since Newton's time. This reliance had that double and circular character that alarms people and even philosophers suspicious of numbers: the instruments not only made the measurement, they also produced the pheonomena to be measured. The more vigorously and successfully the late Enlightenment pressed its quantitative natural philosophy, the more explicitly it recognized that thereby it would not attain knowledge of the true principles of things. The instrumentalist science of the turn of the 19th century made manifest and generalized the complementarity between truth and calculation recognized much earlier by Ptolemaic astronomers.

This recognition demoted natural philosophers to the spiritual level, insofar as numbers were concerned, of numerate merchants and bureaucrats. For all of them, the ostensible purpose of quantifying their material was to control it, to order it, to increase its reach, to guess or make the future from it. This quantifying spirit spread far: we can follow it downwards from the heaven of mathematics, where it presided over the rise of algebra and analysis at the expense of geometry; thence through the spheres of the instrumentalist physical sciences; on to the forests of Germany, where every tree was counted; and at last to the caves of the lexicographer and the classifier, who labored to inventory and control their ever increasing material, and to calculate its amount if not its meaning. A survey of such applications is now available.[5] This book offers further or more detailed examples, drawn from the physical sciences, geodesy, and the reform of weights and measures.

Chapter 1 describes the program of dragging the gravitational theory down from the heavens to help quantify representations of

5. Frängsmyr, Heilbron, and Rider, eds., *The quantifying spirit* (1990).

electrical, magnetic, and thermal phenomena here below. The emphasis falls not on the details of the resultant scheme of imponderable fluids, but on its epistemological and pedagogical status, and on the standards of achievement and precision it helped to impose. Chapter 2 gives illustrations, many of them literal, of representative instrumentation for procuring exact information. Chapter 3 examines the practice of the dominant school in the pursuit of exact physical science around 1800, which took its inspiration primarily from Laplace. Chapter 4 brings exact science literally to earth in an account of trigonometric surveys undertaken largely for military purposes; the deservedly famous expeditions to determine the earth's shape piggybacked on state support of strategic cartographical projects. Chapter 5 rehearses the story of the greatest mobilization of exact science for national purposes that occurred during the 18th century: the design, execution, and implementation of the metric system of weights and measures.

1 THE STANDARD MODEL

The physical science of the late 18th century invoked a set of qualitatively different sorts of matters, which served as carriers of special-purpose forces. These matters divided into ordinary or ponderable stuff and weightless fluids, able to act on ponderable matter and, in certain cases, on one another. Ordinary matter carries, and exerts upon itself, the forces of gravity, cohesion, chemical affinity, and capillarity. Among the imponderables, the particles of light interact with ordinary matter; the fluid(s) of electricity act on ordinary matter and on one another; the fluids of magnetism behave similarly; and the self-repulsive fluid of heat (caloric) counters the various cohesive forces that, without its intervention, would coagulate all terrestrial ponderable matter into one unleavened lump. Borrowing a term from today's physics, we may call this collection of matters the Standard Model of its time. It represented all the physical phenomena known at the end of the 18th century; it had the unity of a common mathematical dress, if not of a coherent ontology; and it was regarded as a model, not as a direct transcription, of God's blueprint for the creation.[1]

The pattern for the Standard Model was the theory of gravitation and the hints at its extension to other phenomena thrown out by Newton in the Queries to his *Opticks*. For most of the 18th century, however, the fit between calculation and observation that made the reputation of the gravitational theory could not be duplicated in any branch of experimental physics. Beginning around 1770, the situation changed rapidly, as electricity, magnetism, and heat began to yield to the sort of analysis that had ordered the motions of the planets; and just after the turn of the 19th century, the phenomena

1. Cf. Crosland and Smith, *HSPS*, 9 (1978), 7–9.

of capillarity and the behavior of light were brought into the scheme, although in a Pickwickian sense. These achievements inspired and exemplified the program described by Laplace in 1796 and brought almost to realization (or so he thought) by Gay-Lussac in 1809: to perfect terrestrial physics by the same techniques as Newton had used to perfect celestial mechanics.[2]

This program had become plausible by the 1790s owing to an increasingly instrumentalist approach to physical theory; to the then recent and remarkable improvement in experimental apparatus and measuring devices; to the availability of accurate measurements of a wide range of physical phenomena; and to the appearance among experimental philosophers of men with talent and training in what the *philosophes* called the language of Enlightenment. This language, which almost met the usual enlightened criteria of universality, naturalness, rationality, and comprehensibility to all but the dullest intellects, was man's ambiguous friend mathematics. Not the rigorous, captiously logical, deadweight mortmain mathematics of the Greeks, however, or the difficult synthetic methods of Newton, but democratic easy-going algebra and creative slip-shod calculus.[3]

The fruitful application of mathematics to wide reaches of physics around 1800 promoted relaxation on both sides. Mathematicians used what worked, without compulsive attention to rigor. Physicists increased the elements of nature—the several sorts of air, the chemically undecomposable substances, the imponderable fluids—without much concern about the old injunction against multiplying essences unnecessarily.

1. THE SCHEME OF IMPONDERABLES

The Standard Model of 1800 supposed the existence of ordinary gravitating matter and six, and perhaps more, special weightless fluids. The matter and the fluids carried the forces—of cohesion and

---

2. Fox, *HSPS*, 4 (1974), 95; Gay-Lussac, *MSA*, 2 (1809), 207–208, announcing his law of the combination of gases in the simplest proportions by volume: "I hope thereby to give a proof of what some very distinguished chemists have advanced, that we are perhaps not far from the time when most chemical phenomena can be calculated." Cf. infra, §1.1.

3. Cf. Rider, in Frängsmyr, *Spirit* (1990), 114–120.

affinity, and of heat, light, electricity, and magnetism—that gave rise to the phenomena of the physical world. The scheme has not received high praise in retrospect. Addressing the Paris Academy of Sciences in 1900, its president, the mathematical physicist Maurice Lévy, identified the doctrine of imponderables of the previous century as the last obstacle to the advance of classical physics. "Physics was then still in limbo, awaiting the savior who would redeem it from the sin of having not yet repudiated the six imponderable fluids."[4] Some modern historians have adopted the same unfriendly attitude: the scheme was empty and arbitrary, also unmathematical, misleading, weak, rigid, and wrong.[5] Others have more properly regarded it as providing the conceptual apparatus for the quantification of branches of physics previously treated only qualitatively. "How perfect the earth," writes Walt Whitman, "and the minutest thing upon it!/What is called good is perfect, and what is called bad is just as perfect,/The vegetables and minerals are all perfect, and the imponderable fluids perfect."[6] Whether positive or negative, the influence of the scheme of imponderables was pervasive: the unquestioned basis of physics at the dominating Ecole polytechnique; the inevitable organizing principle of physics texts; and, hence, an obvious demarcator in the historiography of the physical sciences.[7]

An epoch in physical science

The fashioning of the ingredients of the Standard Model began in earnest in the 1770s. The ultimate causes at play must be referred to whatever brought the quickening of the quantitative spirit in the study of the natural world, which also has been dated to the 1770s.[8] Proximate causes may be readily found, however; and among them the discovery of the simple gases, the recognition of latent heat, and the quantification of electrostatic theory deserve attention.

4. Lévy, *CR*, *131* (1900), 1027.
5. A mixture of Silliman, *HSPS*, 4 (1974), 138, 143–144, 149, and Fox, ibid., 136.
6. Whitman, "To think of time," in *Writings* (1902), 1, 385.
7. Fox, *HSPS*, 4 (1974), 110; Kleinert, in Fabian et al., *Deutschlands Entfaltung* (1980), 109–111, and Haüy, *Traité* (1806), 1, x–xxii; Rosenberger, *Geschichte* (1882), 3, 13ff.
8. Frängsmyr et al., *Spirit* (1990), 1–3, 153–158, 311–312.

Several early historians of physics take the discovery of gases—"the most important scientific event of the eighteenth century"—as an epoch for their subject.[9] Few would dispute the merit of this periodization, which emphasizes the discovery of dephlogisticated air and the gaseous composition of water, for the history of "chemistry." (The quotation marks call attention to an ambiguous conflation of diachronic and synchronic meanings of the word; some clarification of meaning will be offered later.) Why impose the same periodization on "physics"? There are several good reasons.

The discovery of the composition of the atmosphere brought the properties of airs to center stage in the theater of the natural philosopher. They furnished new material for shows: the philosopher burnt hydrogen to make water, asphyxiated small creatures in phlogisticated air, demonstrated the eudiometer and electrical pistol to paying customers, and rose in an aereostat to explore the heavens. Looking back, around 1800, Jean-Henri Hassenfratz, who had worked in Lavoisier's laboratory, spoke about the inspirational airs when describing the lectures he proposed to give as the first professor of physics at the Ecole polytechnique. "At the time of the discovery of gases and aeriform substances, of balloons and their ascension, a ferment broke out in peoples' minds that directed them to the study of physics; some followed their natural bent, others the general fashion, drifting with the crowds."[10]

A joint symbol of these two paths is the balloon ascent made by J.B. Biot and C.L. Gay-Lussac in 1804. The technique of aerostation had become "easy and simple," and almost safe, owing to the efforts of private entrepreneurs, who sold balloon rides and ringside seats at launches, and the military, which foresaw the use of space platforms in observing the enemy.[11] The invention thus developed by fashion and utility was put at the disposal of Biot and Gay-Lussac, two prominent chemist-physicists whom we shall meet later on the ground, by the minister of the interior, the chemist Jean

---

9. E.g., Friedrich Murhard, *Geschichte* (1798/9); Fischer, *Geschichte* (1801), 8, lviii; A. Libes, *Histoire* (1810), 1, 1. The quote is from the engineer Thomé de Gamond, *Account* (1870), vii.

10. Hassenfratz, Ecole poly., *Jl.*, no. 5 (an vi [1799]), 238.

11. Gillispie, *Montgolfiers* (1983), 31–32, 58–62, 95–108, 135; Crouch, *Eagle* (1783), 13, 25, 37, 44, 77, 103; Manley, in Cardwell, *Dalton* (1968), 141; Farrar, ibid., 182.

Antoine Chaptal, on the intervention of Laplace. The aeronauts made measurements of magnetism and electricity in the upper atmosphere, worked a Voltaic pile there, and brought back samples of air from 7000 meters. They learned that air in heaven has the same mixture of oxygen and nitrogen as air in Paris.[12]

The gases also served the natural philosopher as a collection of elementary principles. They form the earth's atmosphere, and, as Biot put it, with understandable exaggeration, they constitute in their combinations all liquids and solids. Studied for themselves, they disclosed universal properties of matter freed from the effects of particular affinities, since, according to the Standard Model, the particles of air stand so far apart as not to feel one another's attraction. In short, as Biot put it, "the properties of gases lend themselves to studies relevant to almost all the natural sciences."[13] "They are the most fruitfull and comprehensive of all the physical discoveries of the 18th century."[14]

This point of view recommended inquiry into the behavior of gases when heated—their capacities, densities, and expansions. We need to know how their volumes change with temperature, says Gay-Lussac, in order to develop meteorology, to improve heat engines, and to calculate tables of atmospheric refraction. Many tried their hands at this difficult business, some by analyzing the readings of barometers taken at different heights and temperatures, others by direct measurement by a pressure gauge and a thermometer enclosed in a heated volume of air. The main lesson learned was modesty. Results ran from an expansion of 1 part in 172 to 1 part in 235 per degree centigrade. Many precautions had to be taken, especially to remove unwanted moisture, until the banner year 1802, when Gay-Lussac gave 1:267, which became the definitive value of the age.[15] Not least among the benefits of the study of gases around 1800 was schooling in the needs and the means of exact measurement in physics.

Much of the progressive work on the Standard Model centered on the study of the relations between gases and the fluid of heat, or

12. Biot, *JP*, 59 (1804), 314 (quote), 317–318; Gay-Lussac, ibid., 458–460; Crosland, *Society* (1967), 262–264.
13. Biot and Arago, *MIF*, 1806, 301–302.
14. Jenisch, *Geist* (1800), 3, 498.
15. Gay-Lussac, *AC*, 43 (1802), 140, 152–156, 158–164; infra, §2.2.

caloric. The cooling of a hot body in the atmosphere bore a close analogy to the rush of a gas from a punctured balloon, excess of temperature being in the first case what excess of pressure was in the second. Therefore, concluded the practiced thermodynamicists Nicolas Clément and C.B. Desormes, after a decade's study of heat and air, "the laws of its [caloric's] motion must be the same as those of gases."[16] As a physical agent, caloric performed as a principle of separation, keeping gases gaseous by holding their particles outside the range of affinity. As a concept widely required in physics and chemistry, it acted as the principle of coherence, and even as the anchor, of the Standard Model by maintaining a close analogy between light on the one hand and electricity and magnetism on the other. And, as the element caloric in Lavoisier's chemistry, it tied the physical sciences so closely together as to make "chemistry" and "physics" indistinguishable (or useless as signifiers) within wide areas around their putative boundary.

Caloric, a precise form of the old element of fire, came into existence in the 1770s and 1780s, almost in parallel with the classical concept of electrical fluid.[17] This point may perhaps best be appreciated in the work of Johan Carl Wilcke, who had collaborated with F.U.T. Aepinus in the 1750s on the design of the air condenser. This device forced the transformation of the electrical atmospheres into the electrical fluid(s); thereafter, excesses of electrical matter remained on or very near the surfaces of "charged" bodies and not in the air between them, and the ascription of distance forces to the agent of electricity gradually gained acceptance.[18] In keeping with Franklin's ideas, the fluid(s) were free to move in some bodies ("conductors") and not in others ("insulators"), and every body when uncharged had its appropriate supply of electrical matter. After some hesitation, Wilcke adopted the theory of two electrical fluids as set out in 1759 by Robert Symmer. He then accepted the consequences about localization and distance force that Aepinus had derived from the experiments with the air condenser.[19]

16. Desormes and Clément, *JP*, *89* (1819), 325–327 (quote), 336, text of 1812.
17. Cf. Fox, *Caloric theory* (1971), 6, 9–11; Fischer, *Geschichte* (1801), 7, 524, 529; Muncke, in Gehler, *Wörterhuch*, 10, 56f.
18. Heilbron, *Electricity* (1979), 384–390, 421–426.
19. Ibid., 418–419, 431–435.

In the early 1770s, Wilcke introduced into heat theory an analogue to the concept of the natural quota of electrical fluids in uncharged bodies. This was the concept of latent heat, which occurred to him as an explanation of the temperature reached by a mixture of equal masses of warm water and melting snow. From his experiments, he knew that this temperature came out less than it would have done had he used ice-cold water rather than melting snow.[20] The difference, he said, represented the heat necessary to constitute water from melting ice. This "latent" heat became undetectable, just like the electrical matter in an uncharged body; heat in excess of this combined heat warms the water and reveals itself to the thermometer just as an electrical charge on an isolated body can be read on an electroscope. Further to the analogy, the fluids of heat and electricity tend to expand, the particles of any single fluid repelling one another and the particles of all the fluids attracting those of common matter.[21]

We lack only an analogy to the different electrical capacities of bodies, that is, to the fact that different objects require unequal charges of electrical fluid to affect an electrometer by the same amount. In the early 1780s, Wilcke came across Joseph Black's idea of specific heat, which he appears to have hit on independently a few years earlier. He defined this quantity by calorimetric measurements and gave values of it, which were not very good, for a dozen substances.[22] Furthermore, and more to the current theme, he emphasized the additional parallel between the theories of heat and electricity that specific heats afforded. Not only did the theories invoke subtle, elastic, and apparently weightless fluids, but also they supposed that each fluid was retained in ponderable bodies by specific forces dependent on the nature of the bodies.[23]

Lavoisier entered into the same order of ideas around 1772, when he also invented his new theory of combustion. A précis of Black's concept of latent heat, delivered at the Paris Academy in

20. McKie and Heathcote, *Discovery* (1935), 78–93; Oseen, *Wilcke* (1939), 156, 174–177.
21. Cf. Sebastiani, *Physis*, 23 (1981), 96–97, and 24 (1982), 208–209.
22. Oseen, *Wilcke* (1939), 232–234, 247–248; McKie and Heathcote, *Discovery* (1935), 31–53, 95–108; Guerlac, *DSB*, s.v. "Black."
23. Wilcke, KVA, *Handl.*, 2 (1781), 143–163. Cf. Seguin, *AC*, 3 (1789), 154–155, 160, 177–179, 187, 215, 217, 219; Fischer, *Geschichte* (1801), 7, 532.

1772, prompted him to disclose that a similar notion had occurred to him the previous year. Like Wilcke, Lavoisier had been surprised that warm water melted less snow than he had anticipated; and, again like Wilcke, the experience had suggested to him that heat can exist in bodies either in a "free" or a "hidden" state.[24] Extension of this insight to the process of vaporization gave Lavoisier the clue for building his new view of combustion into a full competitor to the prevailing theory of phlogiston. The theory referred the heat and light of burning to the release of phlogiston from the combustible. Lavoisier saw that he could supply the heat and light from the air, provided aeriform fluids were understood as compounds of a base with the matter of fire.[25]

To help make his new approach plausible, Lavoisier enlisted the help of his younger colleague, the mathematician P.S. de Laplace, in studying the effects of pressure and temperature on evaporation. They aimed to show that the expansiveness of the free heat opposed the weight of the incumbent atmosphere and that the latent heat of evaporation might suffice to account for the heat of combustion. Their results did not see print.[26] But in 1781 Lavoisier renewed the collaboration under the inspiration of an account of Wilcke's work, in which he found for the first time the concept of specific heat. To measure this quantity, Laplace devised the technique of the ice calorimeter, of which more later. Apparently the main purpose of the measurements was to discover a relationship between the specific heats, ease of dilation, and specific gravities of bodies in further service of Lavoisier's new "chemistry."[27]

Lavoisier first presented his concept of caloric to the Academy in papers delivered in 1777 and published in 1780. He supposed that the very subtle heat fluid ("fluide igné," "matière du feu, de la chaleur et de la lumière") exists everywhere, tends perpetually toward equilibrium, and combines with ordinary matter in various

24. Guerlac, HSPS, 7 (1976), 276, and Lavoisier (1961), 68, 92–94, from Lavoisier, JP, 2 (1772), 428–431.
25. Guerlac, HSPS, 7 (1976), 200–203, 220–221, 256, and Lavoisier (1961), 97–101, 218–221; cf. Lavoisier, MAS, 1777, 592–600, in Oeuvres (1862), 2, 228–231.
26. Guerlac, HSPS, 7 (1976), 207–208; Lavoisier, MAS, 1777, 420–432, in Oeuvres (1862), 2, 216–221; Lavoisier, Elements (1790), 11, 14.
27. Morris, BJHS, 6 (1972), 13–14; Guerlac, HSPS, 7 (1976), 27, 229, 233–234, 250–252; Lavoisier and Laplace, MAS, 1780, 364; infra, §2.2.

degrees. Aeriform fluids and vapors contain the most fire matter (as we know, they are nothing but some more ordinary material, like water, plus heat); solid bodies and liquids oppose the spread of heat through them by different strengths in accordance with their natures.[28] We see again several analogies to electricity.

Lavoisier and Laplace took the parallel further in cooperation with Volta, who passed through Paris as they were beginning their experiments with the ice calorimeter. Together the three tried to find evidence of electrification produced during evaporation—an effect that, if substantiated, might throw light on the origin of lightning as well as on the constitution of vapors. They persuaded themselves that boiling water takes up electrical fluid when vaporizing. They thereby demonstrated "the analogy between electricity and heat" as well as Lavoisier's old idea that "electricity is only a kind of combustion, in which the air furnishes the electrical matter just as...it furnishes the matter of fire and light in ordinary combustion."[29] Volta also made much of the "striking analogy by which the science of electricity throws some light upon the theory of heat." He referred particularly to the doctrine of latent heat.[30]

Lavoisier's great textbook of the new chemistry, published in 1789, opens not with an account of chemical substances and their combinations, but with "the combinations of caloric, and the formation of elastic aeriform fluids." Here the analogy with electricity, although not made explicit, was expressed in terms that no informed contemporary could miss. In the single-fluid theory of electricity, unelectrified bodies have no net effect on one another because the attractions between the electrical matter in each upon the ordinary matter in the other cancel the repulsions between the bodies' electrical matters and the bodies' common matters. (In the two-fluid theory, matter-matter repulsion does not occur.) As for a charged isolated body, it remains together and retains its charge by an interplay between the attraction between electrical and ordinary matter, the self-repulsion of electrical fluids, and, in most theories,

28. Morris, *BJHS*, 6 (1972), 5–7; Lavoisier, *MAS*, 1777, 420–432, in *Oeuvres* (1862), 2, 212, 217.

29. Respectively, Lavoisier, *Oeuvres* (1862), 2, 376, 270, quoted by Guerlac, *HSPS*, 7 (1976), 238, from Lavoisier, *MAS*, 1781, 292, and *MAS*, 1780, 334.

30. Volta, *PT*, 72 (1782), xxxii. On Volta's heat theory, see Sebastiani, *Physis*, 23 (1981), 89–90, 101–102.

the pressure exerted by the atmosphere.[31]

In parallel with these ideas, caloric in Lavoisier's theory holds apart the constituent particles of matter against the force of their mutual attraction; it occupies all available interstices; it is the principle of repulsion. A moment's consideration shows, however, that caloric and common matter do not suffice to constitute bodies: when enough free caloric has been fixed by a body to render the attraction between its particles ineffectual, why do they not fly off to the greatest separation they can attain? Once a body becomes fluid, what prevents it from simultaneously reaching the gaseous state? Lavoisier answered, as did the perceptive electricians who realized that the microscopic forces they assumed could not of themselves retain a macroscopic charge, that the pressure of the atmosphere prevented it.[32]

The ascription of all these forces, as well as gravitational attraction, to caloric, electricity, and common matter did not generate a clear or perhaps even a coherent system. It created a useful system of analogies, however, and maybe more: the quantity of caloric required to liquefy solids and vaporize liquids appeared to be a natural measure of the strength of affinity between material particles in various substances and states. "Perhaps," wrote Lavoisier, "some day the precision of the data will be brought so far that the mathematician will be able to calculate at his desk the outcome of any chemical combination, in the same way, so to speak, as he calculates the motions of celestial bodies." Laplace agreed. "Some experiments already done in this way give reason to hope that one day these laws [of chemical affinity] will be perfectly known; then, by calculation, the physics of terrestrial bodies will be raised to the degree of perfection that the discovery of universal gravity gave to celestial physics."[33]

Once again, the case of electricity gave guidance. Aepinus had assumed that the microscopic forces between particles of the electric fluid diminished inversely as some power of the distance, but he had not been able to find the function. Success came to those who

31. Heilbron, *Electricity* (1979), 379, 396–397, 500.
32. Lavoisier, *Elements* (1790), 2–4, 7–8, 15–19; cf. Morris, *BJHS*, 6 (1972), 17, 22, 25–26.
33. Lavoisier, *MAS*, 1782, 530, in *Oeuvres* (1862), 2, 550–551, and Laplace, *Exposition* (1796), 2, 197, quoted in Guerlac, *HSPS*, 7 (1976), 272, 274.

guessed that the "law" of electricity aped that of gravity and developed the identity with mathematical seriousness. Two testable consequences followed. One, developed by Henry Cavendish in 1771, derived from the theorem that a uniform gravitating shell exerts no force at all on a mass point in the hollow it encloses. Hence a particle of electric fluid *within* a uniformly charged sphere would experience no attraction from the spherical layers above it and a radial repulsion from all layers below it. If nothing hindered its motion, as would be the case in a conductor, the particle—and all such particles—would rush to the surface. Cavendish showed, in a work he did not publish, that no charge could be found within an electrified sphere. Hence it appeared that the law of electricity was inverse-square.[34]

The second, better known, and more influential way to the same result electrified small metal balls and measured the repulsion between them by the twisting of a thread. The essential link to the gravitational theory in these measurements, made by C.A. Coulomb in 1785, was to make the distance between the balls much larger than their radii. In this situation, the effects of "segregation" (induction) could be neglected and, in keeping with the grand theorem of the gravitational theory, the charge on the balls could be supposed to lie entirely at their centers. Coulomb's experiment was afflicted with many sources of errors and difficulties in manipulation. Efforts to duplicate it usually failed. Philosophers outside France at first declined to accept its results as definitive. They began to yield around the turn of the century, just as the new chemistry made its way among them.[35] Accepting Coulomb's law often amounted to accepting, as the goal of physics, the search for mathematical relationships among microscopic entities. Many natural philosophers came to regard electrostatics based on Coulomb's law as the cynosure of "an exact theory," from which the phenomena followed by "rigorous calculation."[36] It immediately gave rise to an analogous theory of magnetic interactions, and, as will appear, it inspired fruitful emulation by theorists of caloric.

34. Heilbron, *Electricity* (1979), 478–480.
35. Ibid., 471–476. Cf. Frankel, *HSPS, 8* (1977), 38–39, who offers a parallel between the torsion balance and the ice calorimeter: each established the quantitative foundations of its subject; and infra, §2.2.
36. E.G. Fischer, *Physique mécanique* (1806), 257–261.

The core six imponderables of the Standard Model had many transient fellow travellers—phlogiston, animal magnetism, animal electricity, radiant heat, to mention the most prominent. The Model easily accommodated them. Phlogiston disappeared into oxygen, animal magnetism vanished altogether, animal electricity became ordinary electricity, and radiant heat was assimilated to light in a lengthy process described below.[37] Suppositious analogies between the mode of action of the imponderables and other forces of nature also expanded the reach of the Model. There were analogies between counterbalancing electrical fluids and the formation of neutral salts; between the interactions of the particles of light and the molecules of transparent bodies, on the one hand, and the forces of chemical affinity on the other; and, with the spurious discovery of magnetic effects of ultraviolet light, between the warming of the earth by the sun and the occurrence of natural magnets. "Iron would then have the same relation to the magnetic fluid [in the sun's rays] as pyrophors have for heat and phosphors have for light."[38] Like today's Standard Model, which has made the big bang a household word, the scheme of imponderables and its richness of analogies spread among the bystanders of science around 1800. The last word on the subject was spoken by the first man of his age. "I believe that man is the product of the [imponderable] fluids of the atmosphere, that the brain pumps these fluids and gives life, that the soul is composed of these fluids, and that after death they return to the ether."[39]

A world of as-if

Acceptance of the scheme of imponderables was eased by the spread of the philosophies of Condorcet and Kant.[40] The bottom

---

37. Infra, §2.2. The identity of animal and ordinary electricity was settled to the satisfaction of most natural philosophers by Volta's battery; cf. Delamétherie, *JP*, 54 (1802), 18.

38. Weiss, *Prevost* (1988), 156; Biot and Arago, *MIF*, 1806, 327; Configliachi, *JP*, 77 (1813), 212–213, 233, criticising experiments adduced by Morichini as demonstrations of the magnetizing power of light; Morichini, ibid., 294–297, 308 (quote).

39. Napoleon, quoted by Fischer, *Napoleon* (1988), 283.

40. Cf. Shimank, in *Lessing* (1968), 67: "Kantian philosophy, experimental-physical and physical-mathematical research, and antiphlogistic chemistry with its striving for quantitative determination are three equally remarkable productions of the age of the Enlightenment."

line of the one, that progress in science amounts to nothing more (or less) than the construction of an apt language to describe natural phenomena, and of the other, that the phenomena themselves contain an unavoidable admixture contributed by the perceiving mind, taught the natural philosopher that knowledge of true principles could never be proved and need not be sought. That left plenty to do. "The true business of the philosopher, though not flattering to his vanity, is merely to ascertain, arrange, and condense the facts."[41] The scheme of imponderable fluids provided a way to realize this modest enterprize. Modesty became a prominent ingredient in it, and so of the Standard Model; its practitioners, when in a negative mood, agreed that it could not lay claim to a higher status than a very convenient fiction. The best scheme, according to the leading German physics text of the 1790s, is the one that makes comprehension the easiest, "however far it may be from the truth that we try to approach through its use."[42]

Some argued that even the parent concepts of the Standard Model were merely useful constructs. Jeremy Bentham arrayed attractions and repulsions in his *Theory of fictions* as conventional causes of "opposite and mutually balancing effects." The abbé Haüy, the author of the first French exposition of the Standard Model, goes even further: attraction and repulsion themselves are phenomena, "properly speaking, the velocities with which bodies tend to approach or withdraw from one another." We single them out because, "knowing the law obeyed by this tendency, and applying mathematics, we can determine all the other phenomena (*faits*)."[43] The first historian of physics, Johann Carl Fischer, nailed to the doorstep of his immense *Physikalisches Wörterbuch* the intelligence that "matter itself is nothing more than a phenomenon and therefore presupposes active causes." We do not know things in themselves, he added, and recommended treating "cohesion" and similar words as expressions concerning phenomena, not causes.[44] It

41. Leslie, *Diss.* (1842), 743. Leslie was an enthusiastic follower of Hume; infra, §2.2.
42. Lichtenberg, in Erxleben, *Anfangsgründe* (1791), xxv, quoted by Kleinert, in Fabian, *Entfaltung* (1980), 103. Cf. Delamétherie, *JP*, 46 (1798), 24: "particular attractions must be viewed only as hypotheses suitable for explaining the phenomena."
43. Bentham, *The theory of fictions* (1789), ed. C.K. Ogden (1959), 110, quoted by Fink, *Archiv f. Begriffsgesch.*, 25 (1982), 71; Haüy, *Traité* (1806), vii.
44. Fischer, *Wört.*, 1 (1798), v, 599 (art. "Cohäsion"); cf. ibid., 6 (1805), 773–774, s.v. "Wärme."

is too early, says Jenisch, and may always be too early, to set down particulars about the fundamental forms of nature. Fortunately we can do without knowledge of true causes. "Knowledge of laws of nature is worth much more than explanations from hypotheses," wrote the textbook writer F.A.C. Gren, who liked to say that "only effects are in nature, their laws lie in our minds."[45]

Although our representations cannot claim truth, we are not altogether free to feign any hypothesis we wish. Or so Ernst Gottfried Fischer insisted in corrections he added to Gren's *Grundriss der Naturlehre* to improve its comformity with the Standard Model. We must associate force with matter, says Fischer; that is a "law of our reason...a requirement of our power of imagination." We must do so even if we must suppose the existence of a hypothetical matter that does not fall directly under our senses. This condition of knowing requires great care in thinking. "There is no cause for criticism for introducing into scientific books such symbolic hypotheses, whose scientific purpose should be only to link the laws of the phenomena to a model (*Bild*); but it is censurable to derive from them perceivable and unperceivable consequences without distinction, and thus to give to science the appearance of uncertainty and fanaticism (*Schwärmerei*)." A matter of heat in principle is most acceptable: "matter is everything in which we imagine mechanical or chemical, in a word, physical forces." But beware. By modeling this heat matter as an expansive fluid, you characterize a nonvisualizable thing with an image from the visualizable world. "As a symbol such a model may be valid, but not as a source of knowledge."[46]

This intricate apologetics for the essential heat matter was doctrine among the clear-minded French as well as among German philosophical physicists. The Fischer we have been following wrote a textbook deemed worthy of translation into French by Laplace's energetic disciple Biot. It treated physics at a higher level of

---

45. Jenisch, *Geist* (1800), *3*, 500–501; Gren, *Grundriss* (1797), 3–5, and (1808), 3–4. Cf. Lorenz, *NTM*, 27 (1990:1), 34, on Gren's Kantianism, and ibid., 27–31, 35–37, on German forerunners of the instrumentalism of the Standard Model; and, for French forerunners, Weiss, *Prevost* (1988), 96–98.

46. Fischer, in Gren, *Grundriss* (1808), 4, 286–287 (quote); cf. Krafft, *Ber. Wiss. Ges.*, 1 (1978), 148.

mathematics, and with a wider supply of facts than Haüy's text, and it emphasized the instrumental character of the caloric theory.[47] In that, it agreed fully with Haüy, who recommended the material theory over the perennial alternative (heat being a consequence of the internal motions of bodies) on the basis not of truth but of convenience. "Without deciding between these two opinions, I will use language comformable to the [material theory], regarding it only as an hypothesis more suitable to aid in comprehending the phenomena, and more convenient for expressing them." In the manner of Fischer, Haüy insisted on a distinction between palpable and true fluids, on the one hand, and constructs like caloric and the matter(s) of electricity and magnetism, on the other. "We do not place the latter in nature, but only in theory, since they have the advantage when well chosen of representing results faithfully, of offering a satisfactory explanation of them, and even of helping us to predict them; so that if [these theoretical fluids] are not the true agents used by nature to produce the phenomena, they may be regarded as taking their place and being their equivalents."[48]

It may be useful to have an English example. William Nicholson wrote textbooks of physics and chemistry that became standard in the 1790s. Both teach that the student need not decide the cause of heat; on either hypothesis, he or she could save the phenomena. "The word quantity applied to heat will therefore denote either motion or matter, according to the opinion made use of, and may be used indefinitely without determining which."[49] Hard on the student, perhaps, but easy on the conscience of the teacher.

So far the writers of textbooks and dictionaries. A similar instrumentalism pervaded the programmatic, if not the substantive, writings of the developers of the caloric theory. The Scottish school that invented the concepts of latent and specific heats independent of Wilcke declined to declare the objective existence of the heat material about which they reasoned. The Genevan experts on radiant heat took the same stand. The Frenchmen who perfected the caloric theory—Lavoisier and Laplace—acknowledged that the

---

47. Biot, in Fischer, *Physique mécanique* (1806), iv; Fischer, ibid., 76–77; cf. Biot, *Traité* (1816), 1, xx–xxi.
48. Haüy, *Traité* (1806), 1, 79–81. Cf. ibid., 2, 138–139, which has a similar argument for preferring the particulate to the wave theory of light.
49. Nicholson, *First principles* (1792^2), 6, and *Introduction* (1790^3), 114 (quote).

kinetic rival could also explain all their results.[50] J.A. Deluc talked at length with Laplace during his collaboration with Lavoisier on the measurement of heat. "He is one of those with whom I have had the greatest pleasure discussing these matters," Deluc wrote, "for he has no prejudices."[51] Lavoisier's *Elements of chemistry* emphasized the hypothetical character of caloric: "we are not obliged to suppose this to be a real substance, it being sufficient...that it be considered as the repulsive cause, whatever that might be, which separates the particles of matter from each other...Circumstances take place exactly as if such a repulsion actually existed; and we have a very good right to conclude, that the particles of caloric mutually repel each other."[52]

This instrumentalism propagated through the French caloricists. Lavoisier's disciple, Armand Seguin, delivered three theories of heat—as motion of ponderable molecules, as a material sui generis, as a compound of light and/or another principle—and advised his readers to take their pick. For himself, he found the second theory the most plausible, and accepted it because it gave a useful unified point of view; but all, he said, were no more than hypotheses, "since it is not possible to demonstrate them rigorously and perhaps it never will be." The representation of the caloric theory at the Ecole polytechnique followed Lavoisier slavishly, from the atmosphere's role in maintaining fluids to the theorist's freedom in choosing models. In his *Dictionnaire de physique* of 1800, M.J. Brisson chose the caloric theory over the kinetic not for its truth, but as "by far the simplest way to account for the heat of bodies."[53]

If the material theory had only convenience in its favor, it might have to yield in whole or part to rival conceptions. Gay-Lussac almost made this view explicit in 1812, when he declared that

50. McKie and Heathcote, *Discovery* (1935), 27–30, and Sebastiani, *Physis*, 23 (1981), 92–94; Sigrist, *Origines* (1990), 86, and Evans and Popp, *AJP*, 53 (1985), 745; Fox, *Caloric theory* (1971), 23–26.

51. Deluc to Blagden, 24 Aug 1783 (Blagden Papers, Royal Society).

52. Lavoisier and Laplace, *MAS*, 1780, 355, in Lavoisier, *Oeuvres* (1862), 2, 286–287; Lavoisier, *Elements* (1790), 5, 23; Cf. Lilley, *Arch. int. hist. sci.*, 1 (1948), 631–634; Morris, *BJHS*, 6 (1972), 31–32; Fox, *Caloric theory* (1971), 28–37; Gillispie, *DSB*, 15, 313, s.v. "Laplace."

53. Seguin, *AC*, 3 (1789), 182–184, 189 (quote), 207, 221–224; Barruel, Ecole poly., *Jl*, no. 4 (an iv), 631–636; Brisson, *Dictionnaire* (an viii), 2, 140, s.v. "chaleur;" Cf. Hutton, *Dictionary* (1815), 1, 265, and Friedman, *HSPS*, 8 (1977), 74–76, 90–96.

progress was blocked by the need to make "many hypotheses, which are the less probable because we know nothing about the nature of heat." In 1828, he justified his continuing adherence to caloric theory solely on the ground that it was better developed than any alternative.[54] We leave the last word in this business to the physicist and encyclopedist Georg Wilhelm Muncke, the author of the article on heat, some 600 pages long, in the second edition of of Gehler's *Physikalisches Wörterbuch*. He offers the truism that we cannot avoid making some hypotheses if we are to reason about natural phenomena. The material theory is the best available. Still, heat exhibits some effects, "which either cannot be referred to a *Wärmestoff* acting by simple ways definable quantitatively, or can be so only by very forced hypotheses." Therefore, says he, anticipating the escape of the quantum physicist, therefore we must ascribe "many phenomena of heat to the increase and motion of [a *Wärmestoff*], and others to wave-like vibrations."[55]

Modesty about the scheme of imponderables was further recommended by the realization that it could not be set out in a unique form. Physicists could not decide in the exemplar case whether electricity came in one fluid or two. "Let's not fight over the two fluids," wrote Georg Lichtenberg to a not-quite enlightened electrician in 1784. "I declare for neither; in speaking or writing, however...I am always a Franklinist, just as I always write 'deutsch' although I readily concede that 'teusch' is also possible, and perhaps even preferable." Lichtenberg's fellow writer of physics and chemistry texts, F.A.C. Gren, decided similarly, not because he thought he knew the truth, but because Franklin's view agreed better with Newton's rules of philosophizing than dualism. Though devoted to the single-fluid theory, Volta acknowledged that the dualist position could not be faulted for anything more than requiring two fluids, that is, two hypotheses, where he made do with one.[56] The French dualists—Coulomb, Laplace, Biot, Poisson—gazed

54. Gay-Lussac, *AC*, *81* (1812), 107, *AC*, *13* (1820), 304, 307, and lectures on physics, 1828, quoted by Crosland, *Gay-Lussac* (1978), 125. Cf. Berzelius to Davy, [1812], in Fox, *BJHS*, *4* (1968), 12: "I am fully convinced that no existing hypothesis sufficiently explains the phenomena of caloric."
55. Muncke, in Gehler, *Wörterbuch*, *10*, 95, 97 (quote), 100 (quote), 103.
56. For Lichtenberg and Volta, Heilbron, *Electricity* (1979), 447–448; for Gren, his *Grundriss* (1797), 7–8, 792–793, and (1808), 8–9, 653, approved by Fischer, *Wört.*, *1* (1798), 922.

in the opposite direction from the same point of view. To them the singlist position had the grave inconvenience of placing on common matter, already sufficiently burdened with distance forces, the additional obligation of carrying the attractions and repulsions that they preferred to ascribe to a second electrical fluid. Or, as Haüy put it, although the case for the existence of two fluids stood on weaker ground than that for electric matter itself, "the adoption of these fluids leads to a simple and plausible way to represent the results of experience."[57]

The other major imponderables suffered from a similar uncertainty. How many magnetic fluids did theory require? Aepinus said one, Coulomb two. "In my opinion," wrote Fischer the encyclopedist, "we should not condemn the assumption of either one or two magnetic matters provided that we treat it purely as a scientific fiction, set out as a guide (*Regulativ*) for experiments and observations, but not as the foundation for explanations and hypotheses. For in the last case we would have said nothing more than we already know, namely that there must be something or other that makes the magnet magnetic." Heat and light again presented the question, one matter or two. Fischer judged that most people inclined to believe them identical; but the main textbook writers, including Gren and Fischer himself, treated them under separate articles. Lavoisier had questioned their distinctiveness, but listed them as equally elementary in his table of simple substances.[58]

Laplace's close colleague, Claude-Louis Berthollet, saw strong evidence for identity in the parallel between latent heat and absorbed light, both apparently constituents of bodies.[59] Radiant heat provided a still stronger analogy. Berthollet commissioned experiments to confirm results reported by William Herschel and William Wollaston, who found, respectively, calorific rays beyond the red end of the spectrum and chemically active ones beyond the violet. The experiments succeeded, as did a demonstration of the polarization of radiant heat.[60] Play with ice and mirrors showed

57. For the French dualists, Heilbron, *Electricity* (1979), 473–474, 498–499, and Poisson, *JP*, 75 (1812), 229; for Haüy, his *Traité* (1806), 1, 366.
58. Fischer, *Wört.*, 3 (1800), 429, 452 (quote), 460, s.v. "Magnet," and 5 (1804), 400, 406, s.v. "Wärme;" Lavoisier, *Elements* (1790), 175.
59. Rosmorduc, *Coll. Gay-Lussac* (1980), 216–219, citing Berthollet, *Essai de statique chimie* (1803), 1, 192–204.
60. Bérard, *MSA*, 3 (1817), 7, 9–10, 28; Gay-Lussac and Thenard, *Recherches*

that cold and heat could be reflected and focused just like visible light.[61] Was radiant heat the middle term between caloric and light, or a tertium quid? One spoke of "rayons de calorique," "calorique rayonnement," "calorique...sous forme rayonnante," and "rayons calorifiques."[62] All of which, and other forms besides, easily found accommodation in the Standard Model.

2. BETWEEN CALCULUS AND CHEMISTRY

A most significant function of the scheme of imponderables was to help define the subject matter of physics. The large changes in meaning and purview of "physics" during the course of the 18th century have been noticed: they amounted in gross to a shrinking from the traditional coverage of the entire natural world, from anatomy through zoology, to the inorganic realm accessible to experiment and amenable to measurement and mathematics. The older usage still occurred occasionally around 1800—for example, in a prize competition in physics for an essay on the nature and uses of the liver offered by the Paris Academy of Sciences in 1792 and 1798—along with a then more recent and less flattering one, deriving primarily from the exploitation of electricity in parlor tricks; but to most *physiciens, Naturfoscher*, and experimental philosophers writing at the end of the 18th century, "physics" meant neither artificial games nor natural history.[63] Their problem of definition was to distinguish their subject from chemistry on one side and from mathematics on the other.

Johann Christian Polykarp Erxleben required six pages to define his subject in the physics text he first published in 1772; his critical editor, Georg Christoph Lichtenberg, extended this already lengthy declaration to no fewer than 38 pages, in a rendition of his lectures published by one of his students.[64] These extravagances betray not

---

physico-chimiques (1811), 2, 186–206, cited in Rosmorduc, *Coll. Gay-Lussac* (1980), 224–226.

61. Infra, §2.2.

62. Culled from Tremery, Soc. phil., *Bull.*, 3 (1813), 323–328, and Delaroche, ibid., 131, and *JP*, 75 (1812), 201, 211.

63. Heilbron, *Electricity* (1979), 1–19, and in Rousseau and Porter, *Ferment* (1980), 361–365; Schimank, in *Lessing* (1968); *MIF*, 1 (1798), ii–iii, viii.

64. Gilles, *Erxleben* (1978), 17, 22–33, 28.

only the compulsion of the classifier but also the elusiveness of his material: physics was undergoing rapid change not only by shedding traditional topics but also by deepening the study of those that remained. "If there is a branch of human knowledge that has made rapid progress in the last few years, it is physics."[65] So Hassenfratz, in his first lectures at the Ecole polytechnique in 1799; he drew the obvious and self-interested conclusion that he could not recommend any of the older standard texts—Nollet, Desaguliers, or Musschenbroek—to his students.

Gren felt obliged both as pedagogue and marketer to recast his *Grundriss der Naturlehre* entirely for its third edition of 1797; he pointed to "the quantity of new views" in circulation since the previous edition of 1793. The most important piece of recasting, apart from dropping the earth sciences and astronomy, brought out primary features of the Standard Model. Gren now opened with an account of the nature of science as taught by Kant: "it is inexcusable to ignore the insights [*Aufklärungen*] that the critical philosophy has provided here." He moved the section on air to the first part of the text, devoted to the general properties of matter. His reason: the recognition of the several types of gas and the discovery of their laws of pressure and equilibrium revealed that characteristics previously supposed to be peculiar to the atmosphere applied to all pneumatic fluids. As for the phenomena of heat and fire, they now all received attention in a chapter on *Wärmestoff*. Gren's competitors also self-consciously attended to the arrangement of their texts. "A system of physics [*Naturlehre*] should not be a dictionary."[66]

Nor should it be a system of chemistry. But how tell one from the other? Lavoisier had made distinctions difficult. He had transformed "chemistry" from a collection of ill-assorted recipes into, yes, "a system," so wrote the literary executor of the 18th century, "a system that simultaneously opens a new world to the thinker and brings a wonderful simplicity to the infinitely complicated interrelationship of natural things."[67] Lavoisier called himself a *physicien* (Volta called him *fisico e chimico*), and advertised his work on combustion as "a revolution in physics and chemistry."[68]

65. Hassenfratz, Ecole poly., *Jl*, no. 6 (1799), 372 (quote)–374.
66. Quotes from Gren, *Grundriss* (1797), v, x–xi, resp.; cf. ibid., xv.
67. Jenisch, *Geist* (1800), *1*, 482, *3*, 499–500 (quote).
68. Cf. Guerlac, *HSPS*, 7 (1976), 193–195, and *Lavoisier* (1961), 228–230; Perrin, *Isis*, 81 (1990), 262; Volta, text of 1783, quoted by Sebastiani, *Physis*, 24 (1982), 20.

Historians spill much vitriol in trying to reckon the obligations of antiphlogistic "chemistry" to experimental "physics." Did "physics" play its part only in Lavoisier's mind, as a model or catalyst to his thinking? Or was it an essential reagent, which, via Lavoisier and his colleagues, precipitated a system of "chemistry" from a mess of empirical recipes?[69] Or did it enter firmly into combination, transforming both "chemistry" and "physics," as Lavoisier supposed, and his younger contemporaries asserted?[70] Much of the historiographical dispute turns on whether "chemistry" existed as an autonomous discipline before Lavoisier, or, otherwise put, whether his revolution took place *in* or *into* a science.[71]

It appears that we need a richer vocabulary. Although participants in this discussion recognize that "chemistry" and "physics" in the 18th century designated fields of study that overlap only partly in subject matter and approach with the disciplines that later bore the same names, they have not drawn much profit from their penetration. The historiographical article on chemistry in the latest general assessment of 18th century "science" provides a particularly good example. As illustrations of the danger of applying 19th-century disciplinary boundaries, it gives heat theory, which, it says, belonged in chemistry in the 18th century and to physics later on, and pneumatics, which stayed with chemistry throughout.[72] But around 1800, as we know, both pneumatics and heat figured prominently in physics texts as well as in chemistry texts. There is much that we would call physics in the *Annales de chimie* and much that we would call chemistry in the *Annalen der Physik*. The editors of these journals later acknowledged their difficulty: the former became the *Annales de chimie et physique* in 1816, the latter the

---

69. The first view seems to be that of Guerlac, *HSPS*, 7 (1976), 194–196, 267–274, who emphasizes the relationship between Lavoisier and Laplace, and of Donovan, *Osiris*, 4 (1988), 221–222, 226. The second view approximates the interpretations of Melhado, *Isis*, 76 (1985), 195, 210, and in Freudenthal, *Etudes* (1989), 125–126; and Lundgren, in Frängsmyr, *Spirit* (1990), 245–246, 257–263.

70. Lavoisier to Franklin, February 1790, quoted by Donovan, *Osiris*, 4 (1988), 227–228.

71. Donovan, *Osiris*, 4 (1988), 215, 226–228; Perrin, ibid., 55–64, and *Isis*, 81 (1990), 262, 265–267; Melhado, in Freudenthal, *Etudes* (1989), 116.

72. Crosland, in Rousseau and Porter, *Ferment* (1980), 391; cf. Donovan, *Osiris*, 4 (1988), 231.

Annalen der Physik und Chemie in 1819. Meanwhile the *Journal de physique, de chimie, d'histoire naturelle, et des arts,* playing it safe, retained its comprehensive title.

Historians would be well advised to use the names of fields of study only as they were used by contemporaries. A history of "physics" therefore will include fields subsequently lost to the discipline as well as new ones added by conquest or invention. When it is necessary, as in the last sentence, to invoke the name of a field diachronically, or to compare the state of a discipline at different times, the various meanings can be set off by quotation marks or other orthographic convention. Proceeding thus, we not only can understand, but also write down, that Lavoisier's revolution brought techniques associated primarily with physics into the laboratory practice of chemistry, but not the subordination of "chemistry" to "physics."

Let us return to the words of contemporaries. In Lavoisier's, of 1790, "chemistry has been brought much closer than heretofore to physics." In M.A. Pictet's, of 1785: "[chemistry] has become the inescapable aide and companion of physics."[73] Encyclopedist Fischer pointed to the discovery of the gas types, the development of caloric theory, and the invention of antiphlogistic chemistry as the forces that brought physics and chemistry together. Hence, says Haüy, "the true physicist is the one who speaks the language of chemistry."[74] An Italian professor writes his correspondent in London for news about "fisica chimica;" a German academician inquires of the same party about "this most interesting part of physics," namely, agriculture; textbook-writer Fischer divides theorists into "physiciens mécanistes," who prefer to refer heat to a mode of motion, and "physiciens chimistes," who plunk for caloric.[75]

Some professed to worry that the trunk of physics, whatever remained after the lopping off of natural history, would be eaten

---

73. Lavoisier to Franklin, Feb 1790, as quoted by Donovan, *Osiris,* 4 (1988), 227; Pictet, *Jl de Genève,* 28 Nov 1789, as quoted by Perrin, *Isis, 81* (1990), 269. Cf. Levere, in Levere and Shea, *Nature* (1990), 207, 210, on Lavoisier's own evaluation of the rapprochement.

74. Fischer, *Wört.,* 3 (1800), 891–893, 899; Haüy, *Traité* (1806), 1, ii–iii.

75. Landriani to Magellan, 12 Feb 1781, and Achard to Magellan, 6 Aug 1784, in Carvalho, *Correspondência* (1952), 51, 107; Fischer, *Physique mécanique* (1806), 76–77.

away by chemistry.[76] The popularity of the new chemistry, and its accessibility, did perhaps divert attention when discoveries beckoned, or so, at least, the astronomer J.B.J. Delambre thought. The leading chemist in the generation after Lavoisier, Berthollet, liked to say that physics "was scarcely cultivated in France." He referred to the years around 1800 and offered as a tonic tighter ties to chemistry.[77]

The trunk of physics, the surviving *physica generalis*, contained the doctrine of the general properties of bodies. Usually five such properties received separate attention. Thus the first lesson on physics at the Ecole polytechnique took up extension (illustrated by the then new metric measure), impenetrability, mobility, inertia, and gravity. Cohesion occupied a less secure place: a general property of bodies, to be sure, but one closely related to the chemical concept of affinity. Hence the very great interest in capillary phenomena, which appeared to make macroscopic and measurable the differential effects of cohesion and affinity.[78] From this point of view, the general part or underlying principles of chemistry could be considered a branch of physics. The course at the Ecole emphasized the theory of affinity over that of any other branch of general physics.[79] The orderly mind inevitably placed capillarity as a species of "chemical attraction" immediately after the obligatory exposition of gravitational action in *physique générale*. The basic principles of chemistry might therefore be taken up as a link between general and special branches of physics, or in their own right as a piece of *physique particulière*.[80]

The other main division of the old "physics," *physica particularis*, had been reduced by 1800 primarily to the range of phenomena

76. Gehler, *Wört* (1798), s.v. "Physik," summarized by Krafft, *Ber. Wiss. Ges.*, 1 (1978), 149–150. Cf. Hassenfratz, Ecole poly., *Jl*, no. 5 (an vi), 239.

77. Delambre, *Rapport* (1810), 31–32, quoted by Silliman, *HSPS*, 4 (1974), 142; Biot, "A monsieur Berthollet," in Fischer, *Physique mécanique* (1806), i.

78. Barruel, Ecole poly., *Jl*, no. 3 (an iii), 337, and no. 4 (an iv), 623–624. Cf. Hassenfratz, ibid., no. 5 (an vi), 238, and no. 6 (an vii), 144, 376–377, and Fischer, in Gren, *Grundriss* (1808), 94–95.

79. Hassenfratz, Ecole poly., *Jl*, no. 6 (an vii), 375: Physics was accorded one day/décade (infra, §5.1) for 10 months, or 30 lessons/year; optics got 5; affinity, caloric, and electricity, 3 each; mobility and inertia, 2 each; all else, 1.

80. A representative text is Libes, *Trattato* (1803), 1, 244. Cf. Gillispie, *Intersezioni*, 9 (1989), 411.

represented by water (the liquid state), air (the gaseous), and the imponderables. Haüy follows his second chapter, on gravity, cohesion, and affinity, with ones on caloric, water, air, electricity, magnetism, and light. William Nicholson adopts a similar pattern in his *Introduction to natural philosophy* of 1790: light, fluids, airs, but then chemistry (including heat) before magnetism and electricity. Fischer's text retains a little more of the ancient scheme of four elements: after bodies in general we have chapters on solids (earth!), heat (fire), liquids (water), and gases (air) before a final three on electricity, magnetism, and light.[81] In keeping with the instrumentalism of the Standard Model, none of these pedagogues claimed the existence of any of the imponderables. Gren treated "besondere Naturlehre" in much the same order as Fischer, but with slightly greater confidence in the reality of his imponderables. Fischer ventured that the attribution of the phenomena of heat, light, and electricity to special matters is "not without probability." "A certain degree of skepticism is certainly advantageous in science," Gren allowed, "but Pyrrhonism is the death of all true Naturforschung." The concession related to the matter of fire; in treating electricity, Gren admitted that he did not know how many, if any, electrical fluids existed, and offered explanations on both the singlist (which he favored) and dualist hypotheses.[82]

As physics mixed with chemistry, chemistry mixed with physics. The textbook writers adopted the definition that physics deals with the general, permanent, and macroscopic properties of bodies, chemistry with permanent changes in the intimate relations among molecules.[83] The chemist Thomas Thomson clarifies: physics deals with sensible motions, chemistry with insensible ones.[84] Taken literally, the definition awarded all theories of the internal behavior of imponderables to chemistry; Nicholson and others followed out this logic by treating electricity and heat as part of chemistry. But

---

81. Haüy, *Traité* (1806^2), 1, 446–447, 2, 402; Nicholson, *Introduction* (1790^3), 1, 253–314, 2, 1–366; Fischer, *Physique mécanique* (1806), ix–xvi. Cf. Arnold, *AHES*, 28 (1983), 250, 263–264.

82. Fischer, *Physique mécanique* (1806), 3; Gren, *Grundriss* (1793), xi (quote), xii-xiv. We have the same in Libes: caloric, air, water, earth, light, electricity, magnetism, galvanism.

83. Haüy, *Traité* (1806^2), 1, ii, iv; Nicholson, *Introduction* (1790^3), 2, 112; Playfair, *Outlines* (1812), 1, 1.

84. Thomson, *System* (1802), 1, 2–3; cf. Crosland and Smith, *HSPS*, 9 (1978), 3–5.

logic also demanded that these subjects be treated among the general macroscopic properties of bodies. Nicholson responded to this challenge with the odd compromise mentioned earlier: his physics text discusses heat under chemistry and electricity as an independent imponderable.[85]

The scheme of organization by imponderables had the disadvantage, to some minds, of radical incoherence. How do they fit together? How many are there? Jenisch, a literary man, wanted to know the nature of light, electricity, and magnetism. He went to a seminar in physics. "[It seemed to me] an elaborate puppet show, where I admired the astounding effort and subtelty with which the men wanted to examine the secrets of nature, and sadly lamented the obstinacy with which nature preserves her secrets undiscovered and unbetrayed." Similarly, Gehler treated "besondere Naturlehre" as a dumping ground for all the parts of physics that eluded quantification and systematization.[86] But these opinions belonged to a minority, and, in respect of quantification, were in error. One of the most important functions of the scheme of imponderables within the Standard Model was to provide a purchase for the application of mathematics to *physique particulière*.

The program of quantifying physics was already a bromide when F.K. Achard, a prominent member of the Berlin Academy of Sciences, announced in 1773 that "everyone now agrees that a physics lacking all connection with mathematics...would only be an historical amusement, fitter for entertaining the idle than for occupying the mind of a philosopher." "Having [thus] learnt to distinguish knowledge from what has only the appearance of it," added J.A. Deluc, a virtuoso performer on the barometer and hygrometer, "we shall be led to seek for exactness in everything."[87] The sobriquet "matematico fisico," mathematical physicist, made its appearance.[88]

85. Nicholson, *First principles of chemistry* (1792^2), 5–31, provides about the same coverage of heat as Nicholson, *Introduction* (1790^3), 2, 73–77, 112–126; and gases are treated similarly in *Principles*, 47–58, and *Introduction*, 2, 153–157. Cf. Miller, *Retrospect* (1803), 1, 443, awarding electricity to chemistry.

86. Jenisch, *Geist* (1800), 3, 497; Krafft, *Ber. Wiss. Ges.*, 1 (1978), 134, 151, with some errors in interpretation.

87. Achard, in Heilbron, *Electricity* (1979), 74; Deluc, *PT, 68* (1778), 493.

88. Used by the astronomer Barnaba Oriani of Pybo Steenstra, a teacher in Amsterdam, in a diary entry of 19 June 1786; Tagliaferri and Tucci, *Incontri*, 4 (1989), 68.

Achard's generation differed from its predecessor by practicing what it preached—by laboring, as he did, to measure accurately, to correct for errors, to tabulate, calculate, and systematize. Gradually, some of Achard's colleagues learned to use their numbers to devise and check quantitative theories. Biot named several of them as counter examples to Berthollet's interested claim that few true physicists existed in France: Haüy's research on crystals, Coulomb's on electricity, Borda's on geodetic instruments, and Laplace's on capillarity.[89]

They were but the most successful. Still according to Biot, mathematics had been the inspiration for all physical knowledge for a generation. "This exchange of knowledge [*lumières*] is the certain proof of the perfection of the sciences." Hassenfratz attributed the great progress in physics he observed around 1800 largely to "a rigorous and almost mathematical method of experimentation," that is, "the scheme [*étude*] and language of the sciences perfected by the union of geometers and savants who cultivate every branch of knowledge."[90] Applied mathematics, says Jenisch, being a mixture of pure reason and physical laws derived from experience, belongs to physics; further, it is the means by which the 18th century has tried to realize "the general character of its scientific culture, i.e., 'practical application.'" As applications Jenisch had in mind not technology, but mechanics, optics, and astronomy, the old physical branches of applied mathematics.[91]

Most prominent German spokesmen for physics around 1800 insisted that in compensation for the loss of its universal coverage of the natural world, physics had a claim to topics previously denominated "applied mathematics." Gehler: "The objects of applied mathematics...belong in themselves to physics, and the fact that they cannot be understood without mathematics is no reason to take them away from true Naturlehre." Lichtenberg: "applied mathematics is made up really only of individual parts of physics." Fischer: the new physics consists of "the choicest parts" of applied mathematics, plus, more recently, "the most necessary topics of chemistry;" "mathematics is indispensable for the Naurforscher," a

89. Biot, in Fischer, *Physique mécanique* (1806), ii.
90. Biot, *Essai* (1803), 22; Hassenfratz, Ecole poly., *Jl*, no. 6 (an vii), 372. Cf. Gillispie, *Intersezioni*, 9 (1989), 412.
91. Jenisch, *Geist*, 1 (1800), 482.

necessity for deriving laws from data, and the only source of true knowledge in physics.[92] Gren, who had been brought up in the older, qualitative, inclusive natural science, came to realize the need for retooling to keep his textbooks competitive. "In the last years of his life he devoted himself with the most creditable zeal to the study of mathematics, which cannot be recommended enough [to physicists]." Or early enough. For despite his zeal, "his mathematical work showed that he had not completely made himself at home with the rigor of mathematical discourse."[93]

The Grens no doubt felt menaced by the calculators. People worked up their data to crowds of irrelevant decimals. They corrected their instrument readings far beyond the accuracy of their measurements. The craze afflicted even so unmathematical a head as Alexander von Humboldt's. "It seems that he can hear the grass grow," sneered a collague unimpressed by the great man's doctoring of instrumental readings beneath the limit of observational accuracy.[94] A good example of this compulsion is a study done in 1806 by Biot and François Arago on the specific gravities and refractive powers of gases. The subject, the forces between common matter and particles of the imponderable fluid light, touched the core of the Standard Model. Biot and Arago worked to six or seven (in)significant figures, corrected in detail for the effects of temperature, pressure, and humidity on their instruments, and checked their results by computing one of the best cultivated numbers of their day. This noble figure, "the barometric coefficient," stands in front of the "law" by which the pressure of the atmosphere decreases with height. After many adjustments, they made the coefficient 18334 meters.

Their number agreed suspiciously well with the then latest reduction of direct measurements to a prodigious formula given by Laplace:

---

92. Gehler, *Wört.*, s.v. "Physik," and Lichtenberg, in Erxleben's *Anfangsgründe* (1794⁶), quoted in Gilles, *Erxleben* (1978), 12, 15, and Gehler, in Krafft, *Ber. Wiss. Ges.*, 1 (1978), 151; Fischer, *Wört.*, 3 (1800), 889–890, 892.
93. Fischer, in Gren, *Grundriss* (1808), xv.
94. F.X. Zach to M.A. David, in Zach, *Briefe* (1938), 138.

$$z = \{\log(h/H)\,18336(1 + 0.0028371\cos2\psi)\} \cdot$$

$$\{1 + 0.002(t + t')\} \cdot \left\{1 + z/a + \frac{0.0868589z}{a\log(h/H)}\right\}.$$

Here z is the vertical distance in meters between two stations at which h, T, t (and h', T', t') are the heights of the barometer, the temperature of the barometer, and the temperature of the air at the top and bottom of the height to be measured; ψ is the latitude, $H = h + h'(T - T')/5412$, and a is the earth's radius, then given over exactly as 6366198 meters.[95] Laplace had taken a strong interest in the barometric measurement of heights around the time that French engineers levelling the Simplon tunnel found perfect agreement between their trigonometrical measurements and an earlier hypsometric formula devised by H.B. de Saussure. Laplace's complicated improvement of the formula made it possible to obtain heights with great fidelity with barometers readable to a small fraction of a ligne (a ligne is 1/12 of an inch). People with a taste for heights and calculation were soon measuring all the mountains they could climb, also church towers, flag poles, and the masts of sailing ships. Reference works of the time, like the *Annuaire* of the French Bureau des Longitudes, published these heights, which at once advertised and routinized the accomplishments of the "physiciens géomètres," the exact experimental physicists, of the early 19th century.[96]

You must know logarithms, as well as like to calculate, to use Laplace's hypsometric formula. Although almost two centuries old in 1800, logarithms had not been applied widely outside astronomy and geodesy. Missionary mathematicians then undertook to spread the blessing. Laplace guessed that a suitable instrument, "simple and cheap," would make their use "extremely popular," and not only in physical science. The assiduous computer who perfected the coefficients in Laplace's formula, the baron Louis Ramond de Carbonnières, who was also a political scientist, botanist, and geologist, wrote rhapsodically about logarithms: "This barême of learned

95. Biot and Arago, *MIF*, 1806, 309–312; Dhombres-Firmas, *JP*, 75 (1812), 265, 273–274.
96. Ramond, *Mémoires* (1811), v, ix; *JP*, 81 (1815), 234–238, an extract of the *Annuaire* for 1815; Truchot, *ACP*, 18 (1879), 308, on reading barometers to 1/100 ligne.

men would certainly not be restricted to mathematicians alone if it were known how easy it is to use, if people knew better the numberless [!] conveniences it can effect in the most ordinary business of civil life."[97] The logarithm and the slide rule did make their way into physics early in the 19th century. They may figure as symbols of its links to applied mathematics just as gases and caloric indicate its connection to chemistry.

---

97. Laplace, *Oeuvres* (1878), *14*, 134; Ramond, quoted by Dhombres-Firmas, *JP*, *75* (1812), 275; Ramond, *Mémoires* (1811), xii.

2 SOME MEANS TO THE END

Laplace's disciple Jean-Baptiste Biot could date the invention of straight thinking in his country almost to the year. "It is to Borda and Coulomb that we owe the renaissance of true physics in France."[1] These men, one a naval officer, the other an engineer, and both academicians, had invented ingenious and precise instruments during the 1770s and 1780s and, what was much more, had shown what could be done with them. Coulomb's painstaking investigations of torsion in wires and threads led to his famous balance for measuring electric and magnetic forces and for calculating how much his body, electrically considered, disturbed his observations. Coulomb's electrical balance will be described presently, in company with other instruments, newly precise around 1800, used to measure the properties of imponderables. First come improvements in Borda's line of invention, precision and practicality in the measurement of angles.

1. NEW ANGLES ON ANGLES

Borda's main contribution to the art of exact measurement stood at the intersection of three trends in 18th-century science. With one of them, the continually increasing prominence of mathematical thinking, we are already sufficiently familiar. Here Borda marched in the van of what his colleagues regarded as true physics. "In all

---

1. Biot and Rossel, *Biographie universelle*, 5 (1812), 60, s.v. "Borda," quoted by Frankel, *HSPS*, 8 (1977), 41; cf. Daumas, in Crombie, *Sci. change* (1963), 428–429. Biot's coauthor, E.P.E. Rossel, was an expert in nautical astronomy, hydrography, and geography, Borda's main subjects.

his inventions we recognize the mathematical physicist [*physicien géomètre*], who knows how to apply calculation to experiment and to attain the greatest precision by the simplest methods."[2] The second trend was the refinement of angular division, which had proceeded apace for a century, since Jean Picard and his colleagues at the Paris Observatory set the foundations of precision astronomy.[3] The third trend was the preoccupation of the maritime powers with finding their way at sea. Astronomy is the most useful of the sciences, said Napoleon, "both to the mind and to commerce." And to other exercises as well. "Spending the night between a beautiful woman and a clear sky, and the day reconciling observations and calculations, seems to me heaven on earth."[4]

As easy as π

Picard's major innovation was to put lenses in the empty pinnules on the alidade or sighting rule of graduated circular sectors. The improvement would not have made an epoch in astronomy, however, had it not been coupled with micrometer eyepieces. The idea of inserting some movable marks in the common focal plane of the objective and ocular of a telescope to measure small angular differences had occurred to several astronomers in Britain and France before the late 1660s, when Picard's colleague, Adrien Auzout, designed a simple and effective arrangement consisting of fixed and movable threads.[5] The introduction of instruments of angular measurement containing lenses by French astronomers can be dated precisely. On the day of the summer solstice in 1667, they drew a meridian line at the site of the future Paris Observatory and proceeded to observe using an instrument with pinnules (open sights). Fifteen weeks later, on October 2, Picard took the altitude of the sun for the first time with a quadrant and a sextant furnished

2. Biot and Rossel, *Biographie universelle*, 5 (1812), 60.
3. This periodization is commonplace: Delambre, *Hist. astr. moderne*, 2 (1821), 598–601, 611–612; Danjon and Couder, *Lunettes* (1935), 626; King, *History* (1955), 99 ("the new school of precision astronomy"); and the collection edited by Guy Picolet, *Jean Picard et les débuts de l'astronomie de précision* (1987), esp. Lévy, ibid., 139 ("le créateur de l'astronomie moderne").
4. Napoleon to Lalande, 5 Dec 1796, quoted by Fischer, *Napoleon* (1988), 39.
5. McKeon, *Physis*, 13 (1971), 246–266, and 14 (1972), 228–230; Repsold, *Astr. Messw.* (1830), 1, 40–43, and King, *History* (1955), 96–99; Chapman, *Dividing the circle* (1990), 26–30, 40–44.

with lenses. At first he and his colleagues merely inserted lenses in the pinnules of existing instruments; but in 1668 they built a much smaller instrument (the quadrant and sextant had radii of 9.5 feet and 6.0 feet, respectively) to carry a telescope with micrometer.[6] In 1679, Robert Hooke installed a zenith sector having a telescope fitted with a micrometer net in his vain search for stellar parallax; by the late 1670s, John Flamsteed's instruments at the Royal Observatory in Greenwich bore telescopic sights.[7]

With Auzout's micrometer, Picard could measure the apparent horizontal diameter of the sun with an accuracy, as determined by modern analysis, of between five and ten seconds of arc. Some three seconds of this error were accidental, that is, arose from errors in reading, while five or more were systematic, the consequences of imperfection in the micrometer screw, of flexure of the mounting of the telescope, and of other fixed features of the instrumentation.[8]

Picard's accuracy of, say, ten seconds in obtaining the size of the sun was at least six times better than the previous standard. In proof whereof can be cited Gian-Domenico Cassini's rationale for building a very precise meridian line in the great church of San Petronio in Bologna in 1656. According to Cassini, no telescope then existing could be relied upon to determine the diameter of the solar disk to within a minute of arc. Astronomers of the time had a special interest in this quantity. The change in the apparent size of the sun during its annual motion was expected to be about one minute if it obeyed Kepler's laws.[9]

Our ball of mud provided further opportunities for the application of Picard's improved astronomical instruments. "We must look in the heavens [he said] for the measure of the earth." He used a ten-foot zenith sector with which he did not do very well, for reasons easily imagined from a picture of the instrument at work (figure 2.1.1).[10] Although the scale could be read to 20 seconds of arc, the insecure mounting of the instrument and the uncomfortable

6. Olmsted, *Isis, 40* (1949), 224–225; McKeon, *Physis, 14* (1972), 228.
7. Chapman, *Dividing the circle* (1990), 38–39, 49–50.
8. Débarbet, in Picolet, *Picard* (1987), 161–168; cf. McKeon, *Physis, 13* (1971), 276–277.
9. Cassini, *Meridiana* (1695), 15–17.
10. Picard, *Mesure* (1671), 165 (quote), 172; Repsold, *Astr. Messw.* (1830), 1, 44, 65.

38 / SOME MEANS TO THE END

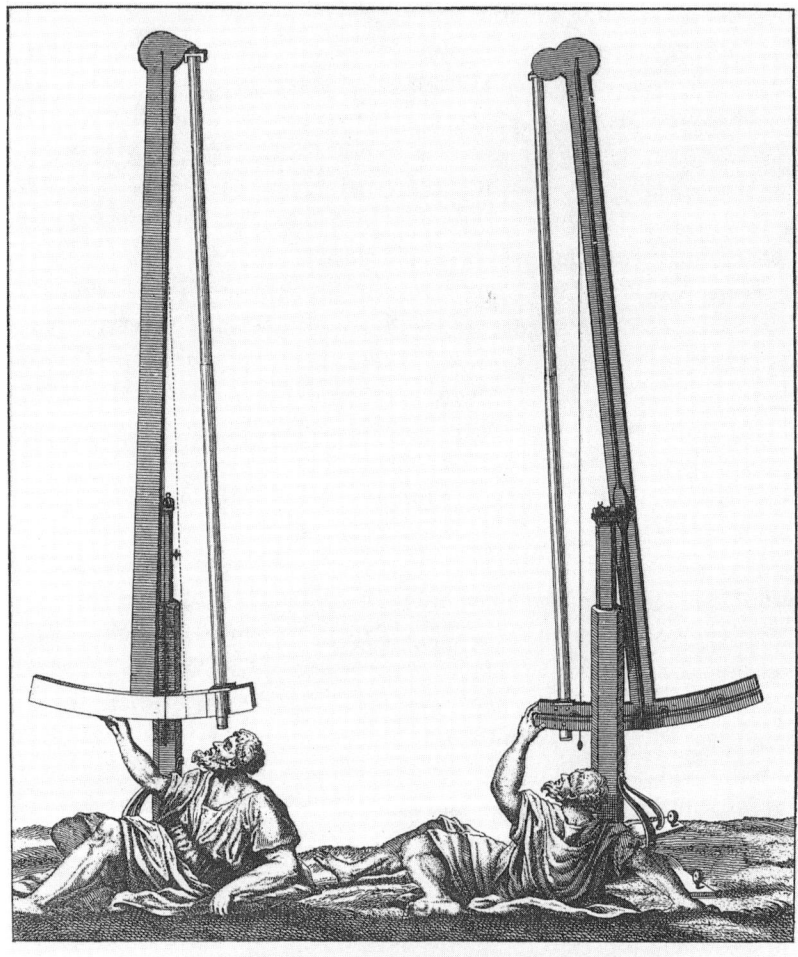

Picard's Sector, 1669,
nach Picard in Mém. Paris avant 1699.

FIG. 2.1.1   Picard's 10-foot zenith sector in use; the angle is read from the division where the plumb line cuts the scale. Note the technique of reversing the instrument to compensate for errors in levelling, alignment, and graduation. Repsold, *Astr. Messw.* (1902), *1*, 44, after Picard.

attitude of the observer helped to throw off Picard's determinations of latitude by as much as 5 minutes. For surveying in the plane, Picard used a quadrant with a diameter of 38 inches (1.02 m) that allowed measurements to 15 or 20 seconds with the help of transversals. Retrospective calculations show that his angles were in fact good to 20 seconds.[11] Even with his errors in latitude, Picard's geodetic work had an accuracy perhaps 20 to 30 times that of his predecessors.[12] His instruments made possible the realization of the ambitious program of geodetic measurements planned for the early years of the Paris Academy of Sciences as a contribution to the advancement of science and the inventorying of the real estate of the realm.[13]

Improvements beyond Picard's came slowly. Two obvious avenues—checking graduation by dividing an entire circle rather than a quadrant or sector and making and mounting instruments more robustly—were opened by Picard's younger colleague Olaus Rømer, but few followed for many decades.[14] The best of the situation around 1700 may be indicated by the qualities of the 7-foot mural quadrant divided to 5 minutes on the limb (and by transversals to seconds) by Abraham Sharp, "the first person that cut accurate and delicate divisions upon astronomical instruments," for Flamsteed in Greenwich. Estimates of its errors range from 11 seconds to over half a minute: Sharp had done his job well, but the quadrant bent under its own weight and when it warmed or cooled its brass limb and iron backing dilated or contracted at different rates. Also, the wall that carried it slowly sank under its weight. By 1715 the wall had descended so far that Flamsteed had to add 14′20″ to all observations made with the instrument. With corrections and inspired adjustments, it provided the information that allowed Newton to develop, and confirm, his theory of the motions of the moon.[15]

The replacement of Sharp's arc in 1725 by an 8-foot quadrant built by George Graham, a clockmaker trained by Thomas Tompion, marked an epoch in the perfection of astronomical

11. Taton, in Picolet, *Picard* (1987), 209, 218; Levallois, in ibid., 234–235, 240–243.
12. Delambre, *Hist. astr. mod.* (1821), 2, 602–603, 612–617, 621–617, 621–624, 627.
13. Olmstead, *Isis*, 40 (1949), 224; infra, §4.1.
14. Repsold, *Astr. Messw.* (1830), 1, 48–54; Lévy, in Picolet, *Picard* (1987), 137; Chapman, *Dividing the circle* (1990), 63–65.
15. Smeaton, *PT*, 76 (1786), 6 (quote); King, *History* (1955), 108; Chapman, *Ann. sci.*, 40 (1983), 463–468, and *Dividing the circle* (1990), 54–60.

instruments. Graham set the design and many of the techniques that would be followed by the most successful circle dividers in the world—John Bird, Jeremiah Sisson, and John Troughton—whose work continued his to the end of the century. Graham's rigid trellis mounting probably increased accuracy by a factor of two, to 7 or 8 seconds (on an 8-foot qudrant readable to 5 seconds); and the switch to solid brass construction, also pioneered by English makers, helped to realize another factor of four, bringing the accuracy of determination of angular separations to 1 or 2 seconds of arc, for the very best fixed instruments. The meticulous John Bird, who made several 6-foot and 8-foot quadrants around the middle of the century, engraved the most delicate lines on his creations only in the morning, in the spring or in the fall, with a compass he had set the night before and allowed to come into full thermal equilibrium with the piece under construction. He would allow but one assistant to work with him, lest the combined body heats of several should expand the limb during division.[16] Having learned that Bird was over 50 and afflicted so much "with the Gravel...that he cannot now bear the motion of a carriage," the director of the Radcliffe Observatory at Oxford urgently petitioned his Vice Chancellor for money to commission two mural quadrants and two sectors from the ill and aging artist. "[He is] the one person living, who is capable of making them with that precision and accuracy, on which the goodness of the observations essentially depends.[17] Figure 2.1.2 illustrates his handiwork.

Another of Graham's important innovations was to drive division by bisection as far as possible. The trick cannot easily be played in a quadrant graduated in degrees; but it can be carried directly to quarter divisions in one graduated into 96 primary parts. Graham would draw an arc a little longer than a quadrant with a beam compass and, picking a spot close to one end, would set one point of his compass there without changing its opening. He could then strike off an angle of two-thirds of a quadrant, and, bisecting that, could add the pieces together to construct a right angle. To divide

16. Shuckburgh, *PT, 83* (1793), 75; King, *History* (1955), 109–118; Daumas, *Instruments* (1953), 234–235, 258; Chapman, *Ann. sci., 40* (1983), 470, and *Dividing the circle* (1990), 70–73.

17. Hornsby, *Proposal* (1771), 1.

FIG. 2.1.2 Bird's 8-foot mural quadrant, 1753, with a mechanism for rotation in azimuth in the style of Le Monnier. Repsold, *Astr. Messw.* (1902), *1*, 61, after Le Monnier.

this angle into 96 parts, he had only to return to the two-thirds quadrant and bisect it six times; the result, an angle 1/64th of two-thirds of a quadrant, was the desired 1/96th of a right angle. Two more bisections of the 1/96th part created an angle smaller than 15 minutes of arc by almost exactly one minute. Graham also divided his right angle into the conventional 90 parts, subdivided both the 90-part and the 96th-part scales to a twelfth of a part (that is, to 5 minutes of arc), added a vernier, and used one scale to check the other. They never disagreed by more than 5 or 6 seconds, which made an important improvement, a factor of two, over Sharp's very best divisions.[18]

Bird extended the principle of successive bisections to the 90-degree scale by beginning with an angle of 85°20′. After 1024 bisections—it took Bird 52 days at least to graduate a quadrant—he arrived at an angle of 5′. But how construct the original 85°20′? Bird computed the lengths of chords that would subtend angles of 4°40′, 10°20′, 15°, 30°, and 42°40′ at the center of the quadrant of radius r he had under construction. He made an auxiliary linear scale of great accuracy, readable to one-thousandth of an inch by vernier, with which he could set the opening of his compass to the calculated lengths of the chords. He made a 60° angle in Graham's manner, then added 30°, subtracted 15°, and added 10°20′, all from the corresponding chords, to generate his 85°20′. He confirmed the accuracy of the work by adding 4°40′ to obtain the full quadrant and by subtracting 42°40′ twice to attain zero. He also inscribed divisions in 96 parts for continual cross-checking in Graham's manner. In another sort of check, the Astronomer Royal, James Bradley, observing in Greenwich, and Tobias Mayer, doing the same in the observatory at the University of Göttingen, arrived at values of the obliquity of the ecliptic that differed only by two seconds of arc. Both used Bird's instruments.[19]

The estimates of accuracy so far presented refer to single, absolute measurements (not small-angle differences) made on quadrants or sextants of 6- to 8-foot radius. Better results could be obtained in special circumstances. A capital example is the 12.5-foot zenith sector made by Graham in 1727 for James Bradley. This instrument was to zenith sectors what Graham's Greenwich instrument was to mural quadrants. Firmly pivoted between two walls, the iron tube of the telescope could be displaced by a fine micrometer screw. The mechanism probably permitted accurate measurements of angular

18. Chapman, *Dividing the circle* (1990), 67–70.
19. Ibid., 71–76, 96.

distances as close as 1.5 seconds. With the help of this sector, Bradley brought to light the annual aberration of stars (which has a maximum value of 20 seconds) and the nutation of the earth (which has a maximum value of 9.2 seconds and a period of 19 years.)[20] By 1750, the precision astronomy built on the innovations of Picard's group could boast some notable discoveries and, in the best instruments of English manufacture, an exactness of measurement of close to one second of arc, an improvement by a factor of ten or twenty in some eighty years. This exactness in manufacture became accuracy in observation through protocols developed by Bradley that used zenith sectors, quadrants, and transit instruments as independent cross checks on one another.[21]

The next forty years saw a similar change, which, as in many other fields of applied mathematics, accelerated around 1770. The limit of accuracy of an absolute measurement made with the equipment of classical astronomy is just under a second of arc.[22] The reliable attainment of this limit rested on the introduction of achromatic lenses; on mechanical improvements owing to the diminution of instruments made possible by the new lenses; and, above all, on the replacement of quadrants by full circles. By graduating over 360°, the maker could check the faithfulness of diametrically opposed marks. The earliest important instruments so divided were the work of Jesse Ramsden, who married into the achromatic Dollond family. He proceeded tediously, by locating approximately the major points of division by compasses and then relocating them under the eyepiece of a micrometer microscope into their correct diametrically opposite positions. It took him 150 days to come full circle. His second major machine in this style, an equatorial completed in 1793 for George Schuckburgh, had a probable error in graduation of 1.5". That would be the case, however, only if Shuckburgh read the scale through one of the six micrometer microscopes with which Ramsden enriched his handiwork. By

20. Bradley, *PT*, 35 (1729), and *PT*, 40 (1748), in Bradley, *Misc. works* (1832), 5, 11, 14, 21–31; Rigaud, "Memoir," ibid., xii–xiii, lxii–lxx; Danjon and Couder, *Lunettes* (1935), 630–631.
21. Chapman, *Dividing the circle* (1990), 83–85, 88–89.
22. Daumas, *Instruments* (1953), 230, 238, 255; Chapman, *Jl hist. astr.*, 14 (1983), 134, and *Dividing the circle* (1990), 109; Danjon and Couder, *Lunettes* (1935), 333, 338, 625.

reading all six and applying appropriate corrections, Shuckburgh could have confidence in the result to 0.5″. Even Bird had had to accept errors of 3″. The tedium of Ramsden's method was very much reduced by Edward Troughton, who introduced a gear-driven roller to find the preliminary divisions and tabulated the errors between these divisions and those required by the microscopic inspection of diametrically opposite points. He then calculated where the definitive divisions should go, and managed to put them there, he claimed, in only 13 8-hour working days.[23]

Ramsden's first full-circle machine was the famous theodolite that entered into operation in 1787 (figure 2.1.3). Commissioned for precise geodetic work, it effectively moved the observatory into the field. With its heavy base plates and sturdy conical struts, it weighed 200 pounds; with its stands, stools, pulleys, and other accoutrements, 400 pounds. Ramsden knew how to protect against the bends of flimsy instruments. The horizontal circle, divided throughout to 15 minutes and readable by micrometer to 1 second, had a diameter of 3 feet. The instrument had a probable error in a single observation of 2 seconds of arc, or under four feet, at 70 miles. Marc Auguste Pictet, a Genevan physist who observed Ramsden's theodolite in action, pronounced it a national treasure; "it is a matter of doubt to me, whether this precious instrument would be allowed to go out of the kingdom."[24] Ramsden made a second grand theodolite, divided to 10 minutes of arc, and so sensitive that a slight breeze, by cooling the limb unequally, could shift the readings of the microscopes differentially.[25]

The works of the great English makers from Graham to Ramsden were the envy of the world. In the 1780s, Jean-Dominique Cassini, the fourth of the name to head the Paris Observatory, sought instruments with which to refurbish the family fief. During a century and more, the Cassinis had not been able to raise up French artisans as able as the English. "Ramsden and the Dollonds are geometers and scientists," he said, "our best artisans are only

23. Shuckburgh, *PT*, *83* (1793), 93–94, 98; Chapman, *Dividing the circle* (1990), 113–116; cf. Westphal, *Zs. f. Instr.*, *4* (1884), 154.
24. Roy, *PT*, *80* (1790), 146–149, 159; Richeson, *English land measuring* (1966), 180–183; Chapman, *Dividing the circle* (1990), 118–119; Pictet, *PT*, *81* (1791), 114; infra, §4.2.
25. Williams et al., *PT*, *85* (1795), 446, 449.

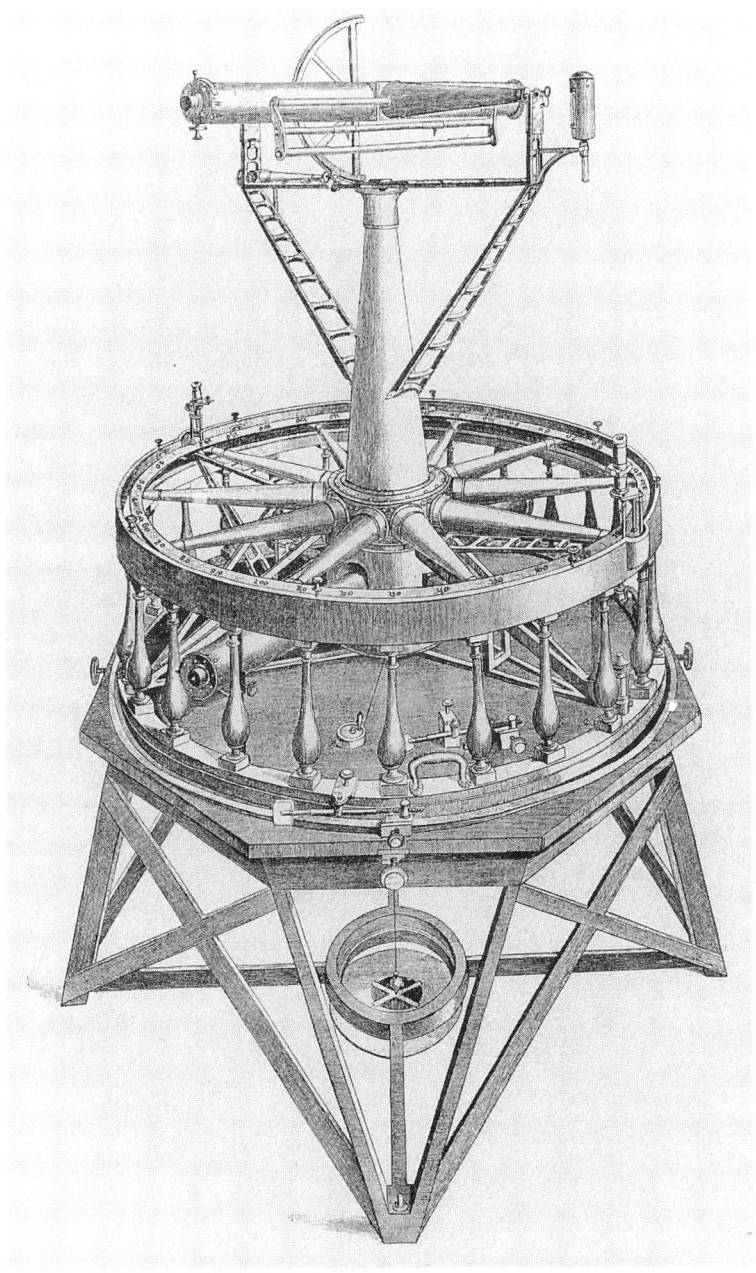

FIG. 2.1.3 Ramsden's 3-foot theodolite, 1787. Repsold, *Astr. Messw.* (1902), *1*, 86, after Ramsden.

workmen." After a visit to Ramsden's shop he wrote the official who oversaw the Observatory's expenditures: "It will be difficult to equal or imitate his work." In 1788, the French reluctantly ordered two instruments for the king's astronomers from the ancient English foe.[26]

They had to wait, until 1803 to be precise. Ramsden took his time and overcommitted his establishment: when the French gave their commission, he had had in hand a meridian transit for the Duke of Saxe-Gotha for five years. The Duke had not sat so idly as Ramsden. When reminders and demands failed, he tried to procure an 8-foot mural quadrant that Bird, then (1786) dead a decade, had made for one Pierre Jacques Onésine Bergeret, who had lent it to the Ecole royale militaire. When Bergeret died, his family put the quadrant up for auction. Several big bidders entered, including the Austrian government of Lombardy (in favor of the Brera Observatory), the Duke of Saxe-Gotha, and the King of France. The bidding went high. "It is the largest and best astronomical instrument we have," the astronomer Joseph-Jérôme Lalande wrote the French minister who put up the money to acquire it. The Duke drove up the price. "The Duke sets the highest price on the possession of this instrument," wrote the Duke's agent to the same minister, pointing out the faithful service that the Duchy of Saxe-Gotha had always rendered the Kingdom of France. The minister decided to retain Bird's masterpiece. The Duke's agent saw to it that the cost of the keeping was very high.[27] Thus the procuring of precision astronomical instruments had become a matter of state, and the French received a reminder they did not need that their inferiority to the English in exact instrumentation exposed them to embarrassment if not to jeopardy.[28] A visit of the director of the Paris Observatory and two leading French astronomers to London early in 1788 provided a further opportunity to measure the disparity. "They saw

26. Cassini, as quoted in King, *History* (1955), 229–231, from Wolf, *Histoire* (1902), 288, 292.

27. David, *Dix-huit. siècle*, 14 (1982), 279–284; Tagliaferri and Tucci, *Incontri*, 4 (1989), 67, 73–4, quoting correspondence of Oriani (for the Brera) and Lalande, 17 June and 3 Jul 1786.

28. For efforts to improve instrument making in France, see *JP*, 6 (1775), 267, announcing competition for a monetary prize put up by the King, requiring a 3-foot quadrant with all the trimmings; Daumas, *Instruments* (1953), 353–360; and Gillispie, *Science and polity* (1980), 117–122.

trained technicians [artistes] that we do not have here [in France] as well as instruments quite different from ours."[29]

Completing the circle

It was Borda's glory to help to arrest the decline that had afflicted France since Picard had led the world in the design of precision instruments. Borda, or rather Borda and one of France's few excellent instrument makers, Etienne Lenoir, invented a worthy competitor to Ramsden's theodolite. During the 1770s, Borda improved the navigational instrument then becoming standard, the octant or sextant, by extending it to a complete circle. The octant, invented in 1731 by an Englishman, John Hadley, requires for successful operation that two mirrors, one of them half-silvered, be set exactly perpendicular to the plane of the instrument (figure 2.1.4a). These requirements at first proved too much even for good English makers; but gradually they learned the necessary tricks and improved the geometry to the simple layout schematized in figure 2.1.4b. Light from the sun at S at altitude α strikes the mirror M, carried on an arm pivoted at O; when $\angle MON = \alpha/2$, the ray sent from M to N is reflected to the eye at E. Only half the surface of N is silvered; through the transparent part, the eye looks to the horizon H. The angle α could be read from the scale around O to a minute of arc.[30] The point of the exercise was to bring the sun (or a star) to the sea, to allow measurement of its altitude without gazing at it and without the uncertainty of trying to hold the horizon on a pitching ship.

In the middle of the 18th century, the calculating astronomer expected to deduce not only latitude but also longitude from sightings of celestial objects. Tobias Mayer, of Göttingen, hoped to win a portion of the prize for a method of finding longitude at sea offered by the British Parliament in 1714. By 1754 he had a theory of the moon's motion accurate to about 30 seconds of arc. Could the octants then in use give sightings with corresponding exactness? Mayer, reflecting French opinion, doubted it. He turned out to be wrong. In order to obtain the elements for computation of the

29. Charles Messier to Magellan, 2 Feb 1788, concerning a visit by Cassini, Legendre, and Méchain, in Carvalho, *Correspondência* (1952), 153.

30. Daumas, *Instruments* (1953), 241–242.

48 / SOME MEANS TO THE END

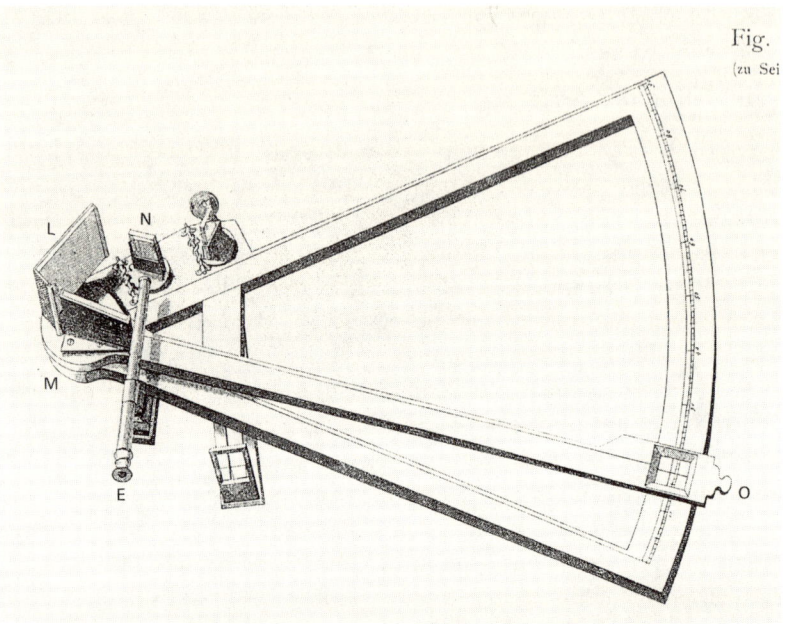

FIG. 2.1.4a Hadley's octant, 1731. M is the mirror that catches the sun's light, N is the half-silvered mirror, L a screen or shade, E the eye of the observer, O the crossing of the scale and the alidade carrying M. Repsold, *Astr. Messw.* (1902), 1, 76, after Hadley.

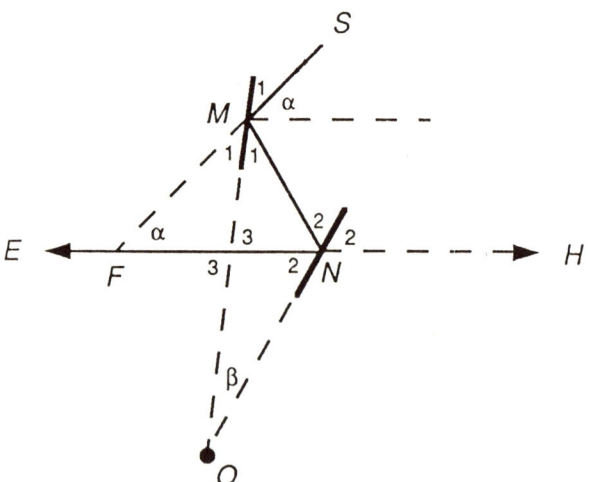

FIG. 2.1.4b The geometry of Hadley's octant. From the equal external angles 3, $\alpha + \angle 1 = \beta + \angle 2$; also, $2\angle 2$ is an external angle to $\triangle FNM$, whence $2\angle 2 = \alpha + 2\angle 1$. Therefore $\beta = \alpha/2$. S signifies the sun, H the horizon; E, M, N, and O have the same meanings as in figure 2.1.4a.

longitude with sufficient accuracy for use in his lunar theory, navigators had to be able to observe the moon's position to within one minute of arc; and they needed an instrument to the purpose at a price they could afford. It appears that Hadley may have been inspired by Edmund Halley's remark of 1731, that the main obstacle to obtaining longitude to within a degree was the want of an inexpensive reliable instrument for measuring the moon's altitude accurately at sea. The first trials of the octant, on which Bradley participated, showed it capable of getting altitude to about one minute—and that with open sights.[31]

The major remaining problem after Bird and others added telescopic sights and expanded the octant to a sextant in the late 1750s was financial. Ramsden solved it. Building on earlier ideas for division of arcs by machine, he devised an engine consisting of a carefully graduated circular plate rotatable through an angle of 10 minutes by turning a worm gear that engaged some of the 2160 teeth on the plate's circumference. He fixed the item to be graduated concentrically on the plate; each time the plate advanced, he made a mark on the limb of the work. A scale attached to the mechanism made possible further divisions to 10 seconds. Ramsden had the first of these "dividing engines" in operation in 1767; seven or eight years later he had an improved version, which easily graduated octants or, because Hadley's instrument by then had expanded by a third, sextants, to an exactness well within a minute of arc and with an expenditure of an hour and a half of time. For this invention, Ramsden received 615 pounds from the Commissioners of Longitude on condition that he make his method public.[32]

In operation, Ramsden's largest sextants had an error of about 6″. Even a small instrument, of six-inch radius, could be read to 30″ and, with skill, judged to 10″ or less. Troughton's 8-inch sextant, turned out with his version of Ramsden's dividing engine, could be read to 5″.[33] These numbers made enthusiasts. Franz Xaver von

31. May, *Naut. mag.*, 145 (1945), 21–23, 26; David, in Zach, *Briefe* (1938), 52, refers to a depreciation of the sextant by N.L. de La Caille, who died in 1762. Martin Alois David (1757–1836) was an assistant at, and later the director of, the Prague observatory.

32. Daumas, *Instruments* (1953), 264–267, and in Crombie, *Sci. change* (1963), 422; Stimson, *Vistas*, 20 (1976), 126–128; Chapman, *Dividing the circle* (1990), 130–132.

33. Ibid., 38–39, 52; Chapman, *Dividing the circle* (1990), 135.

Zach, director of the Giotha observatory, publicist for astronomy and geodesy, and well-nigh a crank for his enthusiasm for English instruments, formed what he called a Hadley Society for the Advancement of Sextants. Zach's correspondents used the instruments on land, to find their latitude. "What can be more striking, convincing, and decisive than your finding [so he wrote a fellow Hadleyan] that the height of the pole at Prague was incorrect!" In Bohemia, a 7-inch sextant of English manufacture did better than a large fixed quadrant. "Amateurs and observers with reflecting sextants are continually increasing." Zach gave a Russian military officer a 5-inch Dollond sextant and instructions for its use. The officer took it proudly to Petersburg, "the first sextant of this terrestrial kind ever in Russia."[34]

As for Zach himself, he never travelled without a sextant and a chronometer ("a portable observatory"), or returned to his headquarters at his stationary observatory on the Seeberg near Gotha without having fixed anew the latitude and longitude of three or four towns. "Travellers should learn the language of the heavens just as they study the language of the countries they must pass through," advised the geodecist Cassini de Thury, who agreed with Zach in little else; "[latitudes and longitudes are] the most important fruit that can be gathered on a trip."[35] Cassini's rival, William Roy, counselled gentlemen and ladies disoriented in London to resort to the marvelous reflector. "[Anyone] desirous of knowing exactly his own situation in this great metropolis may easily satisfy himself, by taking two angles from the top of his house, with a good Hadley's sextant."[36]

Meanwhile, in the summer of 1754, Mayer, despairing of octants, had hit on an improvement based upon a geodetic instrument he had devised two years earlier. In the earlier version, which he recommended for its combination of accuracy with compactness,

34. Zach to David, 8 Jul 1795, 19 Sep 1796, and 3 Oct 1796, in Zach, *Briefe* (1938), 102, 131 ("Hadleysche Gesellschaft"), 132–133. Further expressions of enthusiasm may be found in Lecoq to Gauss, 3 Feb and 22 Mar 1799, in Gerardy, Ak. Wiss., Göttingen, *Nachrichten*, 1959:4, 39, 42.

35. Zach to David, 20 Oct 1791, 16 Jul 1793, and 30 Apr 1796, in Zach, *Briefe* (1938), 62–63, 84, 115; Friedrich, *Peterm. geogr. Mitt.*, 117:2 (1973), 147, 149 (quote), 153; Cassini de Thury, *Relation* (1775), 38 (quote), 45 (quote).

36. Roy, *PT*, 80 (1790), 258; cf. Topping, *PT*, 82 (1792), 99–100.

two similar sticks AB and CD with fine holes bored at their ends pivot from a common axis (figure 2.1.5a). To measure the angle between the distant points P and Q (figure 2.1.5b), line up the holes on the sticks and point them, through the telescope attached to the upper one, at P. Fix the lower stick CD and move the upper to look at Q; rotate the sticks in place until the telescope again points to P; move the telescope back to Q so that the lower stick lies at C'D'. The angle between the sticks is now twice the angle between P and Q. The game may be continued as long as desired, through x circles

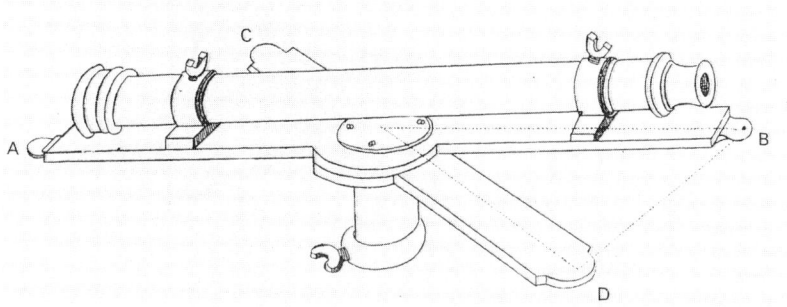

FIG. 2.1.5a Mayer's geodetic instrument, 1752, for multiplying the angle between distant objects sustended at its center. Repsold, *Astr. Messw.* (1902), *1*, 77, after Mayer.

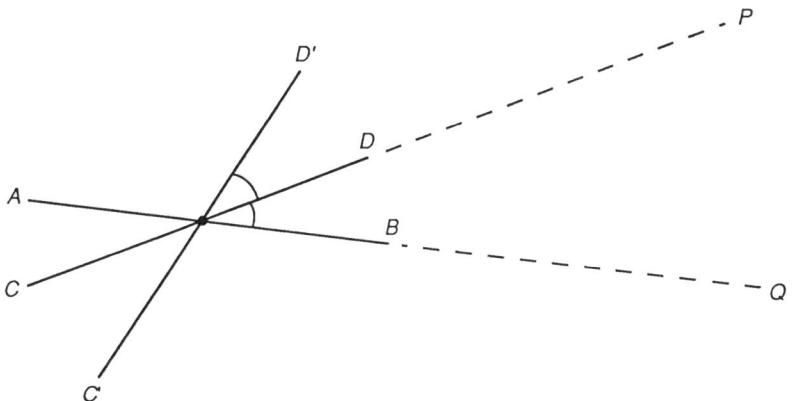

FIG. 2.1.5b The geometry of the duplication of an angle by Mayer's geodetic instrument.

52 / SOME MEANS TO THE END

in n sightings. Then the angle between P and Q is $360° \cdot x/n$. Each complete circle destroys mistakes in graduation; Mayer claimed that it improved accuracy in geodetic measurements from a customary (and perhaps exaggerated) 5 minutes to a most commendable 10 or 15 seconds.[37]

The later marine version of Mayer's repeating circle appears to greatest advantage in the description by its most enthusiastic developer, Charles Borda (figure 2.1.6a). The rule M carried the fixed mirror m as in the movable arm of Hadley's octant; but in contrast to the standard sea-going instruments, the telescope N

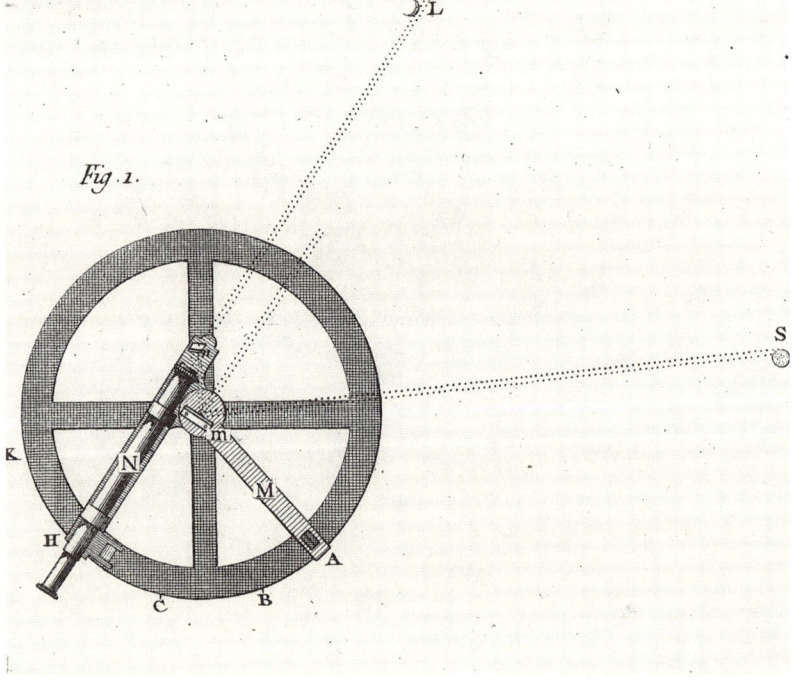

FIG. 2.1.6a Borda's first reflecting circle. The field mirror m rotates on the alidade M; n, which corresponds to the half-silvered mirror of the Hadley octant, rotates with the index carrying the telescope H. Borda, *Description* (1802), pl. 1.

37. Mayer to Euler, 23 June and 7 Sep 1754, in Forbes, *Euler-Mayer corr.* (1971), 88, 92; Mayer, Akad. Wiss., Göttingen, *Comment.*, 2 (1752), 325–331.

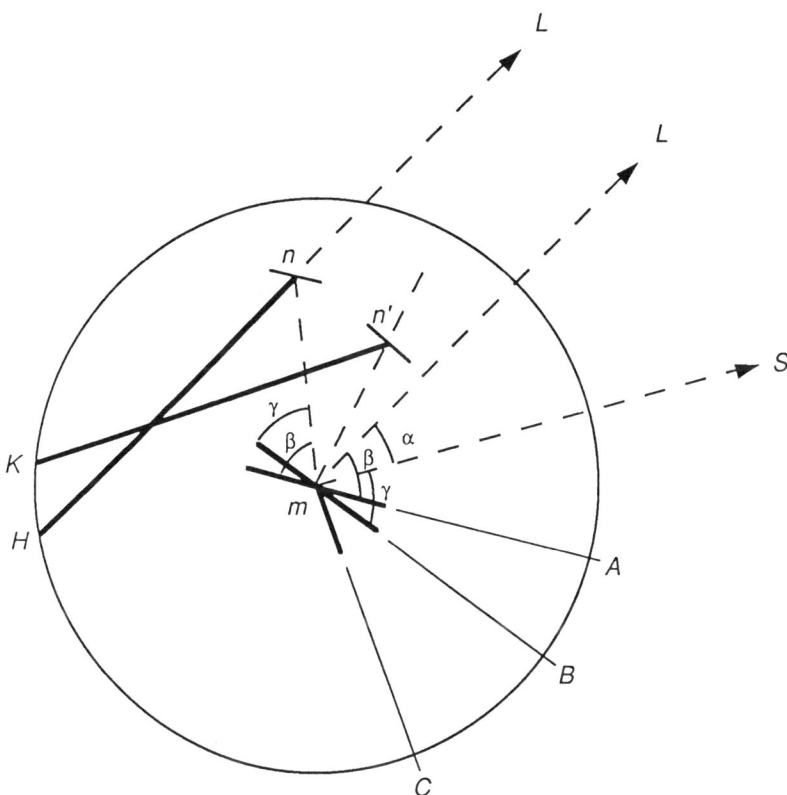

FIG. 2.1.6b Geometry of Borda's first reflecting circle. A, B, C, H, K have the same significance as in figure 2.1.6a; mn is the reflected beam from m when the telescope is at H, mn' the same when the telescope is at K. Since AB is equal to $\beta-\gamma$ and also to $\gamma-(\beta-\alpha)$, $\alpha = 2$ AB.

and the small mirror *n* can also move. The circumference of the circle, 8 inches in radius, is graduated into 720 parts, each divided to 15 minutes and readable by vernier to 2 or 3 minutes. Point the telescope at one of the moon's limbs L and move the rule M until the reflection of L from *m* to *n* falls on the direct image as seen from H. The mirrors *m* and *n* must then be parallel. Clamp N at H and rotate M clockwise until, at B, the reflected image of the star S coincides with the direct image of L. The angular separation α is twice the angle subtended by the arc AB, or, since the circle is graduated into 720 parts, α = AB as marked (figure 2.1.6b). Now reset the instrument by moving the telescope (to K, say) where the direct and reflected images of L coincide. Then move M to C, where the reflected image of S coincides with the direct image of L. We have AC = 2α. Mayer recommended repeating the procedure a half-dozen times, when, he said, the error arising from incorrect graduation even in an 8-inch circle should sink to under half a minute of arc.[38]

Mayer sent a description and model of this device to Britain in 1755 to buttress his application to the Commissioners of Longitude. A sea trial did not reveal that his instrument had any advantage over a standard octant with a large radius. But the captain charged with the test did not follow Mayer's instructions, which were inconvenient to attempt on shipboard. Nor did the repetitions around the circle, to minimize the effect of poor engraving, seem very important to the British; that difficulty, according to Bradley, by then the Astronomer Royal, "might be sufficiently removed by the care and exactness with which Mr Bird is known to execute those [instruments] that he undertakes to make."[39]

Perhaps because France had no Bird in hand, Borda could more readily appreciate the possibilities of Mayer's reflector than the British did. He observed that the greatest inconvenience of the instrument, and a major source of error, was the need to reset the mirrors to parallel after each measurement; and he pointed out that both blemishes could be removed merely by shortening the telescope. If, as in figure 2.1.7, N terminates before the center of the circle, the reflected image of a star to its left as well as to its right can be received in the eyepiece. That obviates the need to reset. With N at H and M at A, the reflected image of S and the direct image of L coincide in the telescope. Turn the entire instrument to direct the telescope to S; move M to B, to send the reflected image of L to H; twice the arc AB equals the angular separation of L and S. The telescope can again be directed at L, and M moved so as to bring the image of S to the eyepiece; the instrument rotated to point the telescope to S; M rotated, to C, to accomplish coincidence; and so on.[40]

38. Mayer, *Tabulae motuum* (1770), 21–27; Borda, *Description* (1802), 7–8.
39. Bradley to Secretary of the Admiralty, 14 Apr 1760, in Mayer, *Tabulae* (1770), cxiii-cxiv; Daumas, *Instruments* (1953), 242–243.
40. Borda, *Description* (1802), 11–12.

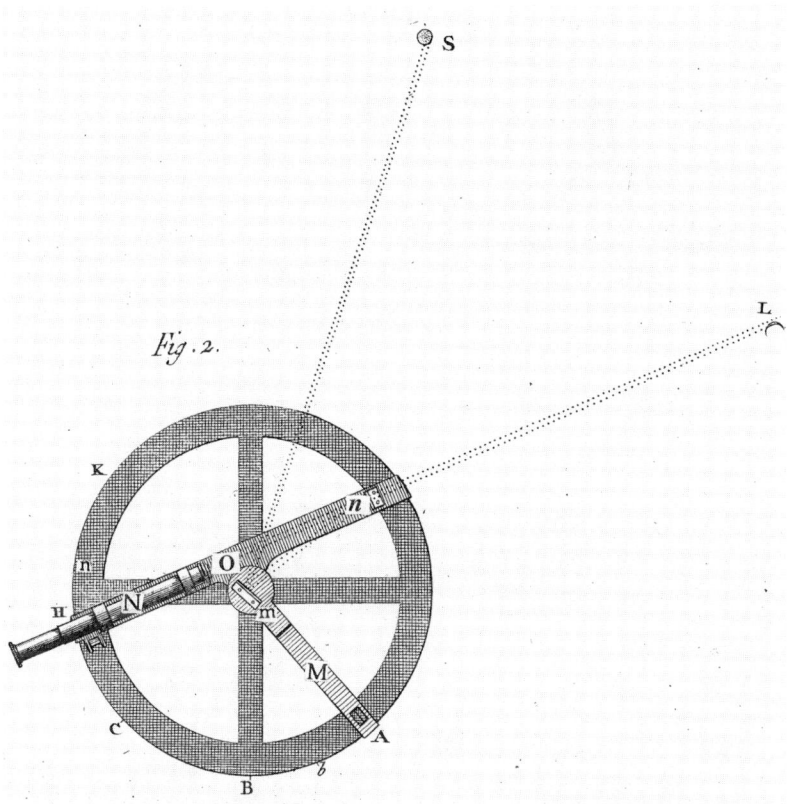

FIG. 2.1.7 Borda's second reflecting circle. Let β designate $\angle$mnO, which is a constant of the instrument; γ, the angle of incidence of the rays from S on m when M is at A; γ', the angle of incidence of the same rays when N points at S and M is at β. Then, as in figure 2.1.6b, AB = $\gamma-(\gamma-\beta)$ and $\gamma'-\alpha = \gamma-\beta$ −AB, whence α = 2 AB. From Borda, *Description* (1802), pl. 1.

Borda perfected the scheme during the late 1770s with the help of his shipmates and a quasi-literate instrument maker named Etienne Lenoir. The instrument (figure 2.1.8) had a diameter of only 10 inches; it was divided into 720 degrees and could be read to minutes by verniers; a handle fixed to the underside of the circle along its axis eased the various maneuvers. It may be operated like

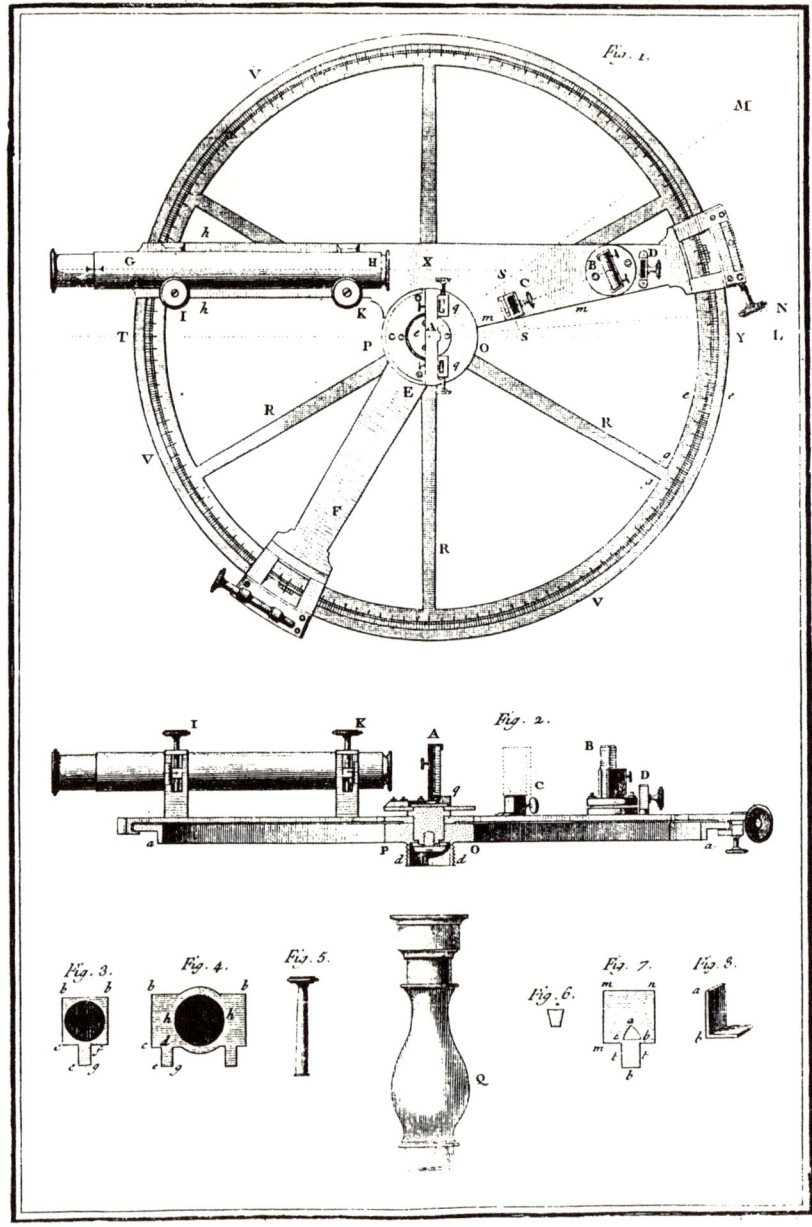

FIG. 2.1.8 Lenoir's version of Borda's reflecting circle. Note the micrometer screws for setting the movable rules and the screen at C to cut down glare. From Borda, *Description* (1802), pl. 2.

Borda's modification of Mayer's device. If, however, the two celestial objects differ markedly in brightness, like the moon and a star, it is better to keep the telescope trained on the dimmer one. Begin by fixing the telescope GH on the star (on the right) and move the ruler EF carrying the large mirror until the image of the moon, M, is received in the telescope. Next, rotate everything around the axis of the telescope, keeping the telescope pointed at the star, until moon and star are again in the plane of the instrument. Then move the rule EF until the images again coincide in the eyepiece. Repeat *ad libitum*, "until you have done the number you think necessary to obtain an acceptable precision."[41] Borda's reflecting repeater did not immediately win the affection of the Navy, but it inspired several French instrument makers to improve its accuracy and reduce its tedium.[42]

Lenoir had the idea how to bring Borda's repetitive reflector to land. Proceeding in the direction opposite to Mayer, he proposed to replace the mirror system by two telescopes, as indicated in figure 2.1.9a. The upper telescope, M, is free to move; the lower, F, is fixed to the circle and marks $0°$. To determine the angle α between two sations, L(eft) and R(ight), point F at R and M at L; clamp M and rotate the circle until M points at R; free M and move it back to spy on L. The angle between the telescopes is now 2α (figure 2.1.9b). Clamp M; move the circle clockwise through 2α until F again faces R, release M and rotate back to L: the angular separation of F and M has become 3α. Repeat thirty or forty times, for a total angle $A = n\alpha$. The quantity A/n gives α very nearly, since the averaging diminishes the effect of errors from sighting and the repeated use of the entire limb all but eliminates the effect of slips in graduation. Preliminary trials measuring all angles of a triangle gave a sum that differed from $180°$ by as little as 1.5 seconds and never by more than 4.5 seconds. The poorest measurements with Ramsden's heavy theodolite closed at just under 3.0 seconds. Both therefore represented a very great advance over earlier instruments, including Mayer's repeater, which seldom produced angles that summed to within 20 seconds of $180°$.[43]

41. Ibid., 13, 16, 31 (quote).
42. Ibid., preface; Marguet, *Histoire* (1917), 156–160; May, *Naut. mag.*, 145 (1945), 26.
43. Berthaut, *Carte* (1898), 1, 101–106, 117–119; Cassini, *MAS*, 1788, 712; Daumas, *Instr.* (1953), 243–245. Repsold, *Astr. Messw.* (1902), 1, 78, is not so enthusiastic about Lenoir's handiwork as were the French.

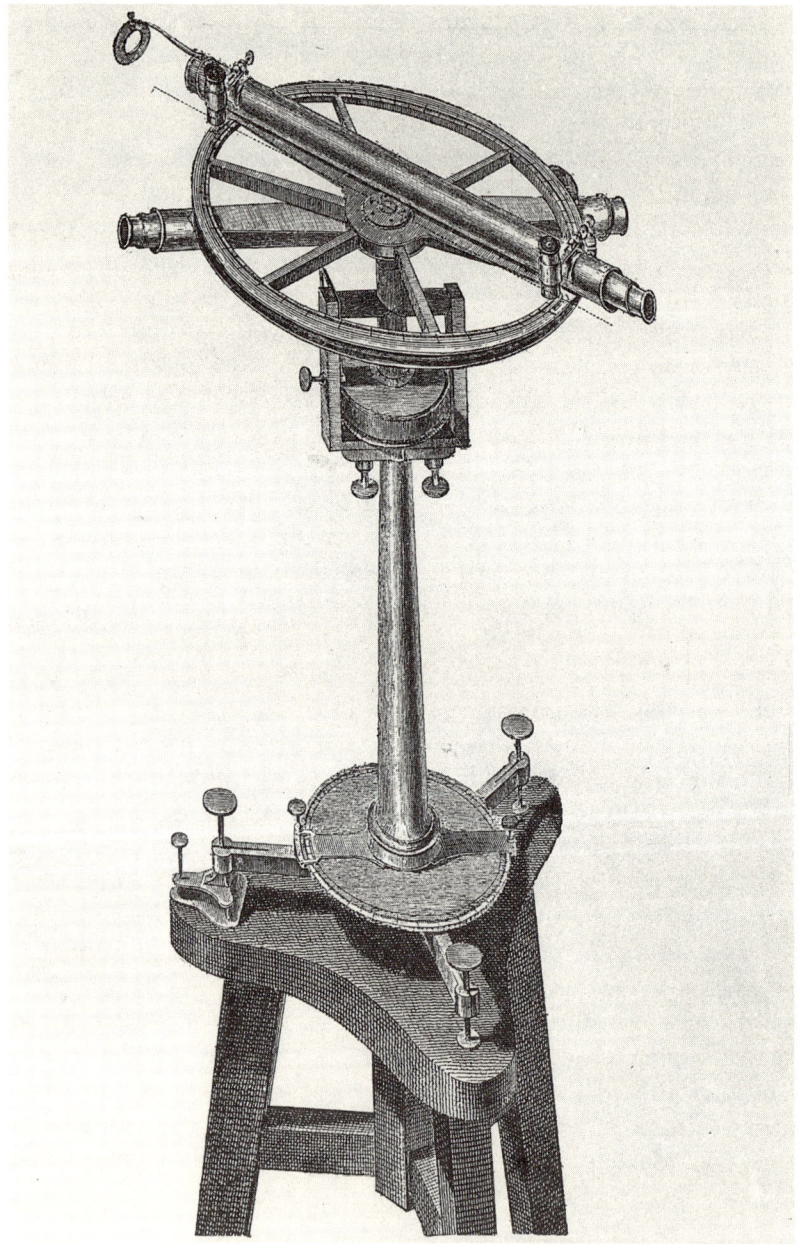

FIG. 2.1.9a A large Borda circle as adapted for geodetic use by Lenoir, ca. 1785. The upper telescope carries microscopes to read the scale divisions. Repsold, *Astr. Messw.* (1902), *1*, 78, after Puissant.

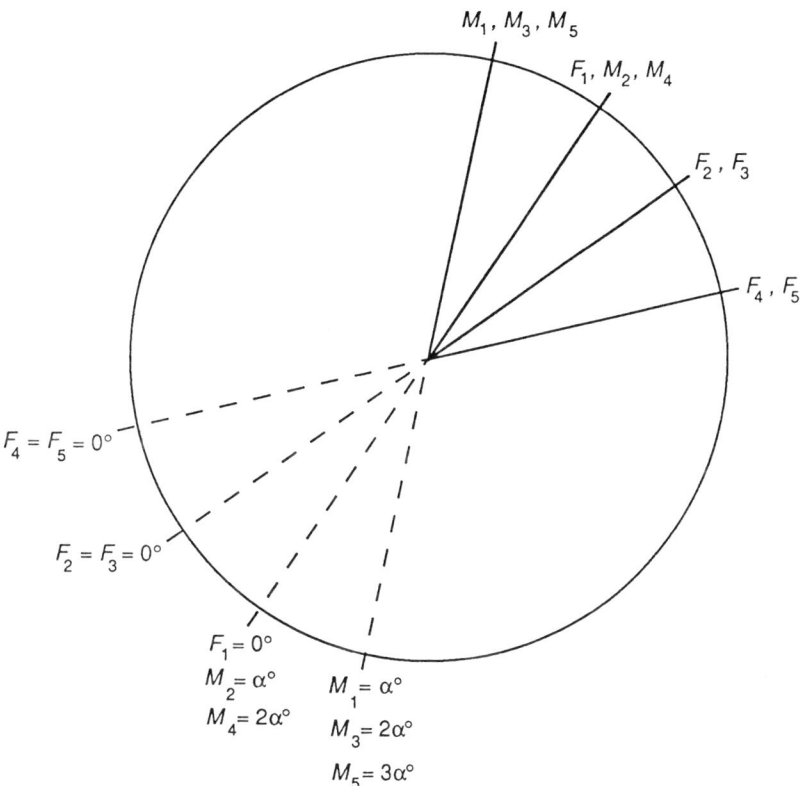

FIG. 2.1.9b The geometry of the Borda-Lenoir circle. The subscripts indicate the positions of the fixed F and movable M telescopes at the first five steps of a measurement.

The Borda-Lenoir circle had a diameter less than half that of Ramsden's theodolite (0.42 m against 0.92 m). It had the further advantages of a moderate cost and easy portability, bought at the expense of fatiguing and demanding manipulations.[44] Nonetheless,

44. Méchain and Delambre, *Base* (1806), *1*, 97–99; Berthaud, *Carte* (1902), *1*, 105. Cf. Pictet, *PT*, *81* (1791), 114, on the theodolite: "too heavy and of too large a bulk

it became the standard high-quality surveying instrument of the following two or three decades.[45] It held its own, more or less, against the theodolite in the trigonometrical survey that linked the observatories of Greenwich and Paris in the late 1780s. The link was proposed by Jean-Dominique Cassini, who understood that, as the leader of the English team put it, the theodolite "may be considered as infallible." "Despite that," says Cassini, "we dared to flatter ourselves that we had on our side an instrument that yielded nothing in the precision of its results to the English apparatus."[46] The French, pointing their repeater from their side of the Channel, found the distance between two flares set out near Dover; the English aimed their theodolite at similar marks near Calais. Unfortunately time and weather did not permit the full circle of French measurements. The competition reached no formal conclusion. It left the French astronomers with a burden of proof they rushed to shoulder when the Revolution provided them, under the guise of a search for a natural basis for weights and measures, an opportunity to use the repeater on a new triangulation of the Paris meridian.[47]

The repeater received mixed reviews. John Leslie, professor of mathematics at the University of Edinburgh, whom we shall meet again soon as a measurer of imponderables, dismissed it in his review of the science of 1800. "This instrument, being complex...in its construction, and tedious and operose in its application, seems after many trials not to be gaining ground in England. Its chief advantage consists in reducing the errors occasioned by imperfect workmanship, which a more skillful execution might nearly preclude." Nonetheless, English makers furnished Borda circles, if only for the Continental market. Von Zach ordered one from Troughton in 1795, with which he proposed to do "things more unbelievable" than he could do with a sextant. The circle arrived many months later. It had been worth the wait. "Nothing outdoes this

---

to be conveniently carried to the top of mountains;" and Gauss to Olbers, 8 Jul 1820, and response, 17 Aug, in Schilling, *Olbers* (1900), 2:2, 20–21, 23, on need for a well-trained assistant to use a repeating circle.

45. Berthaud, *Carte* (1898), *1*, 106.

46. Roy, *PT*, 77 (1787), 214 (quote), 219; J.D. Cassini, *MAS*, 1788, 710–711 (quote). Cf. Blagden to Banks, 25 June 1785 (DTC Papers, 3, 58, Science Museum, London): "They are laboring here [Paris]. Very hard to rival us in instruments."

47. Delambre, *Histoire* (1827), 758; infra, §4.1.

instrument....The exactness is...unbelievable...the most perfect instrument I know." With enough repetitions, a 6-inch circle gave the height of the pole to one second of arc, far better and just as probable as a sextant.[48] And yet in 1809, the officer placed in charge of the triangulation of Thuringia, a project begun under Zach, got poor results with the circle, which he condemned as "inconvenient and unhelpful."[49] Similarly, C.R.T. Krayenhoff, who stretched a geodetic net over Holland early in the 19th century, and J.J. Tranchot, who did the same for four new Rhineland departments annexed by Napoleon, complained of the heaviness and inconvenience of the larger circles they used, some 16 or 17 inches in diameter. Still, by taking two or three series of ten to twelve measurements, they could make their triangles close to within 2 or 3 seconds, and F.C.F. von Müffling, an officer who had worked on the triangulation of Thuringia under Zach, claimed that he could always get his to within one second.[50] Geodesists adopted the repeating circle not because they liked it, but because they prized accuracy over convenience.

Physicists were willing to make the same trade-off. Borda himself may have used his circle to measure angles in his extensive studies of atmospheric refraction. Not a scrap of his results could be found, however, after his death in 1799; and their only memorial was a hollow prism for examining the refractive powers of various gases, which Biot chanced upon in an optician's shop in Paris. On the order of the Académie des Sciences (the first class of the Institut de France) and with the help of François Arago, Biot resumed Borda's work. To take angles, they borrowed a repeating circle from the Observatory; the circle's designer, Nicolas Fortin, perhaps the best instrument maker then active in France, also collaborated in the observations. It was a perfect union of the elements, and a symbol of the ideal, of a precise physical science: the example of Borda, whose "happy combination of observation and calculation is so necessary in the exact determinations of physics;" the presence of the specialist mechanic, Fortin, who helped to attain "the last

---

48. Leslie, *Diss.* (1842), 749; Zach to David, 8 Jul 1795, 30 Apr and 18 May 1796, in Zach, *Briefe* (1938), 99, 116–118.

49. Quoted by R. Schmidt, *Kartenaufnahme* (1973), 1, 38.

50. Ibid., 42 (von Müffling), 84, 88 (Tranchot); Haasbroek, *Investigation* (1972), 28, 31, 35, 156, 161, 163; infra, §4.1.

degree of precision;" and, above all, the implicit reference, via the repeating circle and the problem of refraction, to astronomy, the exemplar of accuracy, the touchstone of human knowledge, "the limit of the efforts of human industry." "We believe that we can flatter ourselves with having achieved a degree of exactness equal to that of astronomical observations, which, in the present state of science, is all that can be required."[51]

The technique is illustrated in figure 2.1.10, drawn for a gas with index of refraction μ less than air's. The direct ray from a distant sight—a lightning rod on the Observatory, 1400 m from the experimental space—was observed through the fixed telescope of a Borda circle, and the refracted ray through the movable one. The determination of the essential angle ϕ required the usual manipulations of the circle; and from ϕ came the measure of the final angle of refraction. The index μ followed from a little algebra applied to the prism angles α, ϵ_1, ϵ_2, and Snel's law of refraction. The results, though expressed to six digits, did not deserve the fanfare with which Biot and Arago presented them. They diverge from modern values by around 2 percent.[52] Nonetheless, at the time they helped make the reputation for precision measurement of the Laplacian school of astronomical physics.

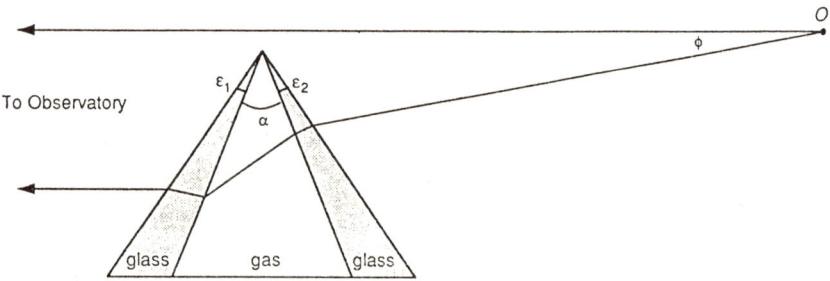

FIG. 2.1.10 Schematic of Biot and Aragot's method of measuring the index of refraction of a gas. O signifies the center of a Borda circle, which delivered, by the usual method, a value for ϕ, the angle between the direct and refracted ray from a distant object.

51. Biot and Arago, *MIF*, 1806, 303–306; "limit of the efforts" comes from Biot, *Mélanges* (1858), 1, 50 (text of 1810).
52. Biot and Arago, *MIF*, 1806, 306, 363–373, and figures 6–7, opp. p. 387; Frankel, *HSPS*, 8 (1977), 60–61.

Another member of the school, Etienne Malus, exploited the Borda circle in an investigation that fully deserved the encomiums for exactness showered upon it. Around 1809, Malus obtained the index of refraction of Iceland spar for its ordinary ray with great precision using a goniometer of his invention schematized in figure 2.1.11. Its operation will be clear from its application to determining the angles of a prism. The prism is placed over the center of an alidade pivoted at O and movable along a graduated scale; initially it points to 0°, and one telescope of a Borda circle receives the light of a signal reflected from BC (figure 2.1.11a). Now turn the alidade clockwise, carrying the prism through an angle β, at which the reflected light from the face AB traverses the telescope. In this situation, the face AB is parallel to the former position of the face BC, and β equals the supplement of $\angle ABC$. The measurement can be repeated, "to astronomical exactness," by clamping the alidade and rotating the entire goniometer counter-clockwise until the reflected light from face BC again enters the telescope (figure 2.1.11b); and the whole process can be multiplied *ad libitum*. With similar but more complicated maneuvers, the physicist can measure the angles of incidence and refraction of light pasing through the prism and thence deduce the index of refraction μ of the crystal. For Iceland spar, Malus obtained, as an average of many trials, the ambitious result $\mu = 1.6563296$.[53]

In another striking application, Malus determined the separation sq of rays travelling in the principal plane of the crystal. Borda's circle fixed the angles as usual and also helped determine sq by the method indicated in figure 2.1.12. The line rp, divided into 100 millimeters, formed one leg of a right triangle with the hypothenuse rt; tp = 1 cm. The crystal sat with its principal plane precisely perpendicular to the plane of the triangle, which was maintained horizontally with the help of a spirit level. Malus moved the triangle until the image of the hypothenuse, as given by the extraordinary ray su, interesected that of the scale, given by the ordinary ray qu: then, as figure 2.1.12 indicates, sq = 0.1 rq, where rq is the scale reading at the place the images intersect. Hence sq is given immediately in tenths of a millimeter, and by visual interpolation to 0.2 mm. This technique also served when the incident ray entered the crystal at

---

53. Ibid., 317–318; Malus, *MSA*, 3 (1817), 122–124, 128 (quote).

64 / SOME MEANS TO THE END

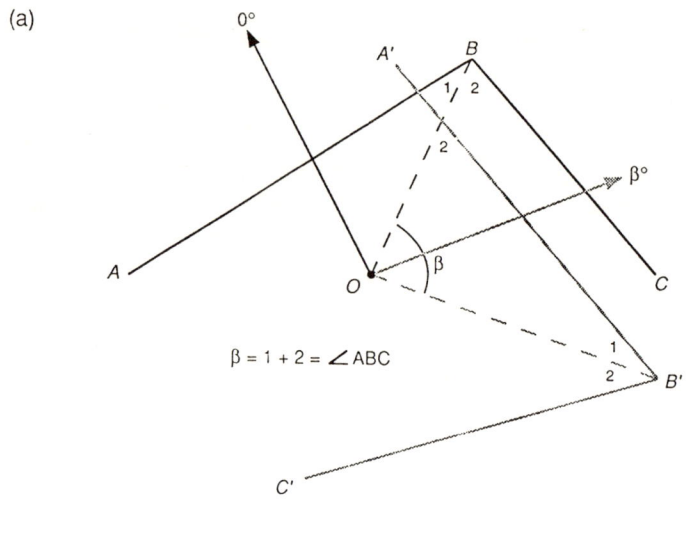

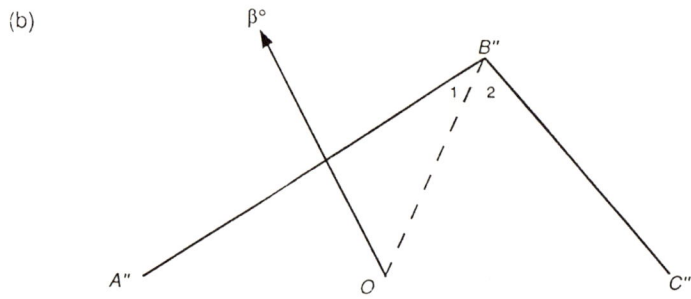

FIG. 2.1.11ab The geometry of Malus' goniometer. The first value of ∠ABC = π − β comes from rotating the crystal through an angle β to bring the face AB to A'B', parallel to the original position of face BC (figure 2.1.11a). This angle can be multiplied by rotating the entire goniometer counter-clockwise through the angle β until the prism has returned to its original position at A"B"C" (figure 2.1.11b).

any orientation to the optic axis: it was basic to Malus' success, and, according to a close student of 19th century physics, it raised experimental optics to a new level of precision. Borda's circle made it possible. "Because of its accuracy this device distinguishes Malus' experiments in double refraction from all previous work."[54]

54. Malus, *Mem. divers sav.*, 2 (1811); Buchwald, *Rise* (1989), 33 (quote), 36.

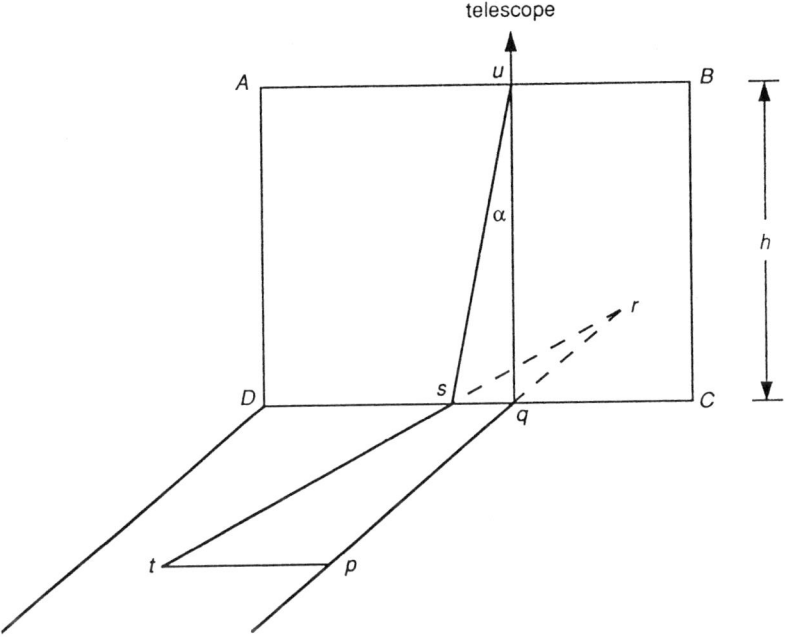

FIG. 2.1.12 Schematic of Malus' measurement of double refraction. The angle α between the extraordinary ray us and the ordinary ray uq in the principal plane ABCD is obtained from sq and the hight h via $\tan\alpha = rq/10h$. The Borda circle made the necessary angular measurements. After Buchwald, *Rise* (1989), figure 1.10.

2. MEASURING IMPONDERABLES

In the case of angles, the measured and the measurer came to the same thing: the operator moved an alidade over a graduated circle or rotated a telescope across a marked scale until the angle he made fit the angle to be measured. Similarly lengths, times, and weights were determined by applications of lengths, times, and weights—the intervals between marks on a measuring rod, the number of ticks of a clock, and the multiples of a standard weight. These quantities had further in common that in most natural philosophies at the beginning of the 18th century they applied universally: all bodies had extension, endured in time, and exhibited weight. An imponderable substance—something that eluded the universal pull of gravity or escaped from the pressure of the terrestrial tourbillon—was as incongruous to the Newtonian as to the Cartesian. Measuring an imponderable world seemed an oxymoron.

When experimental philosophers decided that some of the most active agents in nature had to be considered imponderable, that is, unweighable, they lost the possibility of determining the quantity of these agents directly by the balance. How to proceed? What characteristics of imponderable matter should or could be measured? The first attempts were made on magnetism and electricity.

The first weightless weighings

The difficulty of the problem appears from the flailings of Newton's immediate disciples, who tried to find the "force" of magnets before the concept of imponderable matter had been perfected. They gave no prescription for measuring quantity. Instead, they took an individual lodestone as their subject of experiment and, by analogy to the theory of gravitation, tried to find the "law" by which its "force" on a compass needle diminished with increase of the distance between them. No consistent or general results emerged from these efforts because the law depended upon the particular circumstances of each experimental setup: the shape of the lodestone, the definition of its poles, the distances involved. The measurers had failed to follow out the analogy to the gravitational theory. They did not suppose that the magnetic substance consisted of particles each acting on one other via the same elementary *force* (italicized to indicate the primitive, unobservable, attraction or repulsion between elementary particles), or that the observed "force" on this model is the sum of all the *forces* exerted between all pairs of particles.[55]

The determination of *force* did not enable physicists to measure the quantities they needed to describe everyday experiments. Here the notions of quantity and intensity of magnetism and electricity and, what turned out to be a tertium quid, the capacity of bodies for surplus electrical fluid, played the major parts. We shall survey in turn the methods of measuring *force* and the interrelated quantities charge, capacity, and intensity.

55. Heilbron, *Electricity* (1979), 87–94.

Force. The first to take seriously and successfully the requirement that observed "force" should be computed as the integral of elementary *forces* was Charles Augustin Coulomb, then, in 1785, a senior member of the Paris Academy of Sciences. He understood that the only practical way to achieve the required summation was to use a geometry as close as possible to that which made the world system calculable. He worked not with lodestones of irregular shape but with artificial steel magnets two feet (64.97 cm) long, with well-defined poles 5/6 inch (2.25 cm) from either extremity. In this way he could suppose the magnetic fluids in each magnet to be coagulated at two distinct and distant points—the austral fluid at the south pole, the boreal fluid at the north. He suspended one long thin magnet horizontally in the magnetic meridian by a thread attached to its center of gravity; and he placed a second, similar magnet vertically, with its north pole opposite that of the first magnet, and in the same horizontal plane (figure 2.2.1a). The repulsive force occasioned by all the *forces* exerted between the particles of boreal matter caused the suspended magnet to rotate clockwise as seen from the point of suspension of the thread.

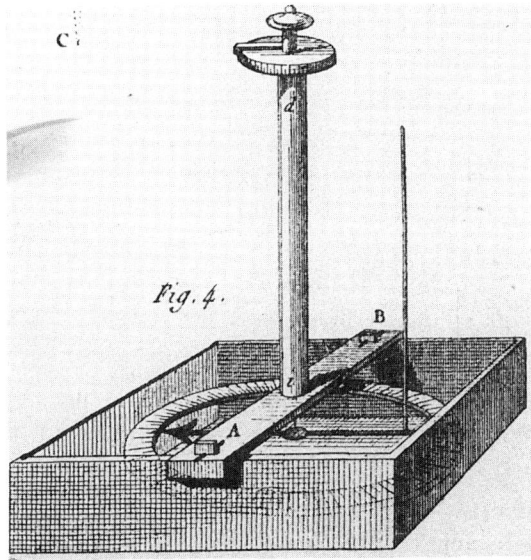

FIG. 2.2.1a   Coulomb's apparatus for measuring the force between magnetic poles. Coulomb, *MAS*, 1785, 578–611.

68 / SOME MEANS TO THE END

Coulomb measured the angle of twist or torsion in the thread when the hanging magnet came to equilibrium (a condition difficult to decide, as will appear). Call θ the angle of twist and ϕ the angle between the hanging magnet and its magnetic meridian. At equilibrium, $\theta_1 = \phi_1$. Coulomb then twisted the thread counter-clockwise, to a new angle $\theta_2 > \theta_1$, thus turning the magnet back to $\phi_2 < \phi_1$. He knew from his own laborious experiments on wires and strings that the twisting force exerted by the thread on the magnet was proportional to the angle of torsion, say $b\phi$. The repulsion between the north poles balances the twisting force plus the restoring force $H\sin\phi$ arising from the earth's magnetic field (figure 2.2.1b).

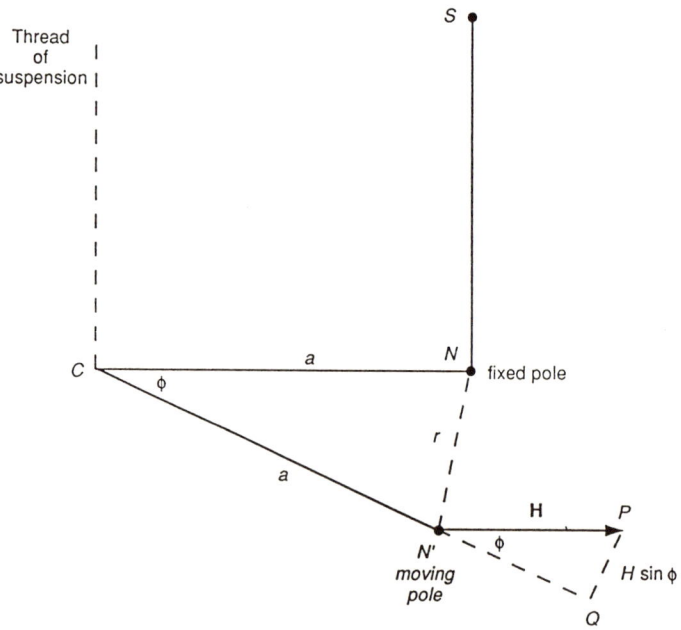

FIG. 2.2.1b Geometry of figure 2.2.1a. The plane of the paper, CNS, lies in the magnetic meridian; C, N, N', P, and Q lie in a plane perpendicular to the paper. Since N'P is parallel to CN (CNS is in the plane of the meridian), $\angle PN'Q = \angle NCN' = \phi$.

Assume that the elementary *force* diminishes as the square of the distance and that the geometry allows the approximation that all the particles of the fluids in the two poles act as if concentrated at two points. Then the force between the poles may be written as $Q/4a^2\sin^2\phi/2$, where a is half the length of the suspended magnet and Q stands for the entirely unknown quantities of magnetic fluid at play. Coulomb worked at angles small enough to allow the further approximation $\sin\phi = \phi$. Consequently he expected to find

$$Q/4a^2\phi_1^2 = (H+b)\theta_1,$$
$$Q/4a^2\phi^2 = H\phi + b\theta, \text{ or} \qquad (2.2.1)$$
$$\alpha/\phi^2 = \beta\phi + \theta,$$

where α and β are constants of the experiment. Coulomb claimed to find these relations to hold very nearly, and to have measured one characteristic of the suppositious imponderable magnetic fluids, namely, the decrease in the repulsion between their elements with increase in distance.[56]

Coulomb gave only one set of measurements. The immediate repulsion between the magnets drove the suspended one clockwise 24° out of the magnetic meridian; hence $\phi_1 = \theta_1 = 24°$. He then twisted the wire of suspension counter-clockwise by $3 \cdot 360°$, which brought the magnet to within 17° of the meridian; hence $\phi_2 = 17°$, $\theta_2 = 1080° + 17° = 1097°$. A further twist through five circles resulted in a deviation of 12°; hence

$$\phi_3 = 12°, \theta_3 = 8 \cdot 360° + 12° = 2892°.$$

In a preliminary measurement, without the vertical magnet, Coulomb had found that it took about 35° of torsion to displace the suspended magnet 1° against the earth's field acting alone. Therefore β in equation 2.2.1 is 35. We have:

	ϕ	ϕ^2	$\beta\phi + \theta$	$\phi^2(\beta\phi + \theta) \cdot 10^{-3}$
ϕ_1	24	576	864	498
ϕ_2	17	289	1692	489
ϕ_3	12	144	3312	477

Or, as Coulomb put it, the numbers in column two are as 4:2:1, those in column 3 as 1:2:4, almost; "whence it follows that the

56. Ibid., 92–6; Coulomb, *MAS*, 1785, 589–90, 601–11.

repulsive action of the magnetic fluid is inversely proportional to the square of the distances."[57]

With his magnetic balance, Coulomb could give a mechanical measure (torsion) of the "force" of an imponderable—or, to be exact, of the repulsion between two parcels of the same magnetic fluid. The apparatus did not work when he opposed unlike poles, since he could not obtain stable equilibrium against their mutual attraction. He therefore had recourse to an indirect method suitable for both attraction and repulsion.

He suspended a short magnetized needle in the magnetic meridian, displaced it from equilibrium, and counted the number n_0 of oscillations it made in a minute. He then placed a long magnet vertically in the plane of the meridian so that one of its poles stood in the needle's horizon. He counted the number of oscillations n_1, n_2, n_3 with the pole of the vertical magnet a distance d_1, d_2, d_3 from the center of the needle. Coulomb chose a thread long and soft enough that he could ignore the effect of its torsion on the motion.

In figure 2.2.2, H indicates the forces parallel to the meridian arising from the horizontal component of the earth's field; f_y and f_z, forces arising from the fixed north pole N. Taking moments around C,

$$d^2\phi/dt^2 \sim -(2H\sin\phi + f_y\sin\chi + f_z\sin\psi).$$

The law of sines applied to $\triangle CS'N$ and $\triangle CN'N$ gives

$$\sin\chi = (d/y)\sin\phi, \quad \sin\psi = (d/z)\sin\phi.$$

Coulomb worked in the approximation $\sin A = A$. Hence

$$d^2\phi/dt^2 = -\text{const.}(2H + df_y/y + df_z/z).$$

In the only set of measurements Coulomb gave, $a = 0.5$ inch and $d > 4$ inches (9.1 cm). He smoothed his way by neglecting a in comparison with d, thus making $y = z = d$ and $f_y = f_z = k/d^2$, k a constant. With this simplification, he had

$$d^2\phi/dt^2 = -2\phi(k/d^2 + H).$$

57. Coulomb, MAS, 1785, 578, in Mémoires (1884), 138–43, 144 (quote).

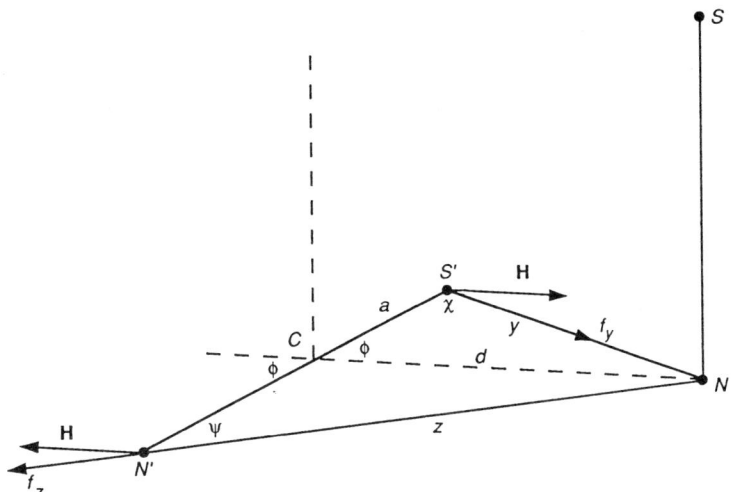

FIG. 2.2.2 Geometry of Coulomb's method for finding the force between like magnetic poles, 1785. C, N, N', and S have the same significance as in figure 2.2.1b; C, N, S lie in the meridian, N', C, S', N in the horizontal plane.

The solution of the preceding equation is $\phi \sim \sin 2\pi nt$, where the number of oscillations in unit time n is given by

$$n^2 = k/2\pi^2 d^2 + H/2\pi^2.$$

In the absence of the pole N, Coulomb found that $n^2 = n_0^2 = H/2\pi^2 = 225$. He hoped to justify the preceding argument by showing that $(n_i^2 - n_0^2) \sim 1/d^2$ for three different distances d_i. Here are his numbers:

	d	n	$n^2 - n_0^2$	$(n^2 - n_0^2)d^2/1000$
d_1	4	41	1456	23.3
d_2	8	24	351	22.5
d_3	16	17	64	16.4 (20.2)

The parenthetic result of the third measurement comes from taking the effect of the distant upper pole of the vertical magnet into account; its power when $d = 16$ inches amounted to about a fifth of that of the lower pole.[58] Thus Coulomb gave an alternative measure of the "force" of imponderables in terms of time rather than torsion, which may have been persuasive to those willing to ignore the difference between 20 and 23.

Measurement of electrical *force* followed the same pattern as the magnetic. At first, failure to follow a strict analogy to the gravitational theory resulted in several different "laws of force" applicable only to the special experimental circumstances from which they derived. The procedures had this in common, however, that they balanced electricity as well as magnetism against gravity. In the most common arrangement, a pair of threads or straws or gold leaves, charged by connection to an electrified conductor, spread and rose until stopped by the torque exerted by their weight. Because the charge distributes in a complicated way along the leaves, their spread does not give a practical measure of force.

In a more favorable geometry, however, the dependence of *force* on distance can be inferred from a direct comparison between electrical repulsion and weight. The first such measurement took place around 1770. It was the work of John Robison, then secretary to Admiral Charles Knowles, who had been sent to Russia to help the Empress Catherine build up her navy. Robison, who had studied natural philosophy under Joseph Black in Glasgow, soon became acquainted with the Petersburg savants, especially F.U.T. Aepinus, the author of a theory of electricity based rigorously on Newton's concept of "force" as a sum of *forces*. Aepinus had guessed, but not demonstrated, that the law of force for electricity diminished as the inverse square of the distance between interacting elements.[59]

In Robison's apparatus (figure 2.2.3a), the repulsion occurs between the brass ball A and the gilded cork ball B, both a fourth of an English inch (0.64 cm) in diameter. The repulsion rotates the waxed silk thread BD in the amber cheek C; the frame AFEL is made of glass to insulate the balls and minimize the effects of induction. The scale GHO in a plane parallel to AFEL allows

---

58. Coulomb, *MAS*, 1785, 578, in *Mémoires* (1884), 131–133.
59. Heilbron, *Electricity* (1974), 396–401, 465–466; Home, "Introductory monograph" (1979), 218, 222–223.

SOME MEANS TO THE END / 73

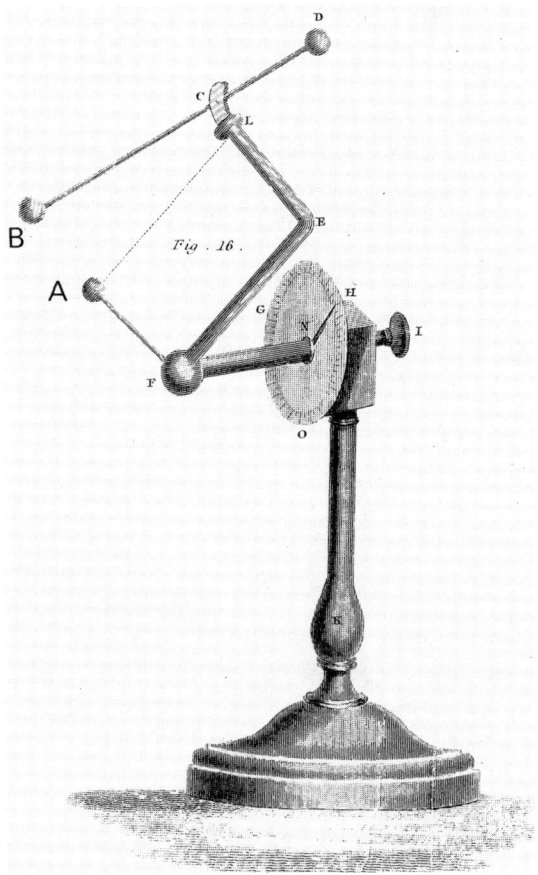

FIG. 2.2.3a Robison's apparatus for measuring the force between small charged spheres A and B, ca. 1770. Heilbron, *Electricity* (1979), figure 19.10.

measurement of the angle α between AL and the horizon and the angle β between BD and the vertical. Robison began with AL horizontal and B resting on A. He electrified the pair but not enough to cause their separation against the weight of B. He next rotated the knob I, and hence FE, weakening the action of gravity perpendicular to AL, until the balls separated and the angle between AL and the horizon rose to α. (If the balls carried unlike charges, Robison twisted the machine so that BC lay below AL.) Since the diameter of the balls was small compared with the gap AB, Robison could calculate as if their entire charges rested at their centers and therefore deduce the law of *force* from the measurement of α and β and the dimensions of the apparatus.

The balance of moments around C (figure 2.2.3b) gives

$$f = \frac{\sin\beta}{\cos\gamma}(W_B - W_D \frac{CD}{BC}),$$

where f signifies the repulsive force between the centers of the balls A and B. To demonstrate the inverse-square law, Robison had to show that $AB^2 \sin\beta/\cos\gamma$ has the same value for several settings of the angle α, on which β, γ, and AB depend. The law of cosines applied to $\triangle ABC$ and $\triangle ALC$ yields

$$AB^2 = a^2 + y^2 + 2ay\cos(\alpha + \beta - 2), \text{ where}$$

$$\cos 2 = (a^2 + y^2 - x^2)/2ay, \text{ and}$$

$$y^2 = a^2 + x^2 + 2ax\cos\alpha,$$

in which a stands for BC = AL, x for CL, and y for AC. The law of sines applied to $\triangle ABC$ and the summation of the angles around A yield

$$\frac{\sin\beta}{\cos\gamma} = \sin\beta(1+K^2)^{1/2}, \text{ where}$$

$$K = (a/y)\sec\phi - \tan\phi, \text{ and } \phi = \alpha + \beta + 2.$$

Robison claimed to have made "many hundreds" of these measurements and to have subjected them to the tedious calculations just outlined, with the spectacular result that the product $(\sin\beta/\cos\gamma)AB^n$ came out to be a constant for all values of α tried if $n = 2.06$. Unfortunately he deprived others of the satisfaction of knowing this extraordinary result and of trying to duplicate it until long after Coulomb had published his deduction of electrical *force* from experiments on the torsion balance.[60]

---

60. Robison, *System* (1822), 4, 68–74.

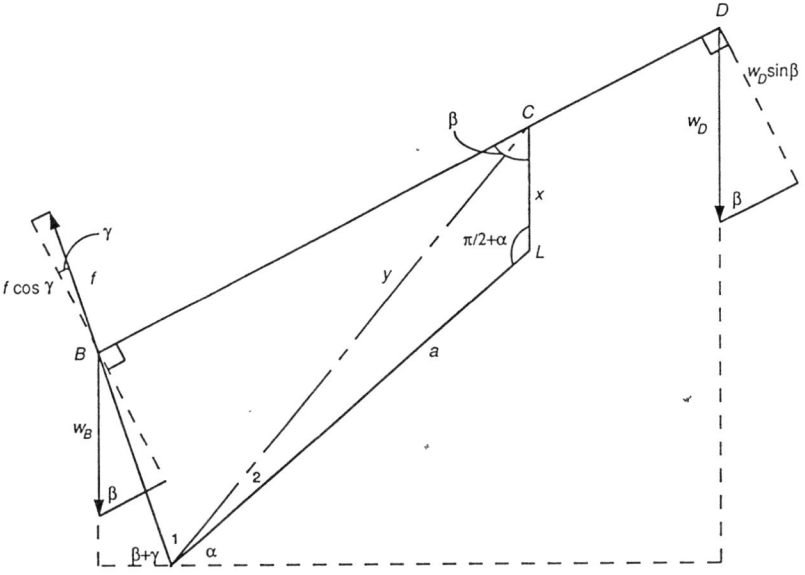

FIG. 2.2.3b Geometry of Robison's measurement, with BC = AL = a, CL = x, AC = y. All lines are in the plane of the paper.

Figure 2.2.4 illustrates Coulomb's familiar apparatus. The waxed silk thread q played the part of Robison's balance arm BC; the paper disk g on one end of q reduced the oscillations set in motion when a charge conveyed to the fixed ball t was shared with the ball a on the other end of q (figures 2.2.6a, b). Coulomb emphasized that owing to the small size of the balls (1/4 inch in diameter) in comparison with the length of the apparatus, he could reason as if their charges were concentrated at their centers. As for the measurement itself, he managed it as he had the corresponding work on magnetism. The initial repulsion produces the situation shown in figure 2.2.5; the balance of torque requires $f \cos\phi_1/2 = \text{const.}\phi_1$ (the subscript signifies "first measurement"). Coulomb turned the knob b (figure 2.2.4c) through an angle θ_2, reducing the divergence to $\phi_2 < \phi_1$, and raising the angle of torque to $\theta_2 + \phi_2$. Since he worked in the region where $\sin\phi \sim \phi$, he could write $ta = 2d \sin \phi/2 \sim d\phi$

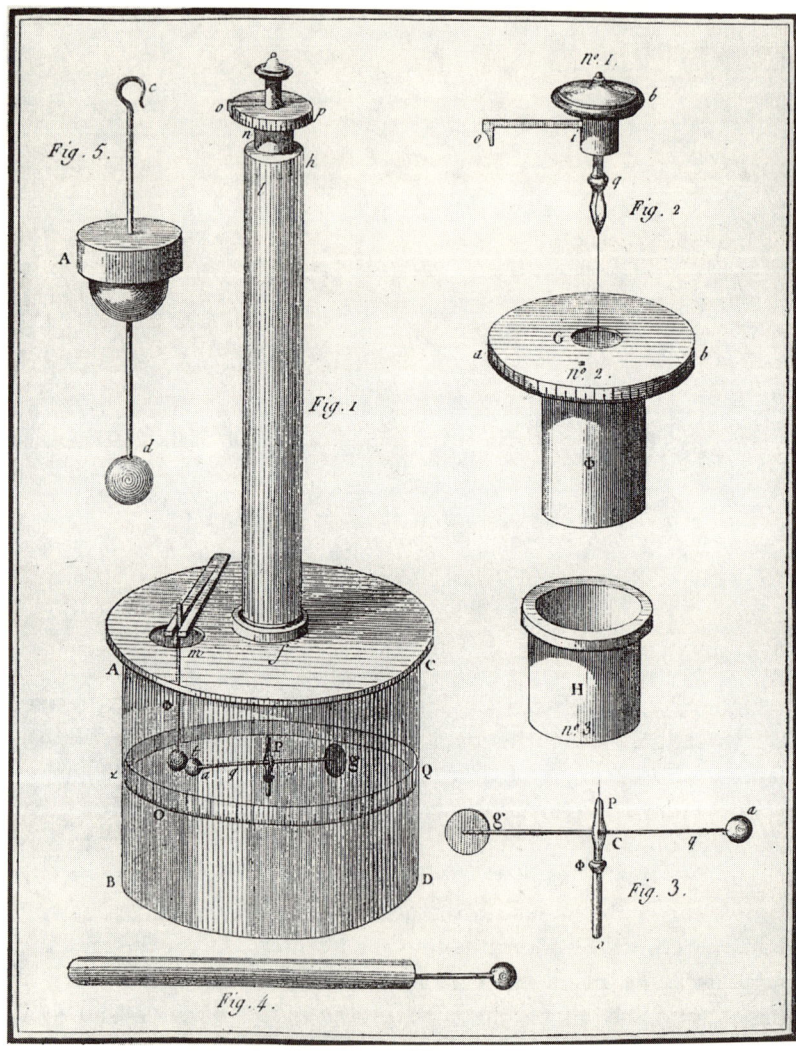

FIG. 2.2.4abc Coulomb's electrical balance, 1785: (a) (fig.1 on plate), the balance as a whole (the charged balls are at a and t) (b); (fig 3 on plate), the balance arm with its damping disk g; (c) (fig 2 on plate), the turning knob and mounting of the torsion wire. After Coulomb, MAS, 1785, 576.

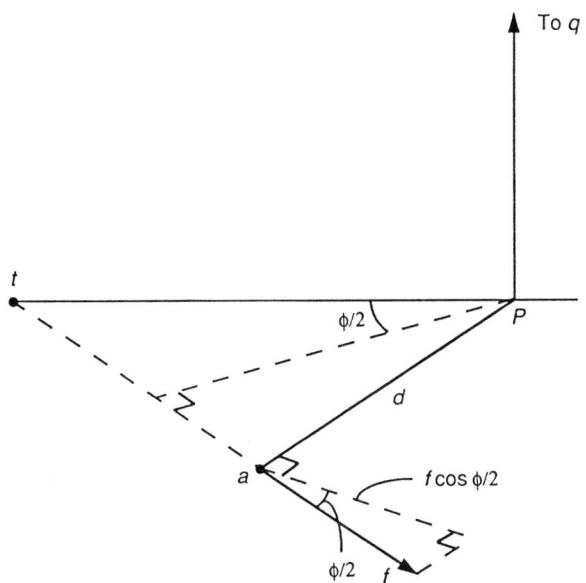

FIG. 2.2.5 The geometry of Coulomb's balance: t, a, and P have the same meanings as in figure 2.2.4a; q signifies the mounting in figure 2.2.4c. The vertical plane contains t, P, and q; the horizontal, t, P, and a.

and $\cos\phi/2 \sim 1$. For the inverse-square law to hold, $\theta \cdot \phi^2$ must be constant. Coulomb gave the following measurements:

	ϕ	θ	$\phi^2\theta/1000$
ϕ_1	36	36	46.7
ϕ_2	18	144	46.7
ϕ_3	8°30′	575°30′	41.6(46.7)

The figure in parentheses results if ϕ_3 is increased to 9°. "It follows," Coulomb wrote, "that the mutual repulsive action between two balls charged with the same kind of electricity follows the inverse ratio of the square of the distances."[61]

And if they carry the same sort of electricity? Coulomb placed an insulated copper globe G, one foot in diameter, with its center in

61. Coulomb, MAS, 1785, 569, in Mémoires (1884), 107–115, quote on 110.

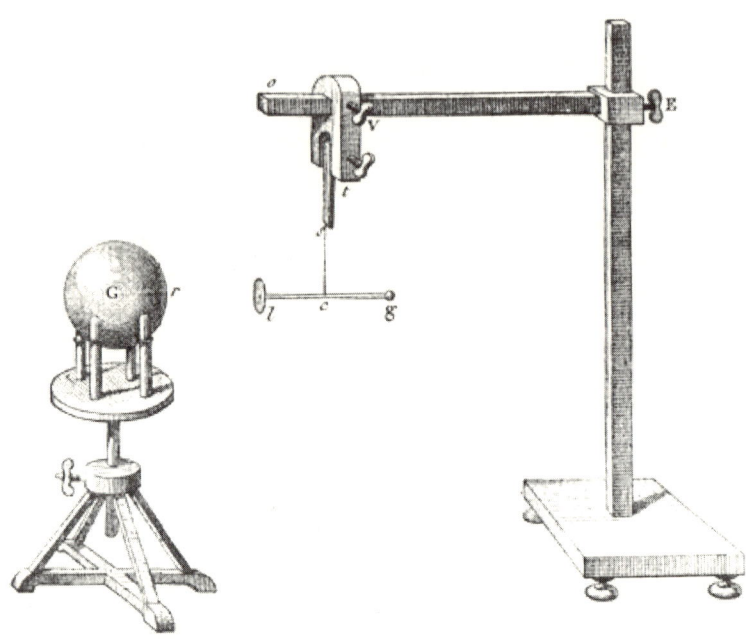

FIG. 2.2.6a Coulomb's apparatus for measuring the pull between like electrical charges, 1785. The block V can be moved along the rod o to change the distance between the charged sphere G and the oscillating rod lg. After Coulomb, *MAS*, 1785, 610.

the plane in which he had suspended a gum-lac rod lg from a silk thread sc attached to a movable block V (figure 2.2.6a). He electrified G and induced an opposite charge on the gilded paper disk l. The geometry allowed the usual approximations about electrified bodies acting at their centers and sines equalling their arguments. Since the suspending thread exerted negligible torsion, lg enjoyed something very close to a simple harmonic motion. Assuming that the inverse-square law held, Coulomb could invoke the law of sines to write (figure 2.2.6b):

$$d^2\phi/dt^2 \sim -(\phi/y^2)(1 + cl/y).$$

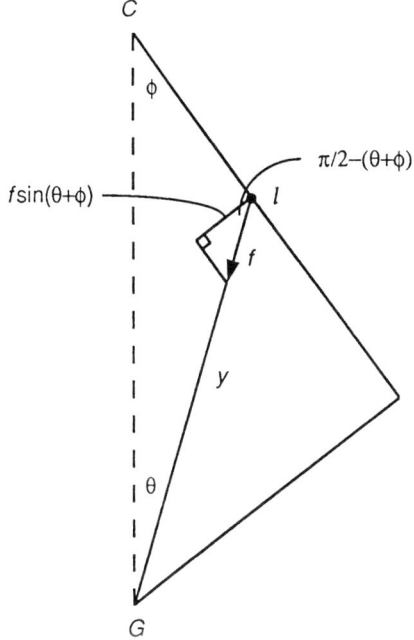

FIG. 2.2.6b Geometry of Coulomb's measurement of the attraction of like electrical charges. The letters have the same significance as in figure 2.2.6a.

Since cl/y did not exceed $1/15$ in the experiments, Coulomb deduced that the frequency n of the motion should diminish linearly with y, or that ny should have the same value no matter where he fixed the block V. His numbers:

	y	n	ny
y_1	9	0.750	6.75
y_2	18	0.366	6.59
y_3	24	0.250	6.00

Ascribing the discrepancy in the third measurement to leakage of the charge during the four minutes the experiment lasted, Coulomb declared that he had recovered his earlier result by an entirely different, and therefore strongly corroborative, method.[62]

The numbers just reported are all that Coulomb offered. They did not compel assent to the law of squares. Not only did some of the measurements deviate seriously from expectation while others agreed precisely and suspiciously with calculation, but also the apparatus behaved badly. Coulomb observed that unlike a gravity balance, the zero point of the torsion balance could drift during experiments since the thread had a habit of untwisting even when unstressed. Some electricians senior to Coulomb, like J.A. Deluc and Alessandro Volta, who did not believe that Newton's concept of *force*, provided the key to electrical theory, felt free to reject Coulomb's conclusions or to regard them, as they rightly did most earlier measurements of the "force" of electricity, as peculiar to the experimental arrangement. German electricians disliked it or dismissed it. P.L. Simon of Berlin and G.F. Parrot of Dorpat, who had no strong aversion to Newtonizing electricity, rejected Coulomb's conclusions because they could not obtain consistent measurements with his "all too unsteady twisting machine."[63] E.G. Fischer scandalized Biot by omitting all reference to Coulomb's balance from his account of electricity in *Physique mécanique*.[64]

The deduction of a law of *force* from direct balance of electrical against mechanical force presents several technical problems. Robison's gravity method could scarcely have yielded a precise result; therefore, perhaps, he did not bother his readers with his numbers. Nor does the Coulomb balance permit exact and consistent measurements. Although the calculations can be corrected for leakage of charge, the drifting of the zero point cannot be compensated without extensive auxiliary investigations. It reflects "elastic after-effect," a phenomenon discovered in the 1830s during experiments directed toward clearing up the ambiguities and inconsistencies of measurements with the torsion balance.[65] Nonetheless,

62. Coulomb, *MAS*, 1785, 578, in *Mémoires* (1884), 116 123.
63. Heilbron, *Electricity* (1979), 475–476; cf. Weiss, *Prevost* (1988), 164–165, re objections to Coulomb's work by Haüy and Louis de Manoël de Vegobre.
64. Biot, in Fischer, *Physique mécanique* (1806), 246n, 257–261.
65. Dörries, *HSPS*, 22:1 (1991), 27–28.

many physicists, particularly in France but also in Britain, accepted "Coulomb's law" and, by 1800, had made it a cornerstone of the Standard Model. Their conviction rested partly on his measurements, partly on the power of analogy (the strengths of electricity and magnetism, like gravity's, diminish with, but remain sensible at, macroscopic distances), and partly on indirect inference from qualitative experiments.

This indirect evidence rested on the theory that a uniform gravitating shell exerts no force on a mass point within it. Applied to the case of electricity, the theory required that all the charge on a spherical conductor (indeed, on any closed conductor) rest at its surface—provided that the elements of the same electrical fluid repel one another in accordance with the law of squares. In 1767 Joseph Priestley reversed this argument. He deduced from the observation that no or little charge could be detected within an almost closed electrified can that the electrical fluid (he was a unitarian in electricity as well as in religion) obeyed the same law of force as gravitating matter. Henry Cavendish demonstrated this null effect by electrifying a spherical surface, enclosing it within another, connecting them together briefly, and showing that the charge on the inner surface moved entirely to the outer when given the opportunity. The demonstration no doubt would have been persuasive, had it been published in its time.[66] Similar arguments offered by Laplace and his coworkers did bring conviction to many.[67] A null method seems especially appropriate to a weightless fluid.

Quantity. The Standard Model provided no way to deduce the absolute quantity of the electrical or magnetic fluids in a body. Relative measurements, however, could be made easily by those able to perform on the torsion balance. In Coulomb's conception, the strength of a magnet indicated the "density" of the fluids at its poles. He could establish relative densities of magnetic needles by hanging them one after another in the plane of the magnetic meridian and twisting the thread of suspension through an angle θ until each stood at some standard angle ϕ_s to the meridian. Then, according to equation 2.2.1, the densities of the fluids in the poles of the needles

66. Heilbron, *Electricity* (1979), 464–465, 479–480.
67. Ibid., 495–500.

were proportional to $35\phi_s + \theta$. The method required the assumption, which Coulomb thought needed no proof, that the magnetic force acting upon the pole of a magnet was proportional to the density of its magnetic fluid.[68]

Coulomb faced a more difficult problem in devising a measure of the "density" of electrical matter. Contrary to the fluid locked up in magnetic poles, electrical fluid is free to move in conductors; indeed, to assure that electrical force vanishes within a conductor, the fluid must in general distribute non-uniformly—with unequal density—across its surface. Coulomb needed a way to find relative densities from point to point. He replied with the "proof plane," a thin small disk of gilded paper attached to an insulating handle, which, when touched to an electrified conductor, picked up a charge proportional to the density of the fluid at the point of contact. He attached a number to the relative density at the contact point by the swing ϕ of the arm of his torsion balance after he electrified it by a kiss from the proof plane. The technique enabled him to confirm calculations of the distribution of electrical fluid on touching charged spherical conductors. The approximate agreement with measurement provided strong indirect support for the law of squares, on which the calculations rested.[69]

A richer approach to the measurement of quantity developed from the primitive thread electroscope. By the early 1770s, it had taken the standard form illustrated in figure 2.2.7, the work of William Henley, a protégé of Priestley's. Electricians used Henley's instrument with permanent apparatus, like the prime conductors of electrical machines; it was screwed into a tap on the conductor, which, when charged, caused its index A to move to an angle α from its stem C. Bottled versions of the diverging threads, like Abraham Bennet's gold-leaf electroscope (figure 2.2.8), served in other situations: charge communicated from a body under examination to the cap of the instrument spread its leaves to an angle α that measured something about the body's electricity.[70]

---

68. Coulomb, MAS, 1785, 578, in Mémoires (1884), 130, 141.
69. Coulomb, MAS, 1787, 421, in Mémoires (1884), 187–190, 200–204, 210–211.
70. Heilbron, Electricity (1975), 449–451.

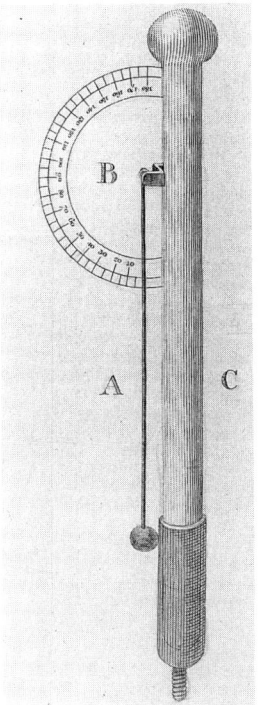

FIG. 2.2.7 Henley's electrometer, 1772. Charge conveyed from a conductor to the screwed fitting at the base is shared with the ball attached to the arm A. Heilbron, *Electricity* (1979), figure 19.1.

What did α measure? Alessandro Volta, who preferred instruments of Bennet's type with straws in place of gold leaves, went far toward answering the question during the 1780s. He knew that α did not measure directly the quantity of charge, Q, on an insulated conductor communicating with the electroscope since, as he had demonstrated brilliantly in 1778, the magnitude of α depends not only on Q, but also on the shape of the conductor. The demonstration: a series of connected metal cylinders (figure 2.2.9), suspended by insulating strings, held a greater quantity of electrical fluid (as estimated roughly by the number of turns of the electrical machine required to charge them or by the shock they delivered when grounded through the body) when hung some distance apart than

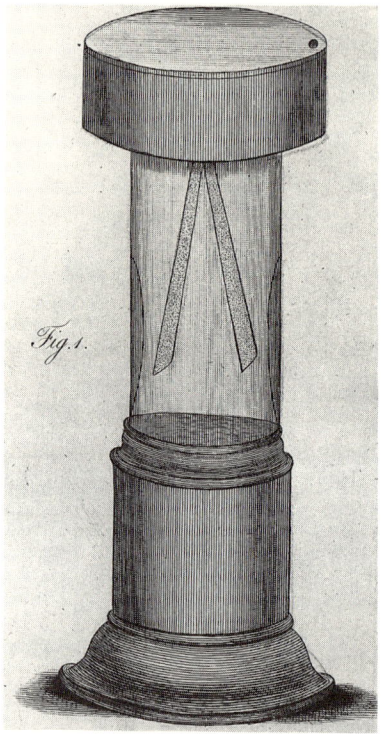

FIG. 2.2.8 Bennett's electroscope, 1786. Note the earthed metal foil on the interior walls to prevent accumulation of charge conveyed by the leaves. Heilbron, *Electricity* (1979), figure 19.2.

when placed close together. Not only the geometry of the conductor, but also the presence of other conducting bodies in its vicinity, affected what Volta called its "electrical capacity." Rather than measuring quantity of fluid, α indicated something more difficult to visualize, its pressure or, what Volta preferred, its "tension." He guessed that quantity depended linearly on tension as measured by α and that "capacity" was their ratio: $Q = CT$.[71]

To demonstrate this relation, Volta required an electrometer that registered equal increments of α for equal increments of charge. He calibrated his bottled straws by giving the electrometer knob a

71. Ibid., 453–456; Volta, *Opere* (1918), 3, 203–206, 212–215, 224–225 (text of 1779), 248–258

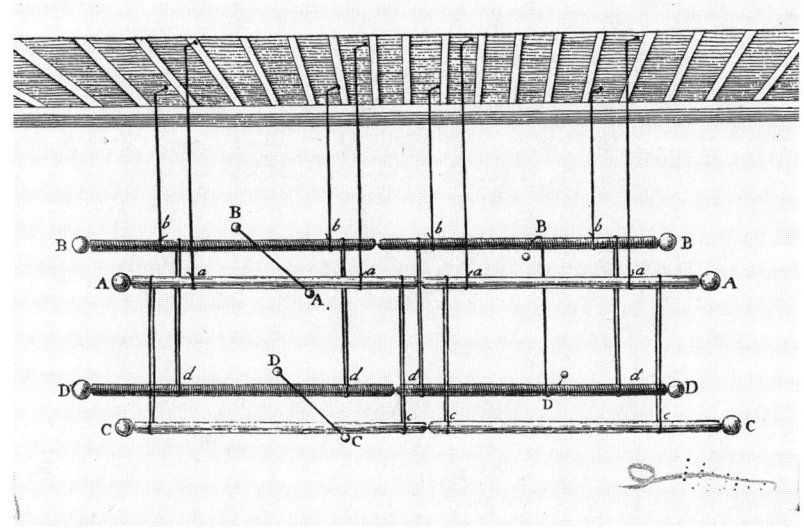

FIG. 2.2.9 Volta's demonstration of the influence of geometry on electrical capacity, 1778. Heilbron, *Electricity* (1979), figure 19.4.

quantity of fluid Q that spread them to their maximum separation α_m; then touching the knob to that of an identical electrometer, thereby reducing the charge to $Q/2$ and the spread to an angle labelled $\alpha_m/2$; and so on. Volta used his calibrated straw instruments to obtain the values of Q/C for various conductors under various circumstances; and, in the case of conductors moved around under the electrical influence of other conductors, to learn how the capacity of any one of them changed with its separation from the others.[72] A conventional measure of the absolute quantity of redundant electrical fluid—fluid in excess of that (or those) natural to a body when uncharged—thus became available.

Volta's way of exploiting the relationship $Q = CT$ in measurement required the intervention of a mechanical quantity, α, the *locum tenens* of T, and thus retained the old approach of balancing gravity and electrical repulsion. Nonetheless, the relationship in principle had no mechanical referent: Q and T described the state of

---

72. Volta, *Opere* (1918), 5, 37–42; Heilbron, *Electricity* (1979), 455–456.

the fluid and C measured the space available for surplus fluid in a conductor in a particular electrical and geometrical environment. By 1800, physicists had grown familiar with self-referring electrical concepts and hence with a self-contained theory of imponderables. They had also begun to develop a similar quantitative description of the heat fluid.

Caloric

The analogy between caloric and the electrical fluid(s) in respect of latency and expansivity, and between both and common matter in respect of capacity, developed further under the impulse of measurement. Although a law of *force* between caloric particles of the form $(1/r)^n$, which many physicists thought might obtain,[73] could not be confirmed, temperature and electrical tension, and heat and quantity of electricity, were found to correlate closely. The discovery of radiant heat, which bore no analogy to electricity but a close one with light, made caloric the middle term in the series of parallels that constituted the strongest case for the Standard Model of 1800.

Like electricity also, the instruments for measuring caloric were themselves affected by the measurement. Whereas in the Coulomb balance the electricity under study induced charges on the apparatus and the investigator, which in turn exerted forces on the balance, so in the thermometer part of the heat under study was consumed in causing the expansion of the instrument, which constituted the measurement. In a still closer analogy, a conscientious student of caloric calibrated his thermometers under glass lest his body heat affect the results.[74] Furthermore, the reliability—or linearity—of the instruments were in both cases sources of recurrent and productive dispute. Coulomb first demonstrated the linearity of the force of the torsion in his wires without involving electricity; but this advantage evaporated when later measurers, confronted as he was by elastic after-effect but less willing to ignore it, rejected his apparatus as uncontrollably inaccurate while accepting the main result he drew from it. The thermometer—that is, the mercury thermo-

73. Leslie, *Diss.* (1842), 760.
74. Evans and Popp, *AJP*, 53 (1985), 752, re Pictet.

meter—also failed the test of linearity. The failure came to light in painstaking comparisons of the expansion of mercury with that of gases.

The measurement of properties of gases under different regimes of temperature became the most fruitful method for advancing understanding of the character of heat and thence of all imponderables. Heat thus succeeded electricity as the bellwether weightless fluid. The onset of the succession was noted in 1772 during the ceremony in which Joseph Priestley received the Royal Society's highest recognition for his isolation of "factitious airs." The Society's President, Sir John Pringle, observed that the new field of gases followed naturally on the investigation of electricity as another route to "the nature...of the subtle fluids of the universe."[75] He spoke presciently.

We shall look at the means of calibrating thermometers, and of measuring the thermal expansion of gases; at the machines used to determine the specific heat capacities of gases and solids; and at the experimental setups for the study and measurement of the properties of radiant heat.

Thermometry, mercurial and gaseous. A good fixed point for the start of modern thermometry is the work of the committee appointed in 1776 by the Royal Society to determine why its thermometers never agreed. Its membership indicated how widely the problem of heat measurement was implicated in the science of the time: the chairman, Henry Cavendish, represented chemistry and natural philosophy; the astronomers Nevil Maskelyne and Alexandre Aubert, astronomy and geodesy; William Heberden, medicine; Jean André Deluc, meteorology; and Samuel Horsley, mathematicians who liked to calculate. They tried the boiling points of several mercury thermometers and found them to differ by over 3°. They traced the variation to differences in the placement of the bulb during calibration—whether immersed in the boiling water or exposed to the steam—and to differences in the prevailing atmospheric pressure. Their recommendation (figure 2.2.10): when marking the boiling point, expose the bulb to the steam with only a short piece of the stem extending above the lid of the pot; heat only the bottom of

---

75. Pringle, *Discourses* (1783), 36, 40 (quote); cf. Kippis, in ibid., xxxvii.

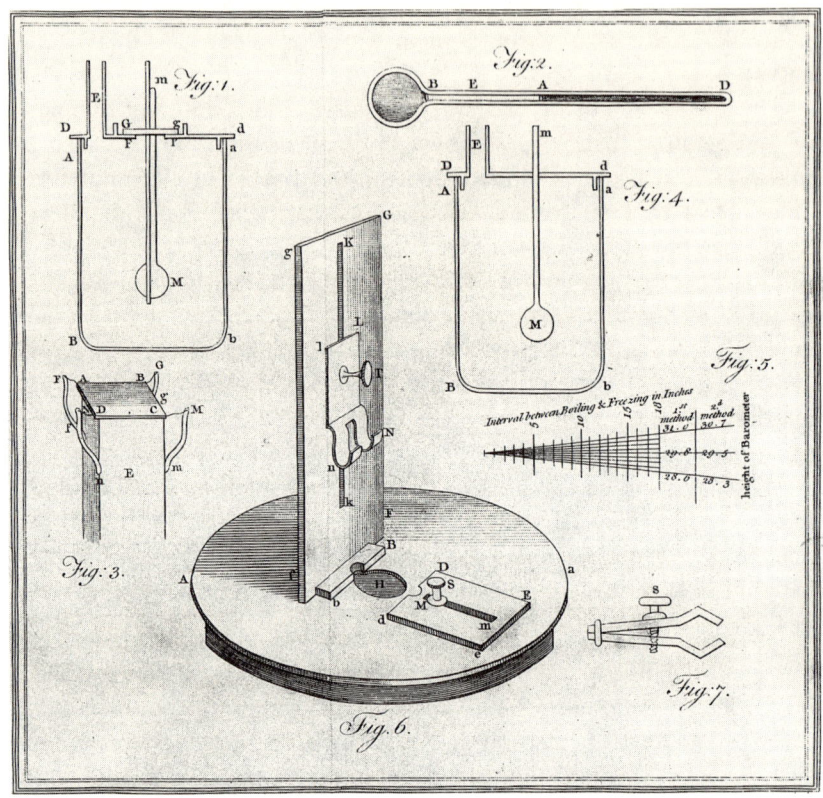

FIG. 2.2.10 Cavendish's procedure for fixing the boiling point of a thermometer, 1777. Here figure 1 is the pot of boiling water; figure 3 gives details about the lid closing the chimney that lets out the steam; figure 6 shows the preferred way to fix the thermometer to the lid of the pot (the instrument is held in the vertical frame NLl and tightened into place by the calimp dDEe); figure 5 displays the change of boiling point with pressure. Cavendish et al., *PT*, 67 (1777), 856.

the pot; and do it all at an atmospheric pressure of 29.8 inches of mercury.[76] Their procedures differed in style as well as rigor from the lackadaisical approach of the preceding generation (figure 2.2.11).

76. Cavendish et al., *PT*, 67:2 (1777), 825, 831, 845–851.

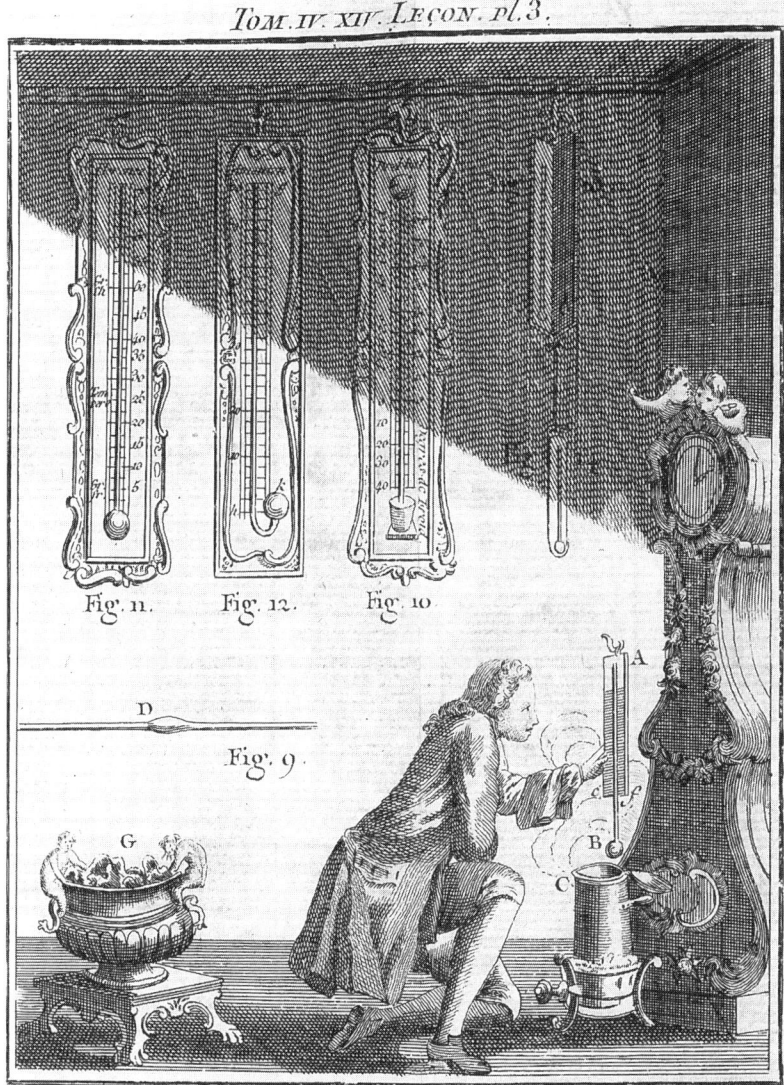

FIG. 2.2.11 Nollet's procedure for fixing the boiling point. In constrast to the elaborate operations recommended by Cavendish et al., Nollet's gentlemanly measurer merely thrusts his thermometer into the steam escaping from hot water. Nollet, *Leçons de physique*, 4 (1764⁴), pl. 3.

Mercury thermometers fixed in Cavendish's manner—with corresponding attention to the lower point—might agree with one another at the temperatures of melting ice and boiling water. To obtain agreement at other temperatures, the maker had to graduate accurately the scale against which the mercury level would be read, and the user had to take into account that he generally applied the thermometer at a temperature and pressure different from those at which its scale was inscribed. Cavendish's committee provided tables and graphs of the small corrections they thought necessary to correct for this last and similar circumstances.[77] Many of their procedures became standard.[78]

While the Cavendish committee steamed its thermometers, Major William Roy, the commander of the Ordnance Survey in Britain, worried about correcting his barometers for the temperatures at which he used them. He could not content himself, in the important matter of choosing high places for fortifications, with the empirical correction to barometric readings devised by Deluc, namely one part in 215 per degree Réaumur above freezing. Roy preferred to measure the coefficient of thermal expansion of air directly. As a preliminary, he examined the boiling points of his best mercury thermometers and, like the Cavendish committee, found them to differ by over 3°F. Picking one as a standard, he inserted it and a tube of air containing a droplet of mercury as a plug into a long vertical tin can filled with water. He then heated the water with lamps brought up to the can; and, when the water was boiling rapidly, marked where the mercury plug stood in the glass tube. He then drained the water, removed the can, and marked the position of the plug in the tube (whose far end was open to the atmosphere) at 20° intervals as defined by his mercury thermometer. He had previously divided the tube into equal volumes by moving a drop of mercury along it. "Notwithstanding every possible precaution...irregularities would occur." Roy ascribed them chiefly to moisture, which gave a large adventitious increase to the measured coefficient of expansion. Nevertheless, he felt confident enough to declare that the expansion of air does not follow precisely that of mercury. For accurate work he preferred a table of expansions at various temperatures derived from his experiments.[79]

77. Ibid., 834, 843, 854, 856.
78. Middleton, *Invention* (1969), 71–72.
79. Roy, *PT*, 67:2 (1777), 689–694 (quote), 696, 704, 713; Feldman, *HSPS*, 15:2 (1985), 162, 173–176.

The determination of ϵ, the coefficient of expansion of gases with temperature, became a small research enterprise. The individuality of thermometers and the difficulty of removing all moisture combined with diverse and unequal experimental technique to yield a crop of divergent numbers. Lavoisier and Laplace settled on Deluc's number, $\epsilon = 1/215$ per °R ($1/269$ per °C), which, with their endorsement, claimed some consensus during the 1780s until challenged by Guyton de Morveau, who wanted a better value to reduce volumes of gas generated in chemical experiments to volumes at a standard temperature. He asigned the work to an officer in the Corps royal du génie, Prieur du Vernois, whom we shall meet again as a revolutionary politician and metric reformer. Prieur worked for two months on several gases and, according to Guyton, "obtained results free from error and with all the precision desirable." In fact, in his revolutionary way, Prieur had neglected to dry his gases and consequently found that they expanded much more rapidly at high temperatures than at low ones and that the rates differed among the various gases. Nonetheless, his numbers carried the field and persuaded natural philosophers who preferred exact measurement to neat theory that uniform laws of heat might be difficult to deduce from the apparently irregular behavior of gases.[80]

Just after the turn of the century a more comfortable view developed under the promptings of theory and practice. On the theoretical side, John Dalton and others who pictured a gas as a set of disconnected particles swimming in a sea of heat fluid inferred that all gases should display the same, presumably simple, mechanics of caloric. Discrepancies in their coefficients of expansion he ascribed to pesky vapors; and, in the one case in which he took the trouble to dry his gas, air, he obtained $\epsilon = 1/483°F = 1/268°C$ over the interval between 55° and 212°. To be sure, this was but an average: like Roy, Dalton made out that gases expand more quickly at higher than at lower temperatures. All gases: "Upon the whole,

---

80. Guyton de Morveau, AC, 1 (1790), 256–261, 265 (quote), 274–275, 284–285; Lavoisier and Laplace, in Lavoisier, *Oeuvres* (1862), 2, 322 (text of 1780); Fox, *Caloric theory* (1971), 63–66.

therefore, I see no reason why we may not conclude that all elastic fluids under the same pressure expand equally by heat."[81] Later, in 1809, Dalton gave an average of $1/479°F$ or $1/266°C$, remarkably, suspiciously, close to the reigning value obtained by Gay-Lussac in 1802.[82]

If we credit his stated rationale, Gay-Lussac undertook to find a good value for ϵ because astronomy (in the calculation of atmospheric refraction), technology (heat engines), physics and chemistry (gas volumes), meteorology (evaporation and humidity), and hypsometry (the heights of mountains, towers, and balloons) all required one. He filled the flask B (figure 2.2.12a) from a bottle M containing the gas under study (and, also, under pressure, from the water in the bucket QS). Having closed the valve R, Gay-Lussac fitted it with the long tube ID, placed the flask and its accoutrements in the iron cage EFGH, and immersed the whole in a water bath (figure 2.2.12b). He heated the bath and, at every $10°R$, opened the valve via the lever LL worked by strings; gas escaped through the mercury bath KX. When the water boiled, he left the valve open and removed the end of ID from the mercury, to allow the gas in the flask to come to equilibrium with the atmosphere. He then closed the valve, allowed the flask to cool, removed it to a bath of known temperature, opened the valve, and measured the amount of water that entered. He thereby obtained the dilation of the air during the heating and deduced a value of $\epsilon = 1/267$ per degree centigrade. In contrast to Dalton, Gay-Lussac dried his gases and confirmed the result for all of them; without, however, taking into account the expansion of the flask.[83]

Later Gay-Lussac showed, or claimed to show, that his ϵ held for all intervals of temperature between freezing and boiling. In these later investigations he used the arrangement of figure 2.2.13: sensitive thermometers placed vertically and horizontally and a tube

81. Dalton, Manchester Lit. Philos. Soc., *Memoirs*, 5:2 (1802), 595–602 (quote); Dalton, in Randall, *Expansion* (1902), 15–21; Fox, *Caloric theory* (1971), 68–69.

82. Dalton, *New system* (1809), 19; Dalton, in Randall, *Expansion* (1902), 22. The absolute readings of Dalton's thermometers were out by perhaps a third of a degree (Farrer, in Cardwell, *Dalton* (1968), 163, 172); but the error might not have afflicted the calculation of ϵ, which depended on a difference of temperatures.

83. Gay-Lussac, *AC*, 43 (1802), 135–140, and in Randall, *Expansion* (1902), 28, 38–44, 48; Gilbert, *Ann. Phys.*, 12 (1802), 396–398, called attention to neglect of the expansion of the vessels.

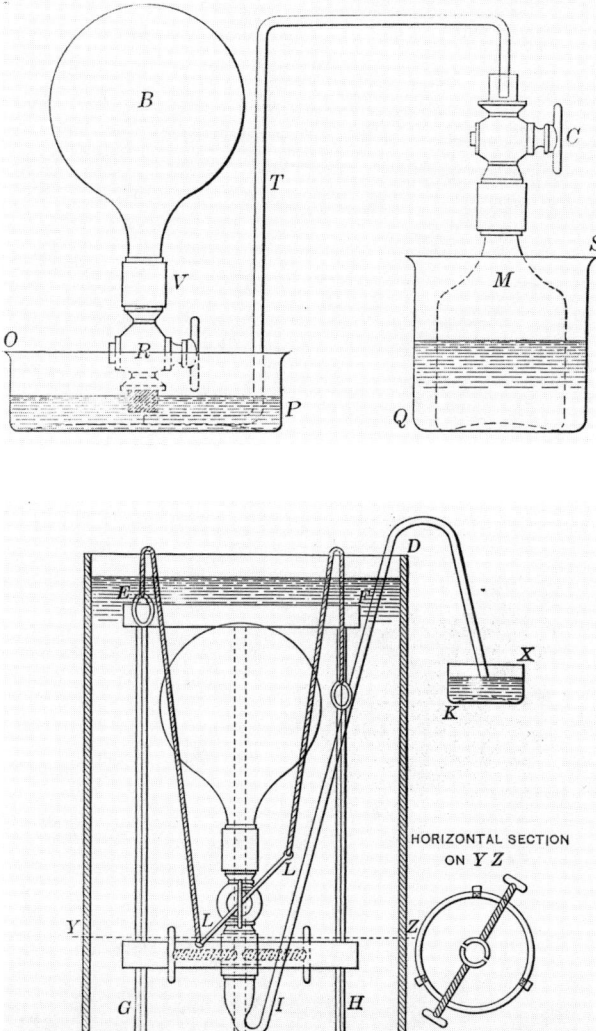

FIG. 2.2.12ab Gay-Lussac's apparatus for determining the coefficient of thermal expansion of gases, 1802. (a) Arrangement for filling the experimental flask B from a gas under pressure in the vessel M (b) flask B in its bath; the gas is expelled through the tube ID as the temperature of the bath increases. Randall, *Expansion* (1902), 38–39.

enclosing a sample of gas stopped by a mercury piston and open to the atmosphere shared a steam bath above water in a stove. Since the water providing the steam was enclosed in a metal vessel AABB, Gay-Lussac had to pull out the thermometer t and the tube GT from time to time to learn the temperature and the corresponding position of the plug M. He confirmed a linear expansion over 100° with the constant coefficient $\epsilon = 1/267\,°C$.[84]

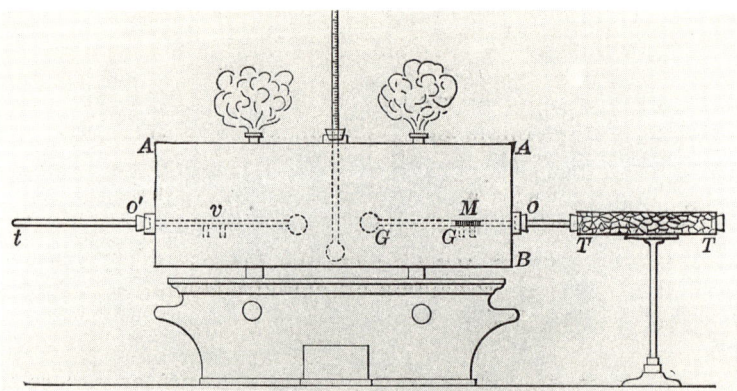

FIG. 2.2.13 Gay-Lussac's second method of obtaining the thermal dilation of gases. The instrument measures the motion of the plug M, which indicates the expansion, against the temperature registered on the thermometer t. TT is a tube for drying the experimental gas. Randall, *Expansion* (1902), 55, after Biot, *Traité* (1815).

Gay-Lussac's value of ϵ has stood up well. The most strenuous efforts by Victor Regnault, the world's authority on the physical properties of gases in the middle of the 19th century, succeeded only in driving the expansion down slightly, to 0.3665 for air (against Gay-Lussac's 0.375) over the hundred degrees between the freezing and boiling points of water. This was the mean result of four different methods all of which agreed in the first three decimals. A fifth method, which Regnault could not control, yielded values from 0.3552 to 0.3647. This unsteady method was Gay-Lussac's own.[85] With gases other than air, Regnault also obtained

---

84. Biot, *Traité* (1815), 1, chapt. 9; Randall, *Expansion* (1902), 54–55, 59–60.
85. Regnault, *ACP*, 4 (1842), 5–63.

agreement in the first three figures in all cases but carbonic acid. Further comparative measurements disclosed that gases differ very slightly in their rates of expansion.[86] Gay-Lussac's conclusion about the uniformity of the gaseous state does not hold exactly. That did not prevent it from inspiring the most advanced work in heat measurement during the heyday of the Standard Model.

Gay-Lussac's ϵ was both product and premise of his measurement, which required the assumption that equal increments of temperature on a mercury thermometer are proportional to equal increases in the volume of the expanding gas. If θ denote the reading of the thermometer and V the corresponding volume of the gas,

$$V_2 - V_1 = V_0 \epsilon (\theta_2 - \theta_1),$$

where V_0 indicates the volume at a reference temperature, say 0°C. This equation may be rewritten as $V = V_0 (1 + \epsilon \theta)$. Since, according to Boyle's law, $pV =$ constant for any particular sample of gas, or, rather, $pV = $ const.n, where n measures the amount of gas in the sample, the equation

$$pV = \text{const.}\, nT, \qquad 2.2.2$$

where $T = \theta + 1/\epsilon$, should describe the behavior of a gas in general. Equation 2.2.2 is known as the law of Gay-Lussac, or of Charles, after the physicist and balloonist Jacques-Alexandre-César Charles, whom Gay-Lussac credited with unpublished experiments that first indicated that all gases might expand equally when heated.[87]

Charles' law liberated thermometry from dependence on the instrument of its discovery, the mercury thermometer, and other materials unique of their kind. Thenceforth, in principle, physicists could employ interchangeably any representative of an entire state of matter.[88] With this liberation and generalization, the quantitative theory of heat attained a parallel to the measurement of magnetic and electral force. Just as the introduction of a geometry consistent with the Newtonian theory of imponderable fluids gave experimenters a way to deduce properties of *force* applicable not only to

86. Regnault, *ACP*, 5 (1842), 52–83.
87. Crosland, *Gay-Lusac* (1978), 25–28.
88. Cf. Barnett, *Osiris*, 12 (1956), 302, 312, 321; Gliozzi, *Cultura e scuola*, 14 (Oct-Dec 1975), 200.

the special apparatus of the immediate investigation, but also to all magnetic or electrical phenomena, so the discovery of the regularity of the dry-gas thermometer gave a universal measure of heat and temperature.

After the acceptance of Dalton's and Gay-Lussac's results, thermometricians reversed their approach and tested mercury and other thermometers for agreement with air thermometers. (To escape the vicious circle, the latter were calibrated by marking equal increments of volume along their stems, a procedure with some analogy to Volta's calibration of straw electrometers.) One of the earliest precise inquiries in this line was prosecuted, appropriately, by a practitioner of the most exact of the sciences, the astronomer Honoré Flaugergues. In 1813, in the service of hypsometry and astronomy (via temperature corrections to estimates of atmospheric refraction), Flaugergues published the results of measurements of ϵ for air from 0° to 80°R (0° to 100°C); "with minute and laborious attention," that is, by meticulously drying his samples, he obtained a value that reproduced his measurements to better than one part in 1000. On this secure base, he found that the mercury thermometer did not dilate equally with equal accessions of heat, but exponentially, as $y = y_0 \cdot \exp(x/a)$, where y is the length of the mercury, x the heat (that is, the temperature as determined by an air thermometer), a a constant. To first approximation, $y = y_0(1 + \epsilon_{Hg} T)$, $\epsilon_{Hg} = 1/a$; the remaining terms may indicate that logarithms played a greater part in the mind of the measurer than they do in the economy of nature.[89]

In the taking of pains, Flaugergues held not a candle to Pierre Louis Dulong, a physician and pharmacist who worked for Laplace's colleague Berthollet, and Alexis-Thérèse Petit, a prodigy from the Ecole polytechnique. In lengthy memoirs published between 1816 and 1818, these compulsive measurers reported their comparisons of the mercury and air thermometers, their determinations of the relative dilations of mercury, iron, copper, and platinum, and their opinions of Dalton and John Leslie. The last is most easily told. The English, like many others, had confined their measurements to too small a range of temperatures and too meager an array of materials to support their generalizations; Dulong and Petit

---

89. Flaugergues, *JP*, 77 (1813), 274–275 (quote), 278–279, 284–287.

ascribed their own strength to their breadth, and also to their pettifogging meticulousness. "Neglect[ing] nothing that could contribute to the exactness of the results," they attained, or so they boasted, "the highest degree of precision compatible with such delicate measurements."[90]

They compared mercury and air thermometers by heating them together in an oil bath; that of mercury placed vertically with its stem extending beyond the bath; that of air placed horizontally and ending, outside the bath, in a point open to the atmosphere. When the oil reached the temperature θ_1, as told by the mercury thermometer, they sealed the air thermometer and removed it to a mercury trough maintained at $\theta_2 < \theta_1$. When the air thermometer had cooled to θ_2, they broke its point under the mercury, which entered until it had filled the volume vacated by the gas during its heating from θ_2 to θ_1. They weighed the mercury that entered (w_2, say) and also the amount of mercury that just filled the tube (w_1). At both temperatures the gas filled the same volume at atmospheric pressure; from the law of Gay-Lussac et al. (equation 2.2.2), we deduce that $n_2/n_1 = (T_1/T_2)_{air}$, where the subscript indicates that the temperature is that registered by as well as in the gas thermometer. But $n_2/n_1 = (w_1 - w_2)/w_1$. Hence Dulong and Petit had the ratio $(T_1/T_2)_{air}$ independently of the mercury thermometer. Comparison with the readings $(T_1/T_2)_{Hg}$ showed that, although the scales ran parallel over the usual range, 0° – 100°C, thereafter the mercury thermometer advanced faster than the air thermometer; when quicksilver showed 300°, air indicated only 292°.[91]

The question naturally arose whether the exemplary behavior of the mercury thermometer indicated that mercury and glass separately expanded uniformly between the freezing and boiling points of water or only that, when combined into a thermometer, they cancelled one another's irregularities. To answer the question, Dulong and Petit had to measure the absolute dilation of one or the other. They chose mercury. They observed that the question, whether mercury expanded by equal increments of heat, bore not

90. Dulong and Petit, *ACP*, 7 (1818), 116, 141 (quote), 114–115, 340–341, 362 (criticism of Dalton and Leslie); Lemay and Oesper, *Chymia*, 1 (1948), 171–172, 177–178.

91. Dulong and Petit, *ACP*, 2 (1816), 244–250.

only on thermometry, but also, and more importantly, on hypsometry: "knowledge of the absolute dilation of mercury became essential as soon as the possibility of measuring mountain heights exactly by means of the barometer was perceived." Their method of pursuing this knowledge was precise and clever.[92]

The method yielded absolute dilations by obtaining the density of mercury as a function of its temperature. The measurement amounted to finding the height of a column of heated mercury that balanced a similar column maintained at the melting point of ice. In figure 2.2.14a, the reference column stands in the tube AB, which is enclosed by a bucket containing melting ice. The heated column A'B' communicates freely with AB via the pipe BB', maintained rigidly horizontal by the levelled iron frame MN. A'B' rises through the cover of a copper cylinder filled with oil; the cylinder also contains the mercury thermometer DE and the air thermometer D'E'. The brick structure represents half of the oven in which the cylinder resides. When the furnace heats up, some oil is expelled to the pot Q; some mercury, to the dish L; and some air, to the graduated tube H'K'. The mercury in the tube A'B' rises to the height h' above D and that in AB settles at the height h. Dulong and Petit read the heights through the telescope of figure 2.2.14b, which, in keeping with the other lavish instrumentation, was mounted on a marble slab T resting on a heavy masonry pillar.[93]

The columns of mercury in the tubes AB and A'B' must be in hydrostatic equilibrium. The heights of the columns h and h' do not depend, however, upon the lengths of the tubes containing them. That is the capital point of the method: the expansion of A'B' during heating does not menace the measurement. The densities δ of the mercury in the two columns goes inversely as their volumes, or, since the tubes have the same cross section, as the heights: $\delta h = \delta' h'$. A little cylinder of mercury in the column AB of volume V would expand to a volume $V' = V(1 + \epsilon_{Hg}\theta)$, where ϵ_{Hg} is the average absolute dilation of mercury over the range $0°$ to $\theta°$. Dulong and Petit determined θ by the mercury thermometer DE; they weighed

---

92. Ibid., 259, 262–263, with the result that the expansion of the two substances compensate; Dulong and Petit, *AC*, 7 (1818), 119–121, 124 (quote). Cf. Barnett, *Osiris*, 12 (1956), 325–327, and Fox, *Caloric theory* (1971), 235–236, for other plaudits.

93. Dulong and Petit, *AC*, 7 (1818), 127–134.

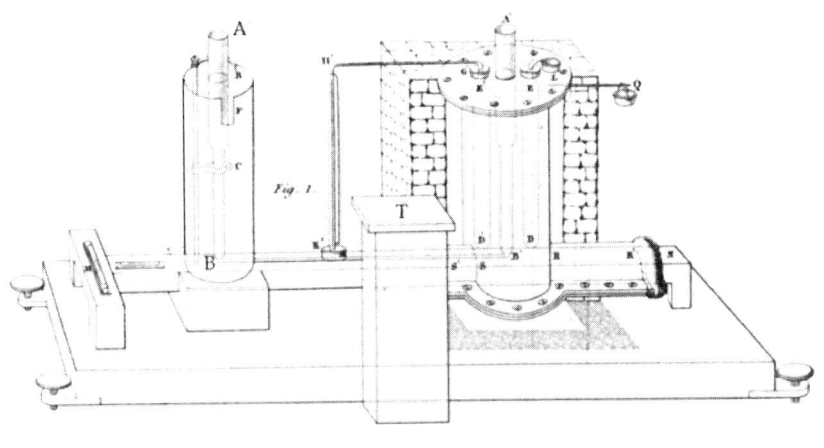

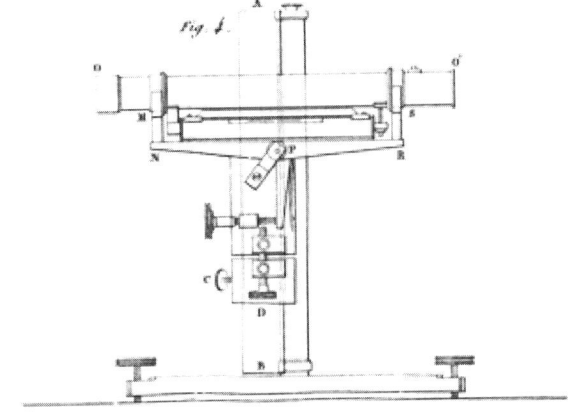

FIG. 2.2.14ab Dulong and Petit's measurement of the expansion of mercury. Upper figure: the apparatus for balancing mercury at the temperature of melting ice (in the tube AB) with mercury at the boiling point of water (in the tube A'B'). Lower figure: the telescope, mounted on the pillar T in figure 2.2.14a, for finding the heights of the mercury in the tubes. Dulong and Petit, *CP*, 7 (1818), pl. 1.

the overflow into the basin L and fixed θ_{Hg} as (w/W) degrees, w being the weight of the overlow and W that of the mercury that just fills DE at 0°. Since the volume V' of the little cylinder goes inversely as its density, $V'/V = h'/h$. Hence

$$\epsilon_{Hg} = \frac{h' - h}{h\, \theta_{Hg}}.$$

Dulong and Petit also obtained the temperature on the air thermometer D'E'. In a preliminary experiment, they brought the oil bath almost to a boil leaving the end of the tube H'K' open to the atmosphere. They then plunged K' into a reservoir of mercury. As the oil cooled, the mercury entered the tube. At a subsequent measurement at a temperature between those of cool and boiling water, Dulong and Petit determined the reading of the air thermometer θ_{air} from the atmospheric pressure p and the height k of the mercury in the tube H'K': the pressure p' in H'K' equals $p - \delta_{Hg} g k$, where g is the acceleration of gravity, and θ_{air} follows from p' and Charles' law. As they remarked, θ_{air} adds nothing to the precision of the experiment; rather, the experiment afforded another chance to relate θ_{Hg} to θ_{air}. Their results, the averages of "a great many measurements" each reliable, in their estimate, to three or four figures:

Absolute dilation of mercury[a]

θ_{air}	θ_{Hg}[b]	Average dilation/degree
0–100°	0–100°	1/5550
0–200°	0–204.61°	1/5425
0–300°	0–314.15°	1/5300

a. After Dulong and Petit, ACP, 7 (1818), 136.
b. Temperature computed on the mercury scale, not the mercury-glass thermometer. From the general supposition, or definition, of temperature via linear expansion, $V = V_0(1+\epsilon\theta)$ or $\Delta V/V_0 = \epsilon\theta$, we have $\Delta_1 V/\Delta_2 V = \epsilon_1\Delta_1\theta/\epsilon_2\Delta_2\theta$. Here the subscripts refer to the intervals 0°–100° and 0°–200° as defined by the air thermometer; measurement disclosed that $\Delta_1\theta = 100°$ on the mercury thermometer as well as on the air thermometer. Therefore $\Delta_2\theta = (100)(\epsilon_1/\epsilon_2)(\Delta_2 V/\Delta_1 V)$. But, by definition, $\Delta_2 V = 2\Delta_1 V$, and $\Delta_2\theta = 200(5550/5425) = 204.61°$.

Later measurers have abundantly confirmed Dulong and Petit's numbers and the conclusion that mercury expands more rapidly at higher than at lower temperatures.[94] The upshot of their inspired petifoggery was to double the range over which their colleagues could measure temperature reliably and thence compute the

94. Middleton, Barometer (1964), 180.

quantities of heat exchanged in their experiments. For example, in 1820 Gay-Lussac, whose measurements had underpinned exact thermometry, joined Dulong and Petit in taking the air thermometer as standard, and employed one that registered 0.06°C/mm over the normal range of temperature.[95]

Quantity and capacity. As mathematicians know very well, imaginary quantities may have real measures. Physicists easily absorbed this truth while dealing with the imponderable fluids. An elegant example is calorimetry as practiced by Lavoisier and Laplace. The instrument they invented—the ice calorimeter—had, as they said, the very great advantage that it could produce numbers about heat "independent of every hypothesis about [its] nature." They represented that their primary purpose in proffering their numbers was to introduce the instrument, and to recommend it for its "precision and generality," not to declare truths about thermodynamics.[96] Nonetheless, their asides favored the (im)material theory. For example, they invoked Lavoisier's conjecture that air supplies both light and heat during combustion (by putting its caloric in play) and they guessed that specific heats should increase with temperature since bodies bind heat (as they might bind a substance) as they expand.[97]

Lavoisier and Laplace intended their method to supersede the adaptation to specific heats of an earlier technique of finding the temperature of a mixture of hot and cold water. In this adaptation, practiced by Wilcke and Black, the measurer dropped an object heated to a temperature T_1 into an equal mass of water at temperature T_2. Then, if θ is the equilibrium temperature and ω the specific heat of the object, $\omega = (\theta - T_2)/(T_1 - \theta)$.[98] As Lavoisier and Laplace observed, the technique suffered from grave inaccuracy for substances with specific heats much different from water's and for all measurements that took any time to come to equilibrium; and it was scarcely applicable to substances that reacted chemically with water or to the heat developed by combustion or respiration. Wilcke had sought to improve the technique by immersing the calorimeter in

---

95. Gay-Lussac, *ACP*, 13 (1820), 305.
96. Lavoisier and Laplace (1780), in Lavoisier, *Oeuvres* (1862), 2, 287 (quote), 283 (quote), 332.
97. Ibid., 315, 319–320. Cf. Fox, *Caloric theory* (1971), 52–53.
98. Cf. McKie and Heathcote, *Discovery* (1935), 43, 98–104, 123–130.

snow and taking the water melted from the snow by the cooling of the test body as a measure of specific heat; but heat exchanged between the snow and the environment vitiated the measurements.[99] Lavoisier and Laplace did better.

Their calorimeter consisted of a vessel divided into three compartments. The innermost consisted of the wire basket LM (number 3 in figure 2.2.15). They packed the basket containing the specimen and its cover, HG, with crushed ice at the freezing point of water and inserted it into the intermediate chamber bbb (number 1 of figure 2.2.15), also charged with crushed ice. Ice melted by the cooling of the specimen ran off through the holes in the basket and the lid into the funnel cd, whence, passing through the stop-cock at k, it collected in the jar P. The quantity of run-off, determined by weighing P, measured the heat lost by the specimen in descending from its original temperature to that of melting ice. So far, the scheme resembled Wilcke's. What differentiated it, what made its "principal advantage," was the external envelope aaa, also filled with ice, but separated completely from the experimental space. The melting of ice in the envelope maintained the basket and the intermediate chamber at the freezing point; its run-off, exiting through the spigot T, could not be confused with the tell-tale drip through k. With this apparatus, Lavoisier and Laplace offered to obtain specific heats "exactly," or, anyway, to within one part in forty, or in sixty, if the day were cold; but, in fact, their fundamental measurement, of the amount of heat required to melt a unit weight of water, erred by one part in sixteen.[100] Some of their numbers for specific heats differ from modern values by 50 per cent.[101]

With the description of the ice calorimeter and the reporting of specific heats to six insignificant figures, Lavoisier and Laplace gave natural philosophers the goal, but not the means, of precision measurements of heat phenomena. The ice calorimeter thus played a part analogous to the torsion balance: adherents of the Standard Model accepted it as appropriate in theory but inexact in practice.

99. Ibid., 96–7, 104; Lavoisier, *Oeuvres* (1862), 2, 291, 300n.
100. Ibid., 296–302. Lavoisier and Laplace found that the amount of heat required to melt a unit weight of ice at the freezing point of water would raise the same weight of water from the freezing point three quarters of the way to the boiling point (p. 301). The correct figure is about four-fifths.
101. Lodwig and Smeaton, *Ann. sci.*, 31 (1974), 10–12.

SOME MEANS TO THE END / 103

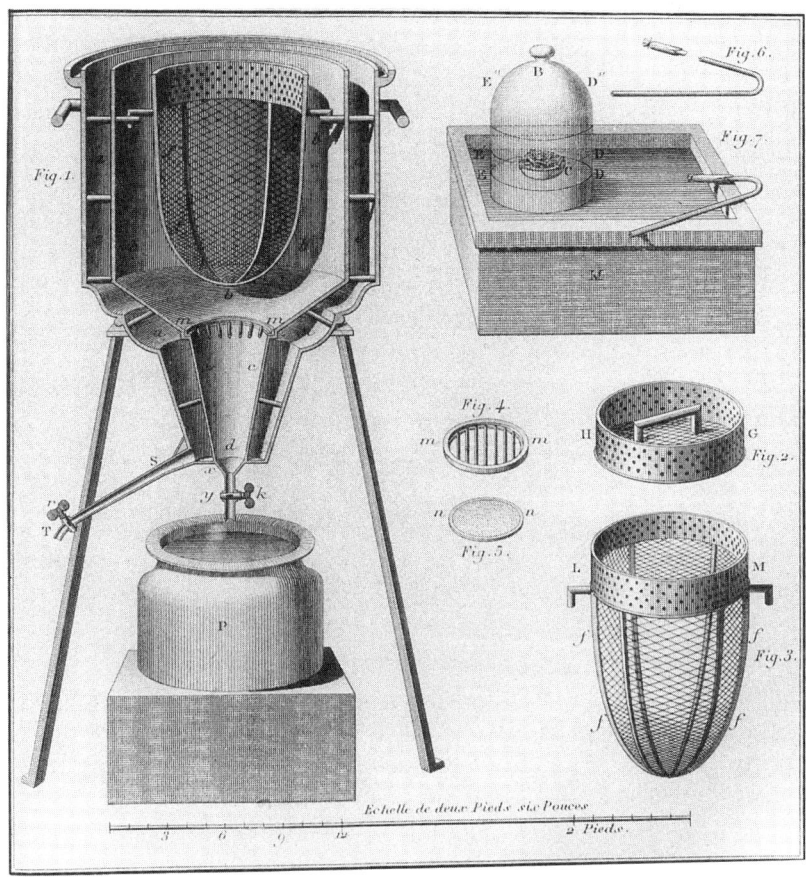

FIG. 2.2.15 Lavoisier and Laplace's ice calorimeter, ca. 1780. The experimental object resides in the innermost basket (figures 2, 3); the entire interior is packed with crushed ice; and the run-off of melted ice through k into the pot P measures the object's specfific heat. From Lavoisier, *Oeuvres* (1862), 2, pl. 2.

The calorimeter failed because not all the melted water made its way into the measuring pot; the rest remained in the ice, retained by capillarity. Lavoisier and Laplace recognized the possibility of this nuisance, but dismissed it as insignificant, since, as they argued, the ice ended the experiment as waterlogged as it began. But the discrepancies between measurements made in their way and ones conducted by the old method of mixtures ruined the reputation of the ice calorimeter, which physicists abandoned the more readily on account of its expense. In this they followed J.H. Hassenfratz, who, despite the advantage of having worked with the ice calorimeter under the direct supervision of Lavoisier and Laplace, could not reproduce their results; and, preferring his numbers to theirs, rejected their machine. "It has failed in the hands of everyone who has attempted it since [Lavoisier and Laplace]," ran the gleeful summary of the British school.[102]

The *Index* to the Royal Society's *Catalogue of scientific papers* lists only one item reporting measurements at the ice calorimeter between 1800 and 1870. Its author had apparently missed the news about the water-retentive ice, which he did not mention in his discussion of the systematic errors of the instrument.[103] In 1870 Robert Bunsen arranged matters so as to avoid collecting water. An inner chamber filled with ice-cold water and surrounded by ice floating on mercury received the hot object under study. The melting of the ice by the cooling object produced water; the loss of volume consequent on the contraction of the melted ice was made good by an inflow of mercury, which registered on a sensitive gauge connected with the mercury in the outer chamber. Bunsen calculated the amount of melted ice from the displacement of the mercury. The method is capable of great accuracy.[104]

With full faith in their calorimeter, Lavoisier and Laplace adapted it to the difficult chore of measuring the specific heats of gases. In order to compensate for the low density of their experimental material, they arranged a continuous flow of gas through the calorimeter. From measurements of the temperature of the gas on entry and exit, the rate of flow, and the quantity of melted ice, they could compute a value for the heat capacity of the specimen under study. In experiments performed during the winter of 1783/4 but not published until 1805, they made the specific heat of oxygen to be 0.65, and the specific heat of air 0.33031 (!), that of an equal weight of water.[105]

102. Ibid., 4–5 (Hassenfratz), 15 (quote from Thomas Thomson, *An outline of the sciences of heat and chemistry* (1830), 67–69).
103. Volpicelli, *Giornale arcadico*, 60 (1833), 70–71.
104. Bunsen, *Phil. mag.*, 41 (1871), 161–163, 178–179, pl. 5 after p. 244.
105. Fox, *Caloric theory* (1971), 35. The modern values for specific heats at constant pressure of oxygen and air at 100°C are 0.218 and 0.240 cal/g.

The year after this belated publication, Gay-Lussac reported measurements on the specific heats of several gases, inspired, he said, by a result he and Alexander von Humboldt had then recently obtained. Examining the classic problem of the proportions in which oxygen and hydrogen combine to form water, they found that the reaction stopped before all the hydrogen had disappeared when oxygen was present in excess. The same thing happened even with the proper amount of oxygen if nitrogen also occupied the experimental space. They supposed that the reaction ceased because the excess gas carried away the heat needed to sustain it; and they conjectured that, because excess oxygen and its substitute nitrogen had about the same effect, the two gases had the same heat capacity. From there the energetic generalizer Gay-Lussac readily sprang to the hypothesis that all gases have the same specific heat.[106] His experiments of 1806 were intended to confirm the hypothesis.

The compulsion to find a "law" about imponderables, which had been rewarded splendidly in the work of Coulomb and Gay-Lussac himself, now plunged Gay-Lussac into absurdity. He took two identical spherical vessels, one filled with air, the other empty, and allowed the air to expand freely into the void. Alcohol thermometers disclosed that, in the process, the full vessel cooled by the same amount as the other warmed. Furthermore, $\Delta\theta$ was exactly proportional to the denisty (or pressure) in the full vessel before the expansion—exactly, to anyone willing to set 61 = 80 and 34 = 40 (the first of these pairs of figures are $100 \cdot \Delta\theta$ as measured, the second what proportionality to density required). A similar result to a similar approximation held for hydrogen, except that $\Delta\theta$ came out larger. Gay-Lussac leapt to another false "law": the specific heats at constant volume of gases under the same pressure and temperature are inversely as their respective densities.[107] In a few years he came to doubt this regularity. In 1812 he replaced it with another, or rather with two. Trying an old-fashioned calorimetric mixing method—two containers of gas at different temperatures put into

---

106. Gay-Lussac, *MSA*, 1 (1807), 180–181; Fox, *Caloric theory* (1971), 129.
107. Gay-Lussac, *MAS*, 1 (1807), 183–201.

thermal contact—he reached the conclusion that all gases have the same specific heat by volume. A little later he reversed, or, better, inverted himself, and found that their specific heats under the same pressure and temperature are as their densities.[108]

Investigators as sloppy as Dalton and as scrupulous as Dulong and Petit shared Gay-Lussac's prejudice that simple laws ruled the world of imponderables. Dalton floated many simple relations between the volume and temperature of a gas and the distance between its particles, for example, that temperature is proportional to distance and hence to (volume)$^{1/3}$. He also fancied that every gas atom at the same temperature had the same capacity for heat, a proposition that did not survive the first exact measurements of the specific heats of gases.[109] Dulong and Petit were luckier in their generalizations. They began a systematic study of the specific heats of metals believing that the specific heat per metal atom c_a might indeed be a constant. Then they measured. Heating their metallic specimens slightly above the temperature of the air, they observed the time Δt each took to cool to its environment; taking $c \propto \Delta t$, they obtained specific heats by weight to three significant figures. Now $c_a = c/n$, where $n =$ the number of atoms in a gram of the substance under consideration; and $n = \text{const.}/A$, A the substance's atomic weight. Measurements on 13 metals gave for $c \cdot A$ values ranging from 0.3685 to 0.3830. "The mere inspection of these numbers shows an approximation too remarkable in its simplicity not to disclose immediately the existence of a physical law capable of being generalized and extended to all elementary substances."[110] They too overplayed their hands, although their law does hold good to a high approximation for metals at room temperatures.

The first trustworthy measurements of specific heats of gases—the measurements that subverted Dalton's guess that c_a is the same for all gases—came as a result of a prize competition of 1812 organized by the First Class of the Institut de France. These gentlemen, who

---

108. Gay-Lussac, AC, 81 (1812), 100–104; Fox, *Caloric theory* (1971), 131–133. Gay-Lussac did often hit when he aimed at generalities; cf. his "law" that gases always combine in simple ratios by volume (MAS, 2 (1809), 207–34).

109. Fox, in Cardwell, *Dalton* (1968), 192–198.

110. Dulong and Petit, ACP, 10 (1819), 398–403, 404–405 (quote); Dulong and Petit took the atomic weight of oxygen as 1. Cf. Lemay and Oesper, *Chymia*, 1 (1948), 178–180.

included Gay-Lussac, called "the attention of physicists" to the need to determine, rather than conjecture, specific heats of gases. Without them, "no exact research can be carried out on the amount of heat given out in various chemical reactions." Two sets of "physicists" responded: François Delaroche, perhaps a medical doctor and certainly a friend of Biot's, and Jacques Etienne Bérard, certainly a medical doctor and perhaps a friend of Gay-Lussac's, who won the prize; and Nicolas Clément and Charles Bernard Desormes, the latter father-in-law to the former and both industrial chemists, who obtained honorable mention. A secondary purpose of the competition—the first being to measure—was to subvert the schismatic caloric sect, represented in 1812 by Dalton and Leslie, who denied the existence of latent or bound heat. In their view, the heat developed in any process arose owing to alterations in the specific heats of the substances undergoing change; consequently, as will appear, they ascribed a finite specific heat to vacuum. Delaroche and Bérard held to the orthodox position of Lavoisier's school; Clément and Desormes betrayed a taint of Dalton's heresy; both sets of investigators obtained excellent values for the specific heats of the common gases.[111]

The apparatus with which the winners worked is schematized in figure 2.2.16. Number 1 shows the calorimeter AB, a copper vessel containing cold water and a spiral tube about a meter long through which the warm gas under study flows. Delaroche and Bérard deduced the relative values of the specific heats from the temperatures $\Delta_i T$ at which the several gases steadily maintained the calorimeter above the ambient temperature. Number 2 schematizes the system for obtaining an even flow of gas. The vessel B contains the gas, the flask A, water, which initially stands at the level GH. In order to have a constant pressure on the gas as it exists through M on its way to the calorimeter, A is firmly sealed apart from the small air-filled tube NO, which plunges to the depth OI. Consequently the pressure at O is that of the atmosphere p_0 and, as long as the water level in A remains above OI, the pressure at D remains constant at $p_0 + \delta_w g \mathrm{IK}$, where δ_w stands for the density of water. The tube QR provides a route to refill B from another gasometer

111. Fox, *Caloric theory* (1971), 134 (quote from prize announcement of 1811)–138; Olson, *Ann. sci.*, 26 (1970), 298–299.

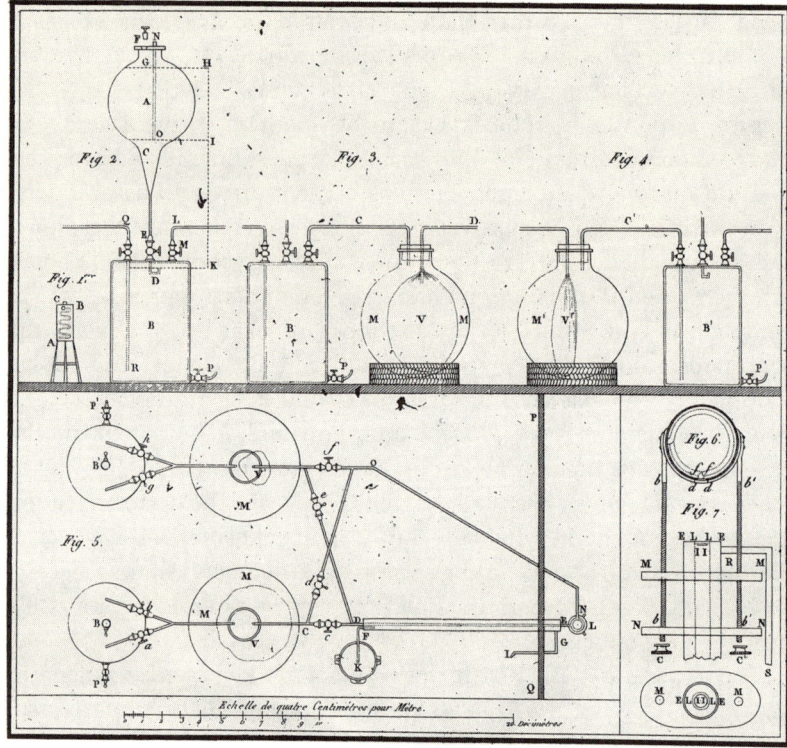

FIG. 2.2.16 Delaroche and Bérard's apparatus for measuring the specific heat of gases, 1813. Figures 2–5 indicate the pressure system that causes the experimental gas to flow; figures 1, 6, 7 give details about the calorimeter. Delaroche and Bérard, *AC*, 86 (1813), pl. 1.

and the tap P allows for drainage. To handle gases soluble in water, Delaroche and Bérard added a flask M between the gasometer and the calorimeter (number 3 of figure 2.2.16). M contains air and the bladder V the soluble gas, say hydrogen. V holds about 40 liters when filled.[112] The duplicate set B', M', V' (number 4) makes possible an almost continuous flow of gas through the calorimeter.

112. Delaroche and Bérard, *AC*, 85 (1813), 86–94, 99n.

Number 5 of figure 2.2.16 shows a horizontal section of the complete apparatus. K represents a stove that sends steam through the long tube FG and out the branch GI; the steam heats the gas almost to 100°C as it passes through DE on its way to the calorimeter L, which dwells in something like thermal isolation from the stove and the experimenters just beyond the wall PQ. An experiment begins with B full of air, B' full of water, and V full of hydrogen; V' is empty. The taps a, c, f, and h are open: air flows from B to M, compressing V, which expels its hydrogen through the heating tube DE and into the calorimeter L; whence the hydrogen, having deposited its heat, flows into V', pressing the air from M' into B', whose water runs out through the drain P'. When all the air has left B and all the hydrogen V, the experimenters (it was good there were two) closed a, c, f, h, and E (see number 2 of figure 2.2.16), and opened P; and also opened g, e, d, b, and E', and closed P', all in the twinkling of an eye (two seconds, to be precise). Thermometers gave the temperature of the gas on entering and leaving the calorimeter. After rendering this description, Delaroche and Bérard remembered their readers. "Perhaps you will find the details we have given about our apparatus a little long; but we thought that we ought not omit them because they will help you to appreciate the exactness of our results."[113] And that also may be sufficient justification for repeating them here.

Let the number of liters of gas passing through the calorimeter in unit time be s (at a standard temperature and pressure), and its drop in temperature during the passage be ΔT. At equilibrium, the gas delivers enough heat to the water to preserve it at a temperature ΔR above that of the surrounding air. If the loss of heat of the water to the air is taken to be proportional to ΔR, then $s \Delta T c_p$ (the rate of delivery of heat by the gas) equals const. ΔR. Here c_p means specific heat at the constant pressure under which the gas moves; Delaroche and Bérard emphasized that the quantity they determined, c_p, differed from specific heat taken at constant volume. They consequently accepted as the relative value of c_p at standard pressure and temperature for the several gases they tried $c_p \propto \Delta R / s \Delta T$. Taking c_p

113. Ibid., 95–103, 109 (quote).

of air as unity, they offered the following values at 0°C and 760 mm Hg:[114]

	Specific heat at constant pressure	
	by volume	by weight
air	1.0000	1.0000
hydrogen	0.9033	12.3401
carbonic acid	1.2583	0.8280
oxygen	0.9765	0.8848
nitrogen	1.0000	1.0318

The numbers have a misleading precision. They are good to two or three figures at most, since Delaroche and Bérard estimated the temperature of the gases at entry to the calorimeter only to three figures and corrected the direct effect of the oven on the calorimeter only to two.[115]

To increase the usefulness of these numbers, Delaroche and Bérard found the ratio of the specific heats of air and water. Their apparatus (number 9 of figure 2.2.17) pushed water through the siphon C at a steady drip. The stove R supplied the steam that warmed the water in its passage through the tube EE on its way to the bottom of the calorimeter at G. The branch EF contained three thermometers, as schematized more fully in the blow-up on the left of the figure; their readings allowed an estimate of the rate of cooling of the water in its travels, and hence of its temperature on entering the calorimeter. It finished its journey dribbling through the capillary point at P, thus making an observable record of its rate of flow. The calculation, which reduced the flow of water to that of air in the earlier experiments, yielded $c_{air} = 0.2669$ that of an equal mass of water.[116]

With this equivalence, Delaroche and Bérard could develop what they rated a devastating rebuttal of the view that the heat liberated in an exothermic reaction derived from the change in specific heats of the reactants. Take the case of water. Let the weights of hydrogen and oxygen that combine to make a gram of water be a and $1-a$, and let their specific heats be c_H and c_O. Hence the difference

114. Ibid., 114–117, 156–170.
115. Ibid., tables opp. 114, 126.
116. Ibid., 140–149, 165.

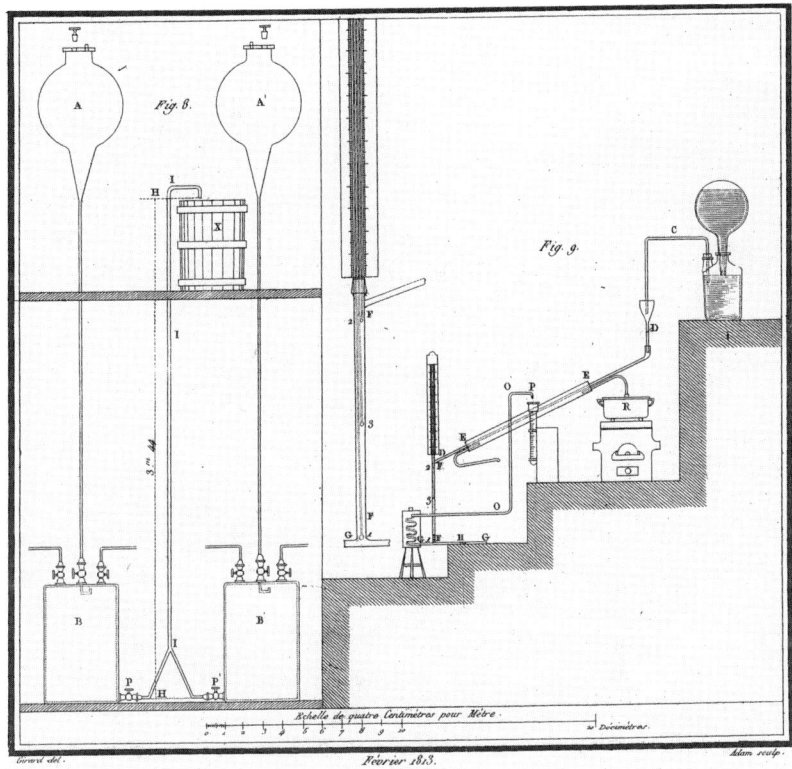

FIG. 2.2.17 Delaroche and Bérard's apparatus for measuring (figure 9) the relative specific heats of gases and water and (figure 8) the specific heats of gases under pressure greater than atmospheric. Delaroche and Bérard, *AC*, 86 (1813), pl. 2.

in specific heats of the ingredients and their combination is $0.2669[(.13)(12.34) + (.87)(.8848)] - 1 = -0.3663$: not only was there nothing to account for the heat produced, but the sum of the capacities of the components fell out far less than the capacity of their compound. Thus did accuracy eradicate heresy, in French caloric theory.[117]

117. Ibid., 74–77, 159–160, 169–175; cf. Fox, *Caloric theory* (1971), 138–142.

One more number. The apparatus at the left of figure 2.2.17 (number 8) pictures the hydraulic arrangement for running gases through the calorimeter at pressures above atmospheric. Delaroche and Bérard attempted only air, and only at one high pressure; their single result, that the specific heat per unit volume increased with pressure, confirmed the expectation with which they did the experiment, and misled caloric theorists for a decade. The increase did not keep pace with the increase in density, however, and so did not confirm one of Gay-Lussac's guesses. On this showing, the specific heat per unit weight would decline with rising temperature.[118] We shall return to this erroneous conclusion.

The experimental work just reviewed required extensive space and what for the time was elaborate apparatus. Delaroche and Bérard used the laboratory provided by Laplace and his colleague Berthollet at their adjoining residences at Arcueil near Paris, "where we had...every facility we could want."[119] Clément and Desormes were thrown on their own resources, which, fortunately, included their chemical plant. Like their competitors, they had bigger game in view than the measurements solicited by the Institute: they aimed to find the specific heat of space and to locate the absolute zero of temperature. They reached these desiderata via the apparatus depicted in figure 2.2.18. In number 2, which shows the method for obtaining specific heats, the large flask A contains the gas under study, the three-necked vessel I does the measuring, and its companion G allows experiments at pressures other than atmospheric. A measurement begins (with all taps closed) by pouring hot water in the box holding A. That causes the gas to expand into the space I and to push the liquid it contains (usually oil) up into the tube BC against the pressure in the vessel G. Clément and Desormes took the time needed for the oil to climb between two marks on the graduated tube BC as proportional to the specific heat of the gas; they obtained another measurement of the same quantity from the time needed for the oil to descend between the marks during the subsequent cooling of the apparatus. Their results, which agreed well with Delaroche and Bérard's, were:

118. Delaroche and Bérard, *AC*, 86 (1813), 132–138.
119. Ibid., 110; infra, §3.1.

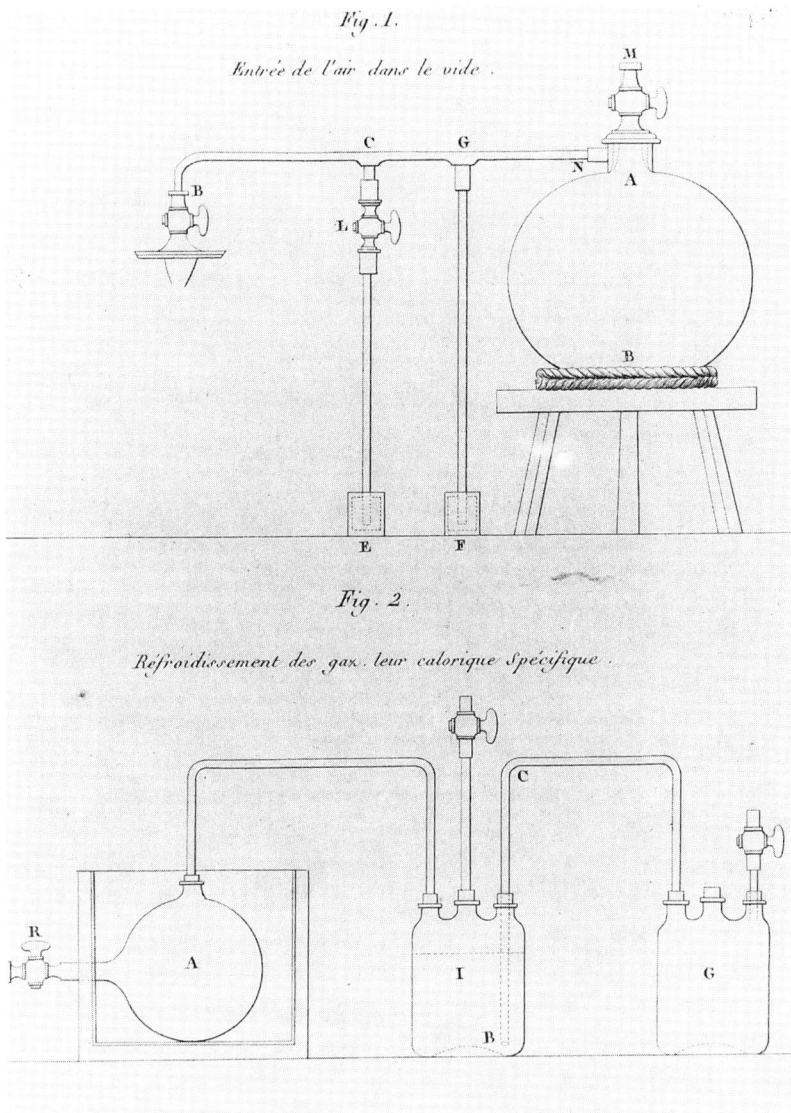

FIG. 2.2.18 Clément and Desormes' method of obtaining specific heats of gases, 1813. The gas in A (figure 2), heated by a water bath, passes into the vessel I, where its heat causes a liquid to rise in the tube BC. The rate of rise of the liquid measures the specific heat of the gas. Figure 1 is an apparatus for finding specific heat of vacuum; the stop cock at B is attached to an air pump not shown. Clément and Desormes, *JP*, 89 (1819), pl.

Volume specific heats, 10°–19°C

Gas	pressure	relative specific heat
Air	762 mm	1000
Air	563	848
Air	510	802
Air	352	679
Nitrogen	762	1012
Hydrogen	762	664
Carbonic acid	762	1500

To convert to the usual scale where water = 1, Clément and Desormes had recourse to the ice calorimeter, and succeeded in obtaining the fair value $c_{air} = 0.250 c_{water}$ by weight.[120] Delaroche and Bérard found 0.267.

Clément and Desormes pictured air at a pressure p less than atmospheric as a mixture of air at atmospheric pressure p_0 and void. They therefore could relate the volume specific heat of the mixture, c, to those of the atmosphere c_0 and of space c_s, through

$$c = (V_n c_0 + V_s c_s)(V_n + V_s), \qquad 2.2.3$$

where the V's indicate the partial volumes of atmospheric air and of space in the mixture. Let n_0 and n signify the number of "molecular units" in a unit volume of air at atmospheric pressure p_0 and at reduced pressure p, respectively. (A molecular unit—a concept Clément and Desormes do not use explicitly—signifies the volume occupied by a gas particle and its surrounding caloric at pressure p_0). Since all measurements were made at the same temperature and volume, $p_0/p = n_0/n$. In Clément and Desormes' vision, the gas at p may be drived from that at p_0 by replacing n_s molecular units by n_s "space units," so that $n_0 = n + n_s$. From $p_0/p = (n + n_s)/n$, $n_s/n = V_s/V_n = p_0/p - 1$. From the last and first entries for air in the preceding table, $p_0 = 352$, $c = 679$, and $p = 762$, $c_0 = 1000$, equation (2.2.3) delivers $c_s = 403$. With the second and third values for air in the table, $c_s = 417$ and 400, respectively; a coincidence that seemed to authorize the value $c_s = 410$ as the specific heat of vacuum.[121]

---

120. Desormes and Clément, *JP, 89* (1819), 335–338 (text of 1812).

The coincidences did not stop there. In number 1 of figure 2.2.18, Clément and Desormes indicated their method of finding the absolute heat contained in a void space. Let the vessel AB be filled with air at barometric pressure p_0. Hook up an airpump to the tap B and diminish the pressure in AB to p according to the mercury and water gauges LE and GF. Close B and disconnect the pump. The air remaining in AB would occupy a volume $V = (p/p_0)V_0$, V_0 being the total volume of AB, if under atmospheric pressure. Now open B for a split second—two-fifths of a second to be precise—which instantly heats the gas in AB and restores its pressure to p_0. When the gas cools to room temperature, note its pressure p'. And reason thus: the pumping introduced an amount of space V_s equal to $V_0 - V$, i.e., to $V_0(1 - p/p_0)$; the instantaneous inrush of air raised the temperature an amount $\Delta\theta$ calculable from Gay-Lussac's law,

$$p_0/p' = [1 + \epsilon(\theta_0 + \Delta\theta)]/(1 + \epsilon\theta_0),$$

$\theta_0 = 12.5°C$ being the ambiant temperature, and left a void space $V_0(1 - p'/p_0)$. Consequently, the heating represented by $\Delta\theta$ resulted from the annihilation of a volume of vacuum (who says *nihil ex nihilo fit*?) expressed by $(p' - p)(V_0/p_0)$.

Clément and Desormes made $V_0 = 28.4$ liters, $p_0 = 966.5$ mm Hg, $p = 752.7$ mm, $p' = 762.9$ mm, $\Delta\theta = $ "1.3212 degrees exactly." Therefore the annihilation of 0.377 liter of vacuum raised the temperature 1.32°C; if the entire V_0 had perished, it could have raised the temperature of an equal volume of air from 12.5° by $(1.32)(28.4)/(3.77) = 99°$. Clément and Desormes guessed that this number would have been something closer to 114° if they could have prevented losses of heat to the atmosphere during their measurement. And this value of 114°, they thought, agreed perfectly with the finding from the theory of the speed of sound (of which much more later) that a compression of ordinary air by 1/116 of its volume raises its temperature 1°. "Couldn't one say that a loss of 116/116 of space, that is, of the entire space, would raise the temperature of an equal volume of atmospheric air by 116 degrees?"[122]

121. Ibid., 341–342.
122. Ibid., 330–334 (quote). It is worth observing that $(p_0 - p)/(p' - p) = 1.35$ is not a bad approximation to c_p/c_v. Gliozzi, *Cult. e scuola*, 14 (Oct–Dec 1975), 205–206.

Further experiments of the same type revealed that vacuum at 18° when annihilated would raise the temperature of an equal volume of air by 102°; and that the comparable figure for vacuum at 98° was 132.24°, the air in both cases having the same density. Let x be the absolute temperature corresponding to 0°C; $c_s(x + 18)$ is then the absolute heat in a unit volume of nothing at 18°C. The heat required to raise a unit volume of atmospheric air by 102° or 102 volumes by 1° is $(102)c_{air}$. On these understandings, $(x + 18)/(x + 98) = 102/132.24$, or $x = 252°$. Alternatively, the value of c_s and the earlier result that at 12.5° an annihilated void can raise an equal volume of air 112° yield

$$(x + 12.5)(400) = (112)(1000),$$

whence $x = 267.5$. A most wonderful coincidence. And there is more. Gay-Lussac had found that all gases gain 1/266.7 of their volume at freezing for each additional degree centigrade. Reversing the process, a gas would contract to nothing, and hence permit no further cooling, at −266.7°C. The direct method gave 267.5, indirect reasoning from entirely different principles, 266.7. "We confess," the lucky experimenters wrote, "we confess that so singular an agreement is for us a powerful reason to believe in the precision of our conclusion."[123]

Two reasons may be offered for dilating on this extraordinary agreement between exact measurement and a theory now (and even then) considered altogether wrong.[124] For one, it serves as a reminder that numerical confirmation of theoretical predictions or expectations does not necessarily indicate a fruitful line of thought. Secondly, it suggests how deeply the set of analogies on which the Standard Model rested had settled. The interaction of vacuum and ponderable matter in the theory of Clément and Desormes resembles that of the two electrical fluids in the standard theory of electri-

---

123. Desormes and Clément, *JP, 89* (1819), 343–346 (quote). Later (ibid., 448–449, text of 1818–19) they added another coincidence: assume that the heat in equal weights of ice water at 0°C are $c_{ice}x$ and $1 \cdot x$, respectively, x being the absolute temperature, and let L be the heat of fusion. Then $x(1 - c_{ice}) = L$; with their own newly measured value for c_{ice}, 0.72, and the standard value for L, 75, they obtained $x = 267$!

124. Fox, *Caloric theory* (1971), 146–153, and Costabel, *Arch. int. hist. sci., 21* (1968), 7–14.

city: a gas changes its density by annihilating vacuum much as a positively charged body loses part of its electricity when touched by a negative one. Clément and Desormes themselves developed similar analogies. Thus, to complete its parallel with magnetism, electricity, and light, caloric must propagate in a space completely void of air. Furthermore, to complete its enrollment in the class of elastic fluids, its laws of motion must be the same as those of gases, temperature being for the one what pressure is for the other. The analogy to gases requires that caloric have mobility and elasticity, extension and impenetrability. And much more to the same purpose.[125]

Radiant heat. The quantitative characteristics of caloric so far examined were inferred from its interactions with matter, especially gases. The inferences required rested on the concepts of latent and specific heat, the expansion of bodies when heated, and the spontaneous flow of heat from warm bodies to cold ones. In none of this did caloric make an appearance apart from ponderable matter. Beginning in the 1770s, however, and thus contemporary with the enrichment of heat theory by the notions of latency and specificity, ways were found to study caloric free from matter. In these early investigations, light acted as the companion of heat. Hence investigators did not have to seek far for conceptual and instrumental apparatus: they adapted the equipment of optics.

A standard parlor trick of the time played with heat as if it were light, by reflecting it from two facing metallic mirrors. At the focus of one mirror the demonstrator placed an inflammable object, which burst into flames when he moved a single hot coal into the focus of the other. Since the ignition could be accomplished over a distance of twenty feet or more by a red hot coal, and over as much as twelve feet by a coal not quite warm enough to glow, the demonstration made an impression.[126] It also raised to a new urgency the old question of the relationship between heat and light. In order not to prejudice the answer unduly, the Swedish chemist Carl Wilhelm Scheele, who had come to the problem through his interest in gases, fire, and the chemical effects of light, introduced the term "strahlende Hitze," or radiant heat. He showed that radiant heat

---

125. Desormes and Clément, *JP, 89* (1819), 325–328, 336.
126. Evans and Popp, *Am. jl phys., 53* (1985), 738, 740, quoting Saussure, *Voyage* (1786), 2, 353.

could be reflected like light from metal mirrors; but he also emphasized the old observation that a glass plate allowed passage to the light from a fire but stopped the heat.[127]

The demonstration of the mirrors and the coal became a measuring apparatus in the 1770s with the substitution of a thermometer for the inflammable body. Using a coal that did not glow as radiator, Saussure learned that it could warm a thermometer whose bulb occupied the focus of a mirror twelve feet distant to 8° above the temperature of the surrounding air.[128] The phenomenon prompted two sorts of queries. First, if dark coals could propagate heat, what might cool coals, or cold coals, or ice, accomplish? Second, do the various colors of the solar spectrum have the same or different heating power, or carry similar amounts of radiant heat? Taking the second question first, several philosophers of the mid-1770s, notably the abbé Rochon, directed sections of the prismatic spectrum onto the bulbs of their thermometers. They learned that rays at the yellow to red end of the spectrum had a stronger "calorific effect" than rays at the blue end; and Rochon thought that he had proved that the maximum effect occurred in the yellow-orange.[129] This promising line of research then stagnated while the few students of radiant heat perplexed themselves over the frigorific emanations of ice.

The question of the heating power of ice, if any, was posed by a mathematician in Saussure's circle in Geneva, Louis Bertrand, to Marc-Auguste Pictet, who had collaborated with Saussure on the double reflection of calorific rays from dark coals. Pictet answered that ice being cold, and cold being the negation of heat, if ice replaced the coal the thermometer would remain at room temperature. The answer satisfied neither Bertrand nor the demands of mathematical continuity. Pressed further, Pictet did the frivolous experiment around 1790 and was nonplussed to discover that the temperature registered by the thermometer sank immediately when he put a chunk of ice in place of the coal. Caught without an explanation, Pictet concocted one that invoked tensions in the caloric

127. Scheele, *Chemische Abhandlung von der Luft und dem Feuer* (1777), quoted by Cornell, *Ann. sci.*, 1 (1936), 219–220, and Weiss, *Prevost* (1988), 222–224.
128. Saussure, *Voyage* (1786), 2, 353–354.
129. Rochon, as excerpted by Evans, *Phil. mag.*, 45 (1815), 404–408; Cornell, *Ann. sci.*, 3 (1938), 119.

held in the air. His friend Pierre Prevost, also of Geneva, invented a better theory: all bodies radiate caloric, warmer bodies more than colder ones; the temperature shown by a thermometer is the net of a dynamic exchange of heat with surrounding objects. To buttress the idea that caloric could fly through space as well as run through bodies, Prevost invoked the obvious analogy to the electric fluid(s), which can burst forth as sparks or flow unperceived in wires.[130]

Prevost prefaced the definitive presentation of his theory with the by-then usual disclaimer of truth. "The word *caloric* was invented to signify the cause of heat, with the intention, formally expressed, of saying nothing about its nature." For himself, Prevost judged the material theory to be the most satisfactory interpretation; it made "a clear and natural language" in contrast to the "obscure and artificial language" of the kinetic approach. Nonetheless, he had no objection to people with different linguistic tastes; "I shall not fight against any system." That understood, let us calculate. Take a and zero to be the heats of two interacting bodies at a time $t = 0$; $1/p$ the decrement of heat in the hot body in time dt; $1/q$, the fraction of the radiation of the one received by the other in dt; $(q-1)/q$, the fraction lost to space. Then, after n intervals dt, the heat H_1 of the body originally hot has become

$$H_1 = \frac{a}{2}\left[\left[\frac{p-1}{p} + \frac{1}{pq}\right]^n + \left[\frac{p-1}{p} - \frac{1}{pq}\right]^n\right],$$

and that of the body originally cold,

$$H_2 = \frac{a}{2}\left[\left[\frac{p-1}{p} + \frac{1}{pq}\right]^n - \left[\frac{p-1}{p} - \frac{1}{pq}\right]^n\right].$$

All this seems a model of a quantitative theory. Unfortunately, Prevost could not confirm it. "I know no experiments made carefully to which these formulas apply directly."[131]

130. Prevost, *JP*, 38 (1791), 317–319, 322n (on electric parallel), and *Calorique* (1809), 14, 28–32, 87–100; Evans and Bopp, *Am. jl phys.*, 53 (1985), 741, 743, 746; Weiss, *Prevost* (1988), 225–227, 246, 251, 255–259.

131. Prevost, *Calorique* (1809), 6, 8 (quotes), 53–54 (quote). The formulas are algebraic re-expressions of the easily deduced states of the bodies after 1, 2, 3...intervals dt: after the first, $H_1 = a(p-1)/p$, $H_2 = a/pq$; after the second, $H_1 = a(p-1)/p - a(p-1)/p^2 + a/p^2q^2$, and $H_2 = a/pq + a(p-1)/p^2q - a/p^2q$; etc.

The concept of dynamic exchange of radiant heat presented difficulties to the physicists of 1800. To the overly literal mind of Count Rumford, for example, exchanging made no sense whatever. It was, he said, "an operation not only incomprehensible, but apparently impossible, and to which there is nothing to be found analogous, to render it plausible."[132] He learned about Prevost's idea, and witnessed Pictet's experiment, during a stay in Edinburgh in 1800, while acting as the Elector of Bavaria's ambassador to Britain. He did not have the leisure to subvert Prevost by experiment until 1802, when, marooned again in Munich, he devoted himself to natural philosophy and to a campaign, at a distance, to win the hand of Lavoisier's widow.[133] Rumford rejected not only Prevost's exchanges, but also the entire caloric theory, against which he had already discharged the heavy artillery of his famous experiments on the grinding of gun barrels. All this made an extraordinary opportunity for Mme Lavoisier. "I think that I shall live to drive *caloric* off the stage [Rumford wrote his English mistress] as the late M. Lavoisier (the author of caloric) drove away *phlogiston*. What a singular destiny for the wife of two *philosophers!!*"[134]

Rumford did not contest the existence of invisible calorific or frigorific rays similar to light rays. Rather, he felt "the difficulty of explaining how, or by what mechanism, it can be possible for the same body to receive and retain, and reject and drive away, the same kind of substance, at one and the same time." His friend Pictet responded with the apt analogy of the prime conductor of an electrical machine, which, when fully charged, emits as much electricical fluid into the air in sparks as it draws from the machine.[135] Rumford answered with an *experimentum crucis* he believed unanswerable. He had invented a thermoscope, which amounted to a differential thermometer (number 2 of figure 2.2.19) consisting of a drop of tinged alcohol free to move under the pressure of opposing columns of air.

132. Rumford, *Works* (1969), 1, 422 (text of 2 Feb 1804).
133. Rumford, *Works* (1969), 1, 477–483 (text of 1804); Brown, *Rumford* (1979), 247–248, 256–263.
134. Rumford to Lady Palmerston, 8 Feb 1804, in ibid., 255, 268; Rumford, *Works* (1969), 1, 3–26 (the canon-boring experiments, text of 1798).
135. Ibid., 1, 422; Evans and Popp, *Am. jl phys.*, 53 (1985), 749; Weiss, *Prevost* (1988), 261.

A measurement began with the droplet in the center of the horizontal graduated tube. Rumford then placed on either side of one bulb of the thermoscope, and at equal distances from it, two empty identical cylindrical vessels of the type shown in number 3 of figure 2.2.19. One vessel he filled with hot, the other with cold water, the temperature of the one being as much above, as that of the other fell below, the temperature of the room and the bulb. The droplet did not move. Rumford then blackened the face of the vessel that was to receive the hot water, knowing from many previous experiments that a metallic object emitted heat rays more copiously when darkened or roughened than when shiny or polished. On repeating the *experimentum crucis* with the hot vessel blackened, he found as he anticipated that the droplet moved; the calorific effect dominated the frigorific, the air in the bulb dilated, and the increased pressure drove the droplet towards the unexposed bulb. Finally Rumford performed with both vessels blackened. The droplet stayed dead center. How could that be explained on Prevost's theory? Should not the blackening of the cold vessel cause it too to send out more calorific rays? And should that not have heated the bulb more, and driven the droplet further, than when only one vessel was blackened?[136]

Rumford took sound, not light, as the correct parallel to radiant heat. The heat in a body derives from the motions of its particles, supposed to be in constant movement; the radiation of heat occurs through a very subtle springy aether set pulsating by the zigs and zags of material particles. A hot body sends forth rapid pulsations, a cold body slower ones; their radiations differ as high notes differ from low notes. Rumford apparently received initiation into the doctrine of the aether in London around 1800, from Humphry Davy and Thomas Young, assistants at the then new Royal Institution, which Rumford had helped to create. Davy taught that radiant heat arose from the jostling of the aether by hot bodies,

---

136. Rumford, *Works* (1969), 1, 348–352, 361–363, 371–372, 422–426. Let α and β be the quantities of radiation received in unit time by the bulb from the hot and cold bodies, respectively, and let γ be the radiation received by the bodies from the bulb. The bulb's net is $\alpha - \beta + \beta - \gamma$, which is zero in the first and third cases ($\alpha = \gamma$) and > 0 in the second ($\alpha > \gamma$).

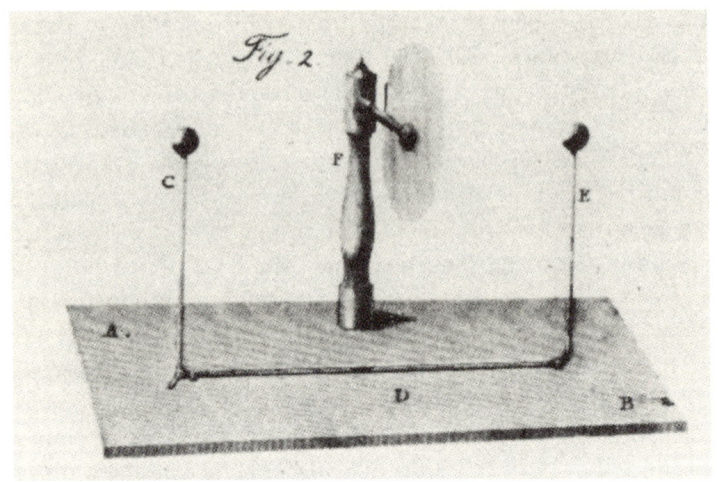

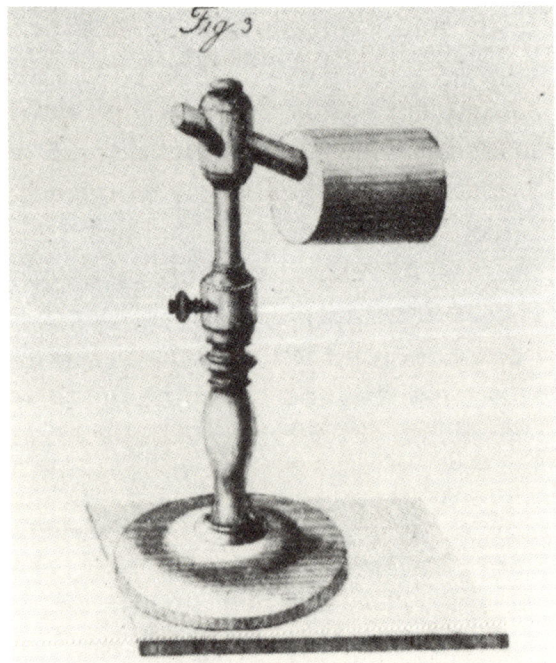

FIG. 2.2.19 Rumford's differential thermoscope, ca. 1802. The droplet of tinged alcohol at D (figure 2) is driven from the hotter of the two bulbs C, E. Cylinders of the type shown in figure 2, containing hot water, were used to warm the bulbs. Rumford, *Works* (1969), *1*, pl. 2.

while Young developed his undulatory theory of light.[137] When Rumford was at his most speculative, he supposed that light and radiant heat differed only in the frequency of their pulsations in the aether; we do not see heat because a benevolent deity, saving us from constant stimulation, made our eyes insensitive to it. Despite his rejection of the fluids of light and heat, Rumford did not escape entirely from the lure of the Standard Model. Throwing a sop to the defenders of imponderables, he allowed that "that eminently rarefied and elastic fluid in which *heat* and *light* are propagated" might properly be called "fire."[138]

Rumford much admired his thermoscope, which taught him that not only blackened surfaces, but also paper, linen, and shellacked metal surfaces, radiate more strongly than shiny ones. He deduced that black skins are advantageous to their possessors in the tropics and that the sensible man wore white clothes in the winter.[139] He took to wearing a very shiny white hat, to preserve the heat in his head. He would have made an odd companion to John Leslie, his overly exact contemporary in heat studies, who used to dye his hair purple. They never met, however, as Rumford was forced to insist. While he was busy publishing the results of the experiments that he had performed in Bavaria during the winter of 1803/4, Leslie's *Experimental inquiry into the nature and propagation of heat* came from the press. Leslie claimed to have begun his experiments in 1801, and by then to have perfected his chief apparatus, his differential thermometer, which was virtually identical with Rumford's thermoscope. In fact, the instruments had the same origin, namely Leslie's differential hygrometer.[140]

Rumford received his copy of Leslie's work from Joseph Banks, who probably expressed the general judgment of the English scientific establishment in saying that the book "contain[ed] many interesting experiments and much bad reasoning."[141] Leslie insisted

---

137. Brown, *Rumford* (1979), 239–240, 257; Rumford, *Works* (1969), 1, 315, 406–408, 414 (texts of 1804); cf. Goldfarb, *Br. jl hist. sci.*, 10 (1977), 27–28.
138. Rumford, *Works* (1969), 1, 309, 416, 440–441 (quote), texts of 1804.
139. Ibid., 330, 337, 435–436 (texts of 1804).
140. Brown, *Rumford* (1979), 260, 269–270, and *Am. jl phys.*, 22 (1954), 13–17; Rumford, *Works* (1969), 1, 348–349, 488–492; Leslie, *Inquiry* (1804), vii, 9.
141. Banks to Rumford, Apr 1804, quoted by Brown, *Rumford* (1979), 269. Cf. Olson, *Scottish philosophy* (1975), 210.

on going his own way: "Throughout the whole, I have freely exercised my reason, unawed by authority and uninfluenced by current opinion." He owed his successes (and also many of his failures) to "the superior delicacy" of the differential thermometer shown in number 1 of figure 2.2.20. It differed from Rumford's design in running the scale vertically and in using tinted sulphuric acid, rather than alcohol (which Leslie found to lose its color too quickly), for the marker. Number 1 of figure 2.2.20 shows the instrument in service: one bulb stands at the focus of the metallic mirror, the other far enough away to register the temperature of the air; the little canister on the left contains hot or cold water and a thermometer to take its temperature; the screen before the canister cuts off its heat rays when experimental protocol so required. Leslie accepted the rise or fall of the acid droplet in the differential thermometer in unit time as his measure of heating effect. Like Rumford but probably earlier, Leslie discovered that the rate of heating or cooling depends sensitively on the character of the surface of the emitting body.[142]

Screens of various materials cut off the transmission of heat to varying degrees. How so? Radiant heat cannot be light, as the effect of a glass screen abundantly shows. Whatever comes from the hot body "bears no analogy to the fluids, real or imaginary, of magnetism and electricity," again because screens affected each differently. What then? Radiant heat consists of pulsations in a medium. What medium? "It is merely the ambient AIR." Heat and light are elastic fluids; indeed, "heat is only light in a state of combination," locked in bodies, striving to free itself. When it gains its liberty it goes off like a gas rushing into a vacuum. The equivalence of the two may be demonstrated by the fact that a body warms in proportion to the light it absorbs and fixes.[143]

---

142. Leslie, *Inquiry* (1804), 9–11, 16–17, 21, 24. Cf. Olson, *Ann. sci.*, 25 (1969), 203–204, and ibid., 26 (1970), 276, 280, 282, 287, 294–295, and Burr, *Isis*, 21 (1934), 174–176.

143. Ibid., 27–29, 31–32 (first two quotes), 41, 143, 150, 160, 162 (last quote), 175, 188. Like Standard Modelists, Leslie drew a strict analogy between heat and electricity, whose propagation he also ascribed to pulsations in the air; Olson, *Scottish philosophy* (1975), 196, 199–200, 213–215.

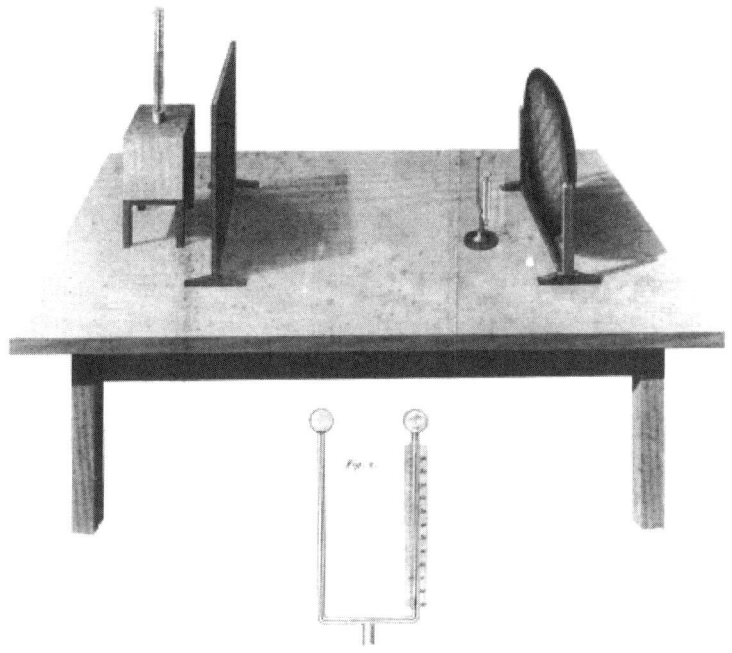

FIG. 2.2.20 John Leslie's differential thermometer, ca. 1801. Figure 1 shows the thermometer, which is presented in detail in figure 2, placed to measure the difference in temperature between the focus of a mirror receiving radiant heat and a neighboring point. Leslie, *Inquiry* (1804), pl. 1.

A body radiates by impelling the igneous fluid in the air. Let AG (figure 2.2.21) be the distance r from a point on the hot body to G, a point on a colder one, and let the air be divided into "primary intervals" s = AB, BC, CD...; just what the value of s might be is "perhaps beyond our penetration to discover." Further, set k = AH = the elevation of the temperature of the hot body above that of the surrounding air, and allow the ordinates at B, C, D...to represent the temperatures of their respective intervals. Now let the first section AB give off a pulse, which drops its temperature to AZ = BI, from which, if it has a continuing source of heat, it rises

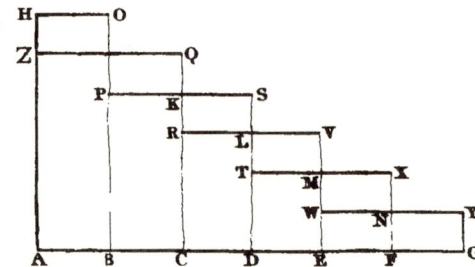

FIG. 2.2.21 Leslie's scheme to show the propagation of radiant heat in air; HA represents the intensity of the heat in the radiating body, YG that at a distant point. Leslie, *Inquiry* (1804), pl. 3.

again to HA = h; meanwhile, the section BC acquires the pulse and the temperature BI, which drops to FB = KC when the pulse passes to the next section, CD, and so on; the pulses require the time $s/e^{1/2}$ to pass each section, where e = the height of the column of air that corresponds to the ambient pressure. According to Leslie, the amount of heat flowing to G from A in unit time is the variation of temperature at each pulse, hs/r, times the power of each section to absorb heat, δas (δ is the air's density, a its "specific attraction for heat"), divided by the time of the pulse, $s/e^{1/2}$, viz., $has\delta e^{1/2}/r$. "Such then is the recondite process, which theory unfolds, of the communication of heat."[144] Since the formula contains unknowables, it could at best be fitted empirically; in fact, it says only that the rate of radiation is proportional to the difference in temperature between the radiation and the environment and, for a cylindrical geometry, inversely proportional to the distance. Leslie's argument may be regarded as a useful indicator of the difficulty felt around 1800 in devising a quantitative theory to fit the ever more precise and multitudinous facts.

Many of the multitudinous facts Leslie acquired concerned the effects of the surfaces of the emitters on the total radiant heat received by the differential thermometer. He varied not only their nature, but also their orientation, and so secured the result that not

144. Leslie, *Inquiry* (1804), 193–200, 201 (quote).

the total surface, but its orthographic projection, determines the heat conveyed to a fixed detector.[145] In this respect, as in many others, radiant heat resembled light. Indeed, despite his independence, Leslie accepted the Standard Model. In his view, light consists of particles whose mutual repellency persists when they are combined in matter; heat is an igneous fluid, composed of mutually repulsive elements, highly elastic and extremely subtle; which, "if it were separately exhibited, would assume a gaseous and expansive form." Although he hedged about the existence of the fluids of electricity and magnetism, he offered no alternatives.[146]

A set of misinterpreted facts inspired Leslie to take on an authority much more formidable than Rumford. This was William Herschel, who lived on a King's pension for his discoveries in astronomy. Herschel did not concern himself with the total radiated heat in the manner of Rumford and Leslie, but with the heat associated with the colors of the Newtonian solar spectrum: violet, indigo, blue, green, yellow, orange, and red. He therefore returned to the problem opened by Rochon and others; not, however, in response to their lead, but because observing the sun brought him in touch with the problem. He literally felt that filters of different colors absorbed different amounts of heat from the sun. To discover which absorbed most, he threw the prismatic colors one at a time onto a thermometer. Preliminary trials indicated that red rays heated more powerfully than violet in the ratio of 3.5:1, and that the heating effect declined monotonically from red to violet. Whence it appeared that "radiant heat, as well as light, whether they be the same or different agents, is not only refrangible, but is also subject to the laws of dispersion arising from its different refrangibility." At this early state of his investigations, he guessed that the maximum of the heating effect might be carried by rays less refrangible than red; "in this case, radiant heat will at least partly, if not chiefly, consist, if I may be permitted the expression, of invisible light; that is to say, of rays arising from the sun, that have such a momentum as to be unfit for vision."[147]

145. Ibid., 69–71, 81, 105.
146. Ibid., 143, 149 (quote), 150, 159, 165; cf. Leslie, *Diss.* (1842), 767: "Every appearance, indeed, seems to indicate that heat is merely light under a latent and combined form."
147. Herschel, *Sci. papers* (1912), 2, 53–56, 60, 63 (quotes), text read Mar 1800. Cf. Cornell, *Ann. sci.*, 3 (1938), 120–121, 126–128, 134–135.

To test the first of these conjectures—that the maximum falls beyond the red—Herschel improved his apparatus to the form indicated in figure 2.2.22. The prism CD mounted in a window shutter spread the colors from violet to red upon the larger table. The leading edge of the smaller table AB bathed in the suppositious invisible heat rays, into which Herschel could advance one or another of the thermometers 1, 2, 3, the others being left as controls. The lines drawn on AB parallel to the long dimension of the visible spectrum allowed easy determination of the distance of the thermometer bulbs from the violet edge. Taking the rate of increase in temperature of the irradiated thermometer above that of the air as shown by its fellows as his measure, Herschel definitively fixed the maximum beyond red, in what he called "the invisible thermometric spectrum."[148] His mapping of the intensity of this spectrum, as well as of the accompanying visible spectrum (figure 2.2.23), is one of the earliest graphical displays of the relations between measured physical quantities. The abcissa is the distance from the violet edge of the spectrum; ordinates indicate relative intensities of the thermometric effect in the case of radiant heat and of luminosity as determined elsewhere by Herschel in the case of light.[149] This mapping should be regarded as no less a means to the end, and no less an invention, than the instruments used to obtain the measurements. The study of radiant heat lent itself particularly well to graphing.

The question remained whether heat and light derived from the same or different agencies. Herschel demonstrated agreement of light and heat obtained from solar and terrestrial sources on six chief points—both sets of rays suffered reflection, refraction, dispersion, absorption, and scattering, and possessed heating power—and, to round out the Newtonian seven, differed on power of illumination.[150] Did that make them the same or different? To respond, Herschel examined the relative transmission of the total spectrum of visible light and radiant heat through various sorts of screens using the apparatus of figure 2.2.24. The box AB, which can be adjusted so as to receive the sun's rays perpendicularly, covers the bulbs of

148. Herschel, *Sci. papers* (1912), 2, 70, 75 (quote), read 24 Apr 1800.
149. Ibid., 99–100 (read 6 Nov 1800).
150. Ibid., 78–79 (read 13 May 1800).

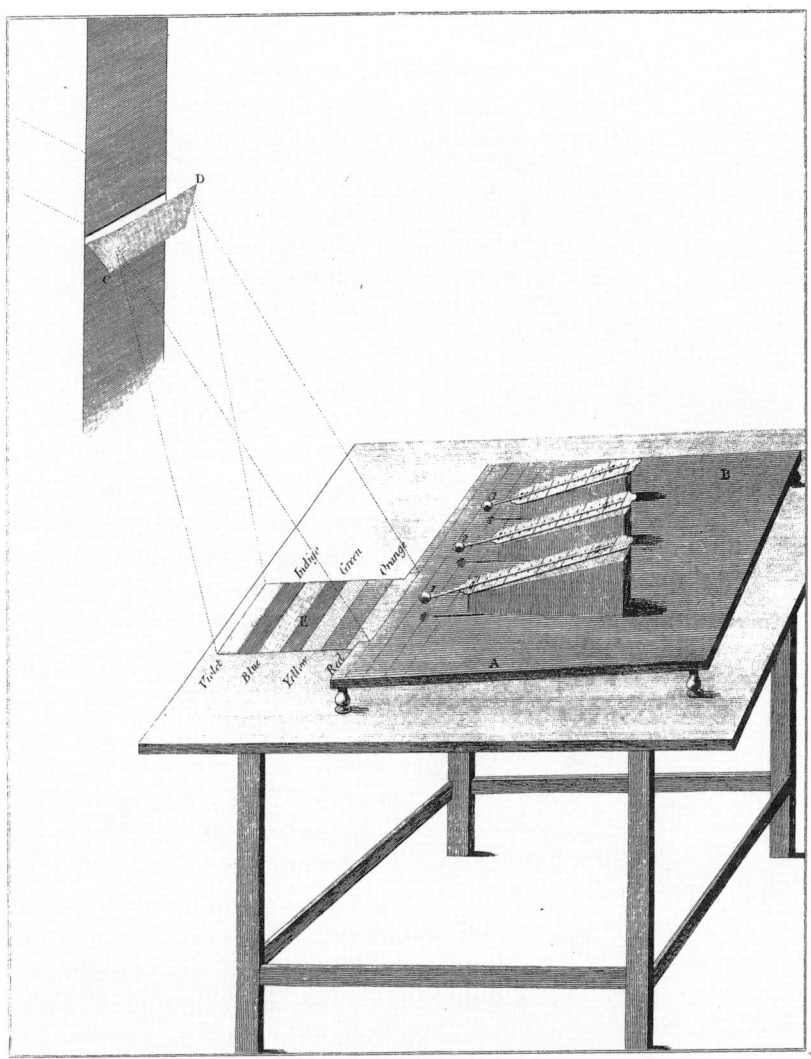

FIG. 2.2.22 William Herschel's set-up for investigating the heating effect of rays beyond the red end of the spectrum, 1800. The prism furnishes the rays, which fall on the bulbs of the thermometers, which lie at different distances from the visible spectrum. Herschel, *Nicholson's jl*, 4 (1801), pl. 14, p. 336.

130 / SOME MEANS TO THE END

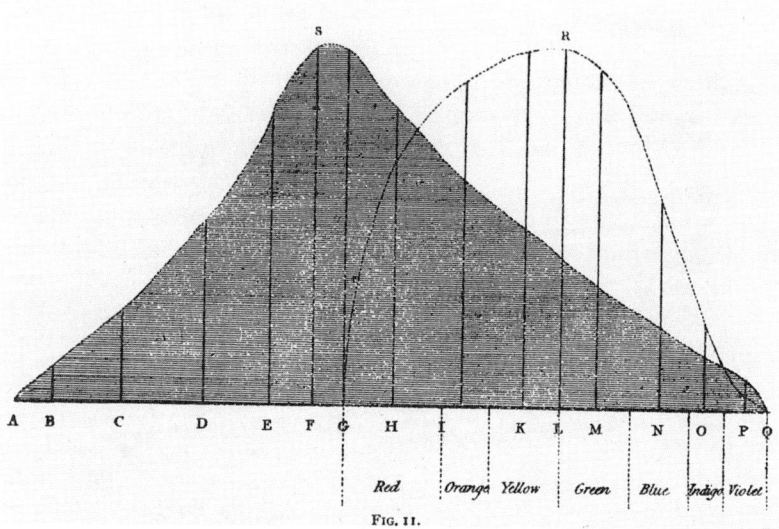

FIG. 2.2.23 Herschel's graph of the relative intensity of the heat and the visible spectrum, 1800. Herschel, *Sci. papers* (1912), 2, 99.

two identical thermometers, apart from two holes. The hole on the right admits solar rays directly, as a control; that on the left admits them after they have passed through the screen under study. Suppose that x percent of the heating effect derives from visible rays and let y be the percent decrease in visual intensity effected by the screen. Herschel found that his many careful measurements of x and y could not be made to agree. Consequently, he ended his inquiry far from where he began it. Heat rays and light rays had to be distinct in nature. Not only did his measurements of transmission oppose their identity, so also, now, did common sense: if heat rays also were light, then the physicist would have to defend the "arbitrary and revolting positions" that, with increasing refrangibility, heat rays suddenly begin to give a little light (at the extreme red edge of the spectrum); that the power of illumination increases as the heating effect declines; and that this power in its turn falls off, to vanish with the heating effect at the far edge of violet.[151]

151. Ibid., 103–107, 129 (quote), 135–137 (read 6 Nov 1800).

FIG. 2.2.24 Herschel's apparatus for studying the effects of screens on the heating power of solar rays, 1800. A screen under test lies over one of the holes in the cover C in front of the bulbs of the thermometers in the box AB; the other hole is left free as a control. Herschel, *Sci. papers* (1912), 2, 104.

Leslie, young, inexperienced, pompous, and bellicose, reproached the famous and accomplished Herschel for experiments "injudiciously contrived, executed without circumspection, and liable to a multitude of inaccuracies," and for "reasoning...still more defective." Leslie had found the truth with the help of a photometer of his own design (figure 2.2.25). The eyes of this robotic face are air-filled bulbs, one transparent and the other black. The instrument functions much as the differential thermometer: heat rays traversing the transparent eye act as a control; those absorbed by the black eye drive down the liquid in the graduated tube by an amount taken as proportional to the heating. Leslie confirmed that blue, green, yellow, and red had different heating effects, which he measured as proportional to 1:4:9:16; he eschewed the Newtonian spectrum of seven colors as a relic of medieval mysticism, one of the great man's "strange and mortifying errors." Returning to the main target, Leslie claimed that the heating effect stopped dead with illuminating power at the red edge of the spectrum. Herschel's experiment had been contaminated by convection, hot air rising from the red patch on the table and flowing over the thermometer mistakenly thought to be recording the effects of invisible rays. "A more objectionable plan of conducting experiments could scarcely have been devised."[152] Later, on realizing that convection currents could scarcely flow up vertically from the table bearing the optical spectrum (figure 2.2.22) and then horizontally to affect the thermometers, Leslie ascribed Herschel's error to failure to take into account the heating effect of diffuse sunlight in the region of invisible rays.[153]

Leslie's objection rested on a misleading inference from his differential technique as well as on his wish to explode "that partiality and timid deference, which the glare of paradox and the weight of authority seldom fail to produce on the bulk of men." The misleading inference: his themoscope would show no net absorption of heat if the transparent and the blackened bulbs absorbed the same amount. But glass transparent to visible rays can be opaque in the infra-red. Apparently Leslie made his bulbs of

---

152. Leslie, *Nic. jl*, 4 (1800), 346–347 (quote against Newton), 348–349, 420 (quotes against Herschel); Leslie also objected to Herschel's seven points of analogy as "mystic" (416). For the photometer, Lelie, *Inquiry* (1804), 407–417, and for a gentler assessment, Olson, *Ann. sci.*, 26 (1970), 283–284.

153. Leslie, *Inquiry* (1804), 457–458, 560.

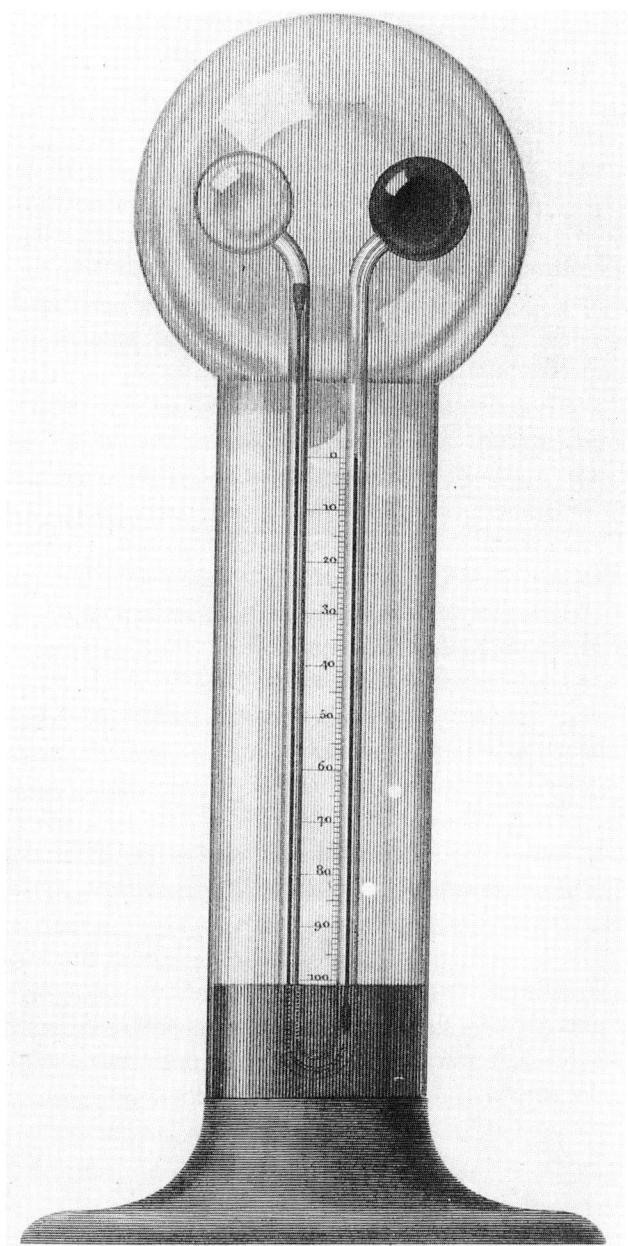

FIG. 2.2.25 Leslie's apparatus for testing the heating effects of the different solar rays. Light falling on the black bulb heats the right branch of the differential thermometer more strongly than the same light falling on the white bulb heats the left branch. Leslie, *Inquiry* (1804), pl. 7, 427.

such glass. Consequently his apparatus was perfectly designed to prove what he assumed at the outset, that no invisible heat rays exist. His "instrument of uncommon delicacy," the photometer he had brought to "the limit of perfection," had two blind eyes beyond the red edge of the spectrum.[154]

The preceding account of the British school of radiant heat exposes an apparently unfettered addiction to modelling. Yet its members felt obliged to do lip service to the positivism of the Standard Model. Herschel declared that his subject was "the rays that occasion heat," not heat itself; "nor do I in any respect engage myself to show in what manner they produce heat." Nor would he declare a preference for rays over waves, or vice versa. Rumford proclaimed "the insufficiency of the powers of the human mind to unfold the mysteries of nature" (a proposition he immediately proved at his own expense) and complemented Lavoisier and Laplace for always speaking about caloric "with that modest reserve which characterizes men of superior excellence." As for Leslie, he used his censure of Newton's errors and "hasty conjectures" as an excuse for a lengthy excursus in praise of Hume's doctrine of cause and effect.[155] For that he was roundly cudgelled by the Scottish clergy, who almost succeeded in defeating his election to the chair of mathematics at the University of Edinburgh in 1805 on the theory that no one who could not tell cause from effect should be given the care of children.[156]

Shortly after the Royal Society of London heard Herschel's first paper on heat radiation, its president, Joseph Banks, rightly judged that "the separation of heat from light [was] pregnant with more additions to science" than Herschel's earlier spectacular discovery of the planet Uranus.[157] Herschel himself contributed largely to establishing his teaching about invisible rays. But Leslie and Rumford muddied the situation so successfully that the French caloricists felt obliged to review everything from the beginning before seeking

154. Quotes from ibid., 559, 403–404, resp.
155. Herschel, *Sci. papers* (1912), 2, 77 (quote), 78; Rumford, *Works* (1969), 1, 405, 303, respectively; Leslie, *Inquiry* (1804), 136, 521–522.
156. Stewart, in Leslie, *Tracts* (1806), 1, 30–35, and Leslie, ibid., 36–39. Cf. Olson, *Scottish philosophy* (1975), 197, 219.
157. Banks to Henry Cavendish, 9 Apr 1800, quoted by Lovell, *Isis*, 59 (1968), 57. Cf. Thomson, *System* (1802), 1, 261–267.

more light on heat. Delaroche set out to clear up the facts and Bérard, on commission from Berthollet, undertook to extend them. Using the usual experimental set-up—parabolic metallic mirrors as reflectors, an incandescent body at the focus of one mirror as radiator, and a thermometer at the focus of the other as the detector— Delaroche defined the thermometer's rise in temperature $\Delta\theta$ in time Δt as the measure of the "thermometric effect" of radiant heat. He established, among much else, that the transmission through a glass plate of the heating power of light increased with the temperature and luminosity of the radiator; also, that passage of the rays through a second plate cut down their heating effect much less than did the first passage. "If caloric is a fluid, this fluid is far from being composed of homogeneous parts." But that only strengthened its analogy to light. "Who will be able to prevent himself from seing in this phenomenon a rather powerful argument to add to those that have led some physicists to regard caloric as a simple modification of light or, rather, light as a special state of caloric?"[158]

To develop further the analogy between heat and light, Bérard and Malus tried to polarize the rays of radiant heat. Malus died during the investigation. Bérard persevered, using a heliostat designed by his late partner. Results: the maximum heating occurs just at, but not beyond, the red end of the spectrum, and heat rays, like "luminous molecules," can be polarized by reflection. Berthollet, Biot, and the chemist Jean-Antoine Chaptal reviewed the findings of Delaroche and Bérard for the Academy. Delaroche's, they said, pointed to "a gradual and progressive transition between caloric and light;" Bérard's suggested further that heat rays, like light rays, have different refrangibilities and that we do not see them only because they do not stimulate our eyes. In all this they were perfectly right. Still, the positivism associated with the Standard Model required a disclaimer. Bérard had not declared himself for any hypothesis about the nature of radiant heat. Berthollet et al.: "We can only approve the wise reserve in which M. Bérard has enclosed himself by not rushing to decide questions about which experiment has not yet made a definitive pronouncement."[159]

158. Delaroche, *JP*, 75 (1812), 202, 205, 214–215 (quote).
159. Berthollet, Chaptal, and Biot, *AC*, 85 (1813), 311 ("a gradual transition"), 313–315, 317 ("luminous molecules "), 323–325 (last quote).

Among the figures of merit Delaroche considered was the elevation E of the temperature of the focal thermometer over that of the air at the instant that the mercury in the thermometer ceased rising. Then the thermometer's rate of loss of heat to the air equalled its rate of gain from the radiator. He tabulated E against the difference D between the temperature of the radiator T and that of the focal thermometer when the thermometer became stationary for several values of T. The display of numbers did not make clear the strength of the effect—the fast rise of E with increase in D. Delaroche therefore had recourse to graphs (figure 2.2.26), on which the letters indicate the numerical entries in the opaque table: thus the ordinate bb gives the value of E corresponding to a value Ab for D. The graph has the advantage over Herschel's of suggesting a simple functional relationship between E and D, and supports the suggestion that the beginnings of the graphical display of experimental data should be located in the study of imponderables.[160]

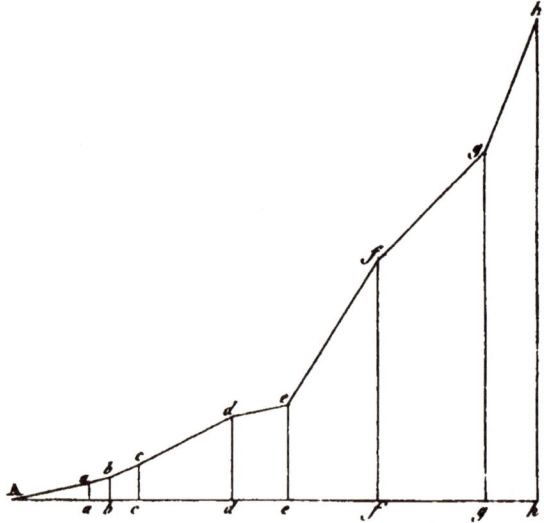

FIG. 2.2.26 Delaroche's plot of E, the difference between the temperature of the focal thermometer and that of the air, against D, the difference between the temperature of the radiator and that of the focal thermometer in the steady state. Values of D are given along the abcisse, of E along the ordinate. Delaroche, JP, 75 (1812), 248.

160. Delaroche, JP, 75 (1812), 217-218 (table), 248f (graph); cf. Trilling, Brit. jl hist. sci., 8 (1975), 200-206.

Delaroche's experimental strategy assumed Newton's law of cooling, according to which the rate of loss of heat of a hot object, $\Delta Q/\Delta t$, is proportional to the difference in temperature between the object and its surroundings, $\Delta Q/\Delta t = -k\Delta\theta$, k a constant.[161] Delaroche recognized that the relation held only for small differences in temperature (under 10°C) and that it lumped together the independent actions of conduction, convection, and radiation. In general, he supposed, a proper cooling law would replace $\Delta\theta$ with $(\Delta\theta)^n$. Following up a suggestion by Biot, he had placed some ice chips at the focus of one mirror and measured their rate of heating without a warm body in the focus of the other. He then brought up the warm body, a copper ingot, first at a temperature of 960°C, then at 427°C. If the "thermometric effect" were proportional to the difference in temperature between the warm body and the ice, the net rate of melting of the ice (the observed melting less the effects of conduction and convection) should have been as 960:427, that is, as 2.25:1. In fact, in two tries he obtained 5.22:1 and 4.61:1; the rate of loss of heat by radiation went more like $(\Delta\theta)^2$ than $(\Delta\theta)$.[162]

In this conclusion and reasoning Delaroche did not agree with Leslie, who had arived at the following expression for a body cooling in air:

$$dh = -(a+h)hpdt,$$

where h is the temperature ("the height of the thermometer") over the ambient medium, a characterizes the surface of the hot body regarded as a radiator, and p is a constant. The term $(a+h)$ represents loss of heat by pulsation of the hot surface against the air, which was Leslie's model of radiation; the term ph lumped convection and conduction. The expression integrates to

$$pt = \frac{1}{a}\left[\log\frac{H}{h} - \log\frac{a+H}{a+h}\right],$$

where H is the original height of the thermometer. A touter of

161. Cf. Grigull, *Physis*, 20 (1978), 213–235.
162. Delaroche, *JP*, 75 (1812), 215. The ratios 5.22:1 and 4.61:1 seem too small. Insofar as the radiator can be likened to a black body, the ratio of its loss rates at the two temperatures would be $(1233/700)^4 = 9.6:1$. Delaroche may have erred in determining the temperatures of the radiators by calorimetry.

logarithms, Leslie recommended this last formula as "abundantly simple."[163] His approximation, $\Delta Q/\Delta t \propto -(k_1 \Delta\theta + k_2 \Delta\theta^2)$, in effect added the second term of a power series to Newton's law. His formula may be reconciled superficially with Delaroche's by interpreting the term in $\Delta\theta^2$ as radiation loss. But that would not agree with Leslie's physical interpretation.

The problem of distinguishing the contributions of radiation, conduction, and convection to heat loss eluded the physical theory of the time. The best that could be done by the best measurers was to seek to find $\Delta Q/\Delta t$ for radiation empirically, by examining a hot body maintained in an evacuated space. The best measurers, Dulong and Petit, confirmed that the rate depended on the nature of the surface but not on shape or volume. From observation of the drop in temperature θ of a hot thermometer radiating into a void bounded by walls at temperature τ, they deduced the empirical formula $d\theta/dt = m(a^\theta - a^\tau)$. Here m is a constant characterizing the object under study and $a = 1.0075$ is a universal constant, at least as judged by comparisons between air and mercury thermometers. Dulong and Petit observed that their formula, "represent[ing] rigorously the progress of cooling in a void, at all temperatures and for all bodies," reduced to Newton's for small values of $\theta - \tau$.[164] They also offered an empirical formula for cooling in a gas-filled enclosure. But in this case, as in the former, they provided formulas applicable only to the experimental set-up from which they derived them. Thus measurement outran theory. Caloricists were reduced to multiplying special cases as students of electricity and magnetism had done before the time of Coulomb.

---

163. Leslie, *Inquiry* (1804), 263–266, 276, 313–314, 319–320, 324 (quote). Leslie's method of determining the effect of convection in air may appeal to hands-on experimenters: he measured $\Delta Q/\Delta t$ as he paraded a hot body around a room and when he whirled it in a sling around his head (ibid., 281–286).

164. Dulong and Petit, *ACP*, 7 (1818), 225–226 (including a mandatory criticism of Leslie), 233–238, 255 (quote), 259, 262. Let $a = 1 + \epsilon$, $\epsilon = 0.0075 \ll 1$; to first approximation in ϵ, $a^\theta - a^\tau = \epsilon(\theta - \tau)$, which gives Newton's law. Cf. Fox, *Brit. jl hist. sci.*, 4 (1968), 4.

3   LAPLACE'S SCHOOL

Buy low and sell high. French savants who cleaved to the watchword of Wall Street in the years around 1800 found their fortunes enhanced beyond their fondest fantasies. Selling high meant leaving posts in abolished institutions gracefully and, perhaps, withdrawing from Paris; buying low meant entering new institutions as quickly as their longevity seemed assured, or, best of all, clutching fast to Napoleon's coat tails when, as a mere general, he began to patronize science. The adroit investor might find himself repossessed of the same portfolio—or, to speak plainly, returned to his old position, suitably increased in value—when an institution closed by the Revolution reopened with new flourishes during the Empire. Such an institution was the Paris Academy of Sciences, suppressed in 1793 and resurrected two years later as the First Class of the Institut de France; its members, who suffered humiliation and an occasional decapitation under the Convention, received from Napoleon the right to wear a special dark green coat embroidered in yellow lace.[1]

The caparisoned academician stood at the top of the institutions for science and near science of the Napoleonic era. Members of the First Class had a significant say in appointments of teachers to the higher civil and military schools and of officials to government technical bureaus like the Observatory, the Bureau des longitudes, and the Museum of Natural History. They themselves held many of the senior appointments in these institutions. To complete the chain, the most successful of the protégés would enter the First Class, whence, with the help of others at lower ranks, they would identify candidates for patronage among the best students in the higher

---

1. Crosland, *Society* (1967), 155. For the death and transfiguration of the Academy, see Hahn, *Anatomy* (1971), 310–312.

schools. The main feeder in the system was the Ecole polytechnique; the main filter, ability at mathematics; the great patrons, the mathematicians Joseph-Louis Lagrange, Gaspard Monge, and Pierre Simon Laplace, and the chemist Claude-Louis Berthollet.

Berthollet and Laplace occupied a special place in the patronage system of Napoleonic science. Berthollet (R.A.S., 1780, M.F.C., 1795)[2] had served in Napoleon's immediate entourage during the Egyptian campaign and taught him chemistry in off-hours and in Italian (as a native of Savoy, Berthollet's mother tongue, like Napoleon's, was a version of Italian); Napoleon named Berthollet a Senator, a position that brought some responsibility and the wherewithal to establish a private research laboratory at his country house in Arcueil just outside Paris. Laplace (R.A.S, 1773, M.F.C., 1795), also a Napoleonic favorite (as an examiner at the Ecole militaire in 1785 he had had the foresight to pass the future general) and likewise a Senator, bought a property adjoining Berthollet's and added his support to the laboratory.[3] Laplace and Berthollet gave scientific direction as well as career opportunities to their protégés. "The subject of all our studies was indicated in their works; their conversation and advice gave us the means of pursuing and concluding our researches."[4] They thus reproduced on the intellectual level the union between experimental chemistry and mathematical physics that brought them together on the societal and professional level.

Laplace and Berthollet furthered their collaboration by setting up a private club, the Société d'Arcueil, so named after the location of their adjoining houses, and by basing their research program on a very few principles of physics and rules of procedure. The Society maintained its doctrine by the strength of small numbers—its total membership during its decade or so of activity was fifteen—and by the dependence of its junior members on their seniors.[5] We shall

---

2. As a contribution to historical notation, R.A.S. = regular member of the Paris Academy of Sciences and M.F.C. = member of the First Class, are here introduced in analogy to F.R.S., fellow of the Royal Society of London.

3. Crosland, *Society* (1967), 57–65, 279–280.

4. Biot and Arago, *MIF*, 1806, 304.

5. The juniors whose work will concern us were Bérard, Biot, Dulong, Gay-Lussac, Malus, and Poisson. Membership lists are given in *MSA*, 1 (1807), iii–iv; 2 (1809), [3]; 3 (1817), vi–vii.

first examine the doctrine in general and then study its application to three representative problems: optical refraction, capillary rise, and the speed of sound.

1. THE SPIRIT OF ARCUEIL

The school tie

The intellectual unity of the circle around Berthollet and Laplace rested on the conviction that the phenomena studied by physical science could be mastered by mathematics if it were assumed that the particles of matter and of imponderable fluids interact only by distance forces. The conviction had as support some of Newton's queries; the great power of the gravitational theory; the then recent successes of Coulomb's approach to electricity and magnetism; apparently quantifiable chemical affinities; and Laplace's assurances. As early as 1783, in his collaboration with Lavoisier on calorimetric measurements, Laplace had expressed the belief that refraction, capillarity, cohesion, crystalline properties, and chemical reactions arose from attractive forces identical in essence with the force of gravitation. He repeated the conjecture in 1796, in the first volume of his *Exposition du système du monde*, where, as we know, he expressed confidence that physics could be brought to the state of perfection to which the theory of gravity had raised celestial mechanics.[6]

The doctrine of reduction to distance forces was more than the goal of Laplacian physics: it was also its definition. On this understanding, apparently conflicting statements by members of Laplace's group about the epistemological status of their work may be harmonized. Laplace himself often spoke as if he believed in the physical reality of his distance forces; in the declaration with which the Société d'Arcueil brought itself to public atention in 1807, we read that "progress in physics has the purpose of attaining to the true causes of phenomena," and that the method of the Society, that is, precision, comparison, and criticism, could "lead to unshakeable theories, to uncontestable truths."[7] But Laplace also taught that

---

6. Lavoisier and Laplace, *MAS*, 1780, 355–408 (Laplace, *Oeuvres* (1878), 10, 149–200); Laplace, quoted by Fox, *DSB*, 15, 358–360, and *HSPS*, 4 (1974), 95; supra, §1.1.
7. Fox, *HSPS*, 4 (1974), 105; *MSA*, 1 (1807), i (quote), iii–iv (quote).

human science must always begin with, and may never shed, ideas that seem obvious and simple to us. Experience may confirm the ideas, but they do not therefore become true: the very simplicity of the laws of mechanics alerts us to their fundamentally subjective character. The only way to penetrate deeper is to generalize laws mathematically, thus removing some contingency at the price, perhaps, of unmanageable complexity.[8] The principal applications of the doctrine of forces made by Laplace and his associates jumbled simplifying physical assumptions with attempts at generalizations; but in no case would any particular solution warrant any higher presumption of truth than that of meeting the definition of true physics, that is, of invoking distance forces.

A similar attitude may be extracted from the writings of Laplace's protégés. Biot (Ecole polytechnique,[9] 1794, M.F.C., 1803), who wrote the textbook of Laplacian physics, laid it down that imponderables were no more than "convenient hypotheses, to which physicists are careful not to attach any idea of reality;" and he took many opportunities to claim ignorance of the true causes of physical phenomena.[10] Yet in his work on optics he wrote that Newton had proved that refraction derives from an attractive force between matter and particles of light, sensible only at short distances ("and, in that, altogether analogous to chemical affinities"); that he, Biot, had demonstrated that double refraction involves a force that puts a torque on light particles by an amount that depends upon their colors ("the theory enlightens us about the true cause," "the principles [are] simple and infallible"); and that no more can be known about physics than the doctrine of forces provides ("the highest goal to which human science can attain.")[11]

Gay-Lussac (Ecole polytechnique, 1797, M.F.C. 1806) is said to have been as much a positivist as Comte.[12] To be sure, he often

8. Vuillemin, Thalès, 9 (1958), 63. Cf., on Laplace's instrumentalism, Guerlac, HSPS, 7 (1967), 249.

9. This designation means student at the Ecole, followed by date of matriculation; Crosland, Society (1967), 120–121, 255–257, for Laplace's sponsorship of Biot.

10. Fox, HSPS, 4 (1974), 102, 120–122; the quote is from 1809. For Biot's Traité de physique (1816), see Frankel, HSPS, 8 (1977), 69–71. Cf. supra, §1.1.

11. Quotes from, resp., Biot and Arago, MIF, 1806, 302; Biot, MSA, 3 (1817), 133, 145 (texts of 1813); and Biot, Traité (1816), 1, xv, as quoted by Frankel, Centaurus, 18 (1974), 242. Biot talks about colored light particles in MSA, 3 (1817), 371–372.

12. Sadoun-Goupil, Colloque Gay-Lussac (1980), 65–66; Crosland, Society (1967), 129–130, 250–254, for Berthollet's sponsorship of Gay-Lussac.

represented his objective as the discovery of descriptive laws or principles, and put forward his results with many qualifications. Hence his attraction to gases, whose "simple and regular laws," in respect of dilation by heat and combination by volumes, he uncovered and established without explicit reliance on molecular models. Hence also his artful mixture of modesty, self-protection, and self-assertion: "I believe that I should stress again that I present these consequences [the law of dilation of gases] only with the greatest reserve, being aware how much I have yet to vary my experiments and how easy it is to go astray in interpreting the results: but although the new researches that engage me are immense, I will not be put off by their difficulty."[13] Yet Gay-Lussac included among "laws" items that transcend positive knowledge, for example, that all chemical combination involves the coming together of "elementary molecules" and that the properties of bodies express "the attractions of molecules."[14]

The same usage occurs in the work of Malus (Ecole polytechnique, 1794, M.F.C., 1810). He devoted endless hours to establishing descriptive laws, both experimental and analytical. It was he who confirmed Huygens' empirical rules for the behavior of light passed through bi-refringent crystals, rules previously doubted in France owing to poor experimental technique and to distaste for the wave theory in which Huygens expressed his results. On the analytic side, Malus developed a quantitative account of light rays, considered solely as geometrical entities; in this mood he characteristically attached no physical significance to his striking result that, after a single reflection or refraction, all the rays in the same bundle stand perpendicular to the same surface or front.[15] Huygens' rules and the geometry of light rays are model positivistic laws.

Not so the "laws" from which, in another mood, Malus proposed to deduce them. From Huygens' "laws," he said, "it follows necessarily not only that light is a substance subject to the forces that give activity to other bodies, but also that the form and disposition of its molecules have a considerable influence on the phenomena."

13. Gay-Lussac, *MSA*, 2 (1809), 207–208, and *MSA*, 1 (1807), 202–203, resp.; cf., Gay-Lussac, *MSA*, 1 (1807), [379].
14. Gay-Lussac, *MSA*, 2 (1809), 227; Sadoun-Goupil, *Colloque Gay-Lussac* (1980), quoting from text of 1808. Cf. Crosland, *Ann. sci.*, 17 (1961), 11–12, 25.
15. Frankel, *Centaurus*, 18 (1974), 227–231; Buchwald, *Rise* (1989), 27.

The phenomena all follow from "this unique law," viz., that the forces at work are repulsive and depend on the sine of the angle between the direction of the light ray and a certain axis in the birefringent crystal. The creative ambiguity of Laplacian law appears to full advantage in a programmatic statement by Malus concerning his grand discovery of the polarization of light by reflection. After reciting the details of his experiments, he wrote: "The study of these diverse circumstances will lead us to the law of these phenomena, which depends on a general property of the repulsive forces that act on light."[16] Here the law of the phenomena is almost entirely assimilated with the law of force. Malus' simultaneous enunciation of positivist and reductionist epistemologies, and his easy assimilation of both, may be the reason that the closest modern student of his work calls it "two-faced."[17]

A better characterization can be found. The play between renunciation of truth claims and assertion of true knowledge occurs often in physics. On the psychological level, it helps to square a fashionable philosophy with the conviction that grows from, and may be prerequisite to, intense scientific work. On the epistemological level, its apparent contradiction may be removed by interpreting the reigning theoretical structure not as a true statement about the world, but as a definition of correct physics.

The doctrine of forces was constitutive of Laplacian science much as circular motion defined Ptolemaic and Copernican astronomy and mechanistic reduction certified British physics of the later 19th century.[18] Laplace and his followers did intend to pursue truth in physics, that is, faithfulness to force representations; but they did not worry about the fit between physics and reality beyond showing that experimental results were compatible with the general scheme. An excellent statement of the position, by Alexis-Thérèse Petit (Ecole polytechnique, 1807), a collaborator of Pierre Louis Dulong (Ecole polytechnique, 1801, R.A.S. 1823), both of whom were associated with the Arcueil group, appeared in the *Journal* of the Ecole

16. Quotes from, respectively, Malus, *MSA*, 2 (1809), 260, and *MSA*, 2 (1809), 158; further to the double usage of "loi," see ibid., 256-257 (positivistic), 261, 263 (reductionist).
17. Chappert, *Malus* (1977), 204-211 ("les deux visages de la physique de Malus").
18. Heilbron, in Westman, *Copernican achievement* (1975), 278-281.

polytechnique for 1813. Petit accepted that cohesion, affinity, refraction, capillarity—the standard phenomena handled by Laplace's circle—derived from distance forces. "None of these phenomena [however] is suitable for revealing the law of action of this molecular attraction." What is significant, what gives power to Laplace's approach, is that all these phenomena can be *represented* by distance forces. "The remarkable conformity of observed and calculated results guarantees beyond question the legitimacy of the hypothesis that serves as the basis of this theory."[19] That is the point: epistemological legitimacy, not ontological truth.

From this point of view, what is perhaps the oddest feature of Laplacian physics may appear less peculiar. In contrast to forces of gravitation, electricity, and magnetism, the dependence on distance of the suppositious forces of optical refraction, capillary rise, cohesion, affinity, and so on, was never made explicit. Laplacian physicists did not hide their inability to provide the sort of information that astronomers and electricians had found essential. "I do not claim to indicate the cause of this general property of the repulsive forces that act on light," says Malus, or, for that matter, the real shapes and forms of light particles.[20] Laplace followed Malus' work closely, and sponsored some of it, in order to find a basis in the system of forces and particles for the rules that Huygens had interpreted in favor of a wave theory. Both Laplace and Malus gave accounts—they can scarcely be called derivations—of double refraction in terms of distance forces and the principle of least action. By showing the compatibility of Huygens' empirical rules with the requirements of current physics, the accounts made the rules into law, and the findings into true knowledge, in accordance with Laplacian epistemology; they did so "although the nature of the force that makes the light rebound at the surface of bodies is not known."[21] Indeed, it appears that Laplace's system was so constructed that the forces could not be determined—their "laws" could not be found—in principle.[22]

19. Petit, Ecole poly., *Jl*, 3:16 (1813), 1–2, 40, emphasis added. For the relations between Dulong, Petit, and the Arcueil group, Crosland, *Society* (1967), 131–132, 145, 167–168, 226.
20. Malus, *MSA*, 2 (1809), 267.
21. Laplace, *MSA*, 2 (1809), 138 (quote); Frankel, *Centaurus*, 18 (1974), 232–236, 239–240; Fox, *HSPS*, 4 (1974), 103–105.
22. Chappert, *Malus* (1977), 206–208; Buchwald, *Rise* (1989), 37–40; infra, §3.2.

The unknowability of the forces was not a weakness in Laplace's system, but rather a strength. On the assumption that they did not act over sensible distances, they could figure in calculations much as the Dirac delta function does in quantum-mechanical computations. They left their traces as constants to be fixed by experiment; the fixing necessarily brought agreement between calculation and measurement; and thus was the compatibility of the system of unknown forces with the unknowable system of nature fully demonstrated. In addition, the unknown forces gave the analysis generality; and, by freeing the analyst from the need to specify or simplify, to that extent purged it of anthropomorphic elements. Laplace ended his study and speculation about the forces of affinity, capillarity, cohesion, and refraction about where he had begun, by supposing their identity.[23] The supposition might seem to do little more than set unknown and unknowable things equal to one another. More generously interpreted, it transcribed the mathematician's insight that the more general the treatment or reach of a theory, the more faithful to its principles it is likely to be.

The dead hand

Physics as practiced by the Société d'Arcueil and the Parisian establishment that it dominated has been criticized for the inhibitory effect of its patronage, the emptiness of its doctrine of forces, and the artificiality of its mathematics. No doubt patronage could and can be stifling: the protégé usually cannot choose his problems, or interpret his results, without consulting the interests and positions of his boss. Among the more benign consequences of these constraints was Gay-Lussac's conclusion from his discovery of the law of combination of gases by volume. He acknowledged that the law favored systems, like Dalton's atoms and Proust's laws of combination, opposed by Berthollet.[24] Like a diplomat or a quantum physicist, Gay-Lussac reconciled the opposition by inventing "this great chemical law: whenever two substances come into one another's presence, within their sphere of activity, they act according to their masses, and generally give compositions in very variable

---

23. Fox, *HSPS*, 4 (1974), 98–102, and *DSB*, 15, 360; Mauskopf, *Ambix*, 17 (1970), 185–187.

24. Crosland, in Cardwell, *Dalton* (1968), 276–280.

proportions [as Berthollet taught], unless these proportions are determined by particular circumstances [as manifested in the results of Proust, Dalton, and Gay-Lussac]."[25] Dulong and Petit, whose law of specific heats of metals favored the atomic theory, had to break from the legacy of Arcueil.[26]

Others could not. Biot and Poisson, for example, held to the doctrine of distance forces long after the development of the wave theory of light by Augustin Fresnel had bankrupted the concept in optics. What remained of Laplace's world picture after the removal of the forces was the positivistic background. And thus historiography has punished him twice, once for the scheme of microphysics that led nowhere and again for making French physics safe for the positivism that was to stunt its growth.[27]

This historiography, which is British, echoes attacks on Laplace made by his contemporaries, the British physicists Thomas Young and John Leslie. Young condemned Laplace's (and Malus') theory of double refraction as arbitrary, insufficient, and physically meaningless, a mathematical artifice and nothing more. By refusing to give an explicit form to the suppositious force of refraction, Laplace was not playing the game of physics as Young understood its rules. "He contents himself with saying that the velocity within the crystal must depend only on the situation of the ray with respect to the axis, and that this is a necessary 'condition' of the refraction....[But] the deduction of this 'condition' from any assignable laws of attraction is *the only difficulty* in question." Young's difficulty ran deeper, however. It was not only that Laplace did not give the physics, he used his prowess as a mathematician to obscure it. And so did all those great European analysts who appropriated and mystified the quantifiable parts of physics. "The first mathematicians on the Continent have exerted great ingenuity in involving the plainest truths of mechanics in the intricacies of algebraical formulas, and in some instances have even lost sight of the real state of the investigation,

---

25. Gay-Lussac, MSA, 2 (1809), 233–234; Crosland, *Society* (1967), gives several examples of the constraints of discipleship.

26. Crosland, in Cardwell, *Dalton* (1968), 283–284; Fox, HSPS, 4 (1974), 1–6, 9–18.

27. The opinions of, resp., Fox, HSPS, 4 (1974), 108–112, 123–125, 130–134; and Herivel, BJHS, 3, (1966), 121–129.

by attending only to the symbols, which they have employed for expressing its steps."[28]

Leslie shared Young's sentiments but could not match his civility. Laplace's formula for correcting the barometer was "extremely complex and inelegant;" Poisson, in developing electrostatics, "exercised profound skill in the play of analysis...without, however, arriving at any conclusion that is not obvious or of no value;" and the whole lot, invoking "the higher calculus to unfold the motions of gaseous fluids," expended their skill and ingenuity without gaining "any material result."[29] It is worth emphasizing that this attitude was not confined to that "damned ignorant set in everything which relates to mathematics," as Charles Babbage characterized the British scientific establishment. Prevost wrote for many continental natural philosophers when he expressed irritation at the effort mastering French mathematical physics had cost him. "So great a piling up of calculations (hardly necessary in my opinion)...[when the author] could and should have set forth his interesting results more simply."[30]

The exaltation of mathematics was peculiar to France, where, according to a recent essay, it prospered under the Empire in part because Napoleon could neutralize squabbles among mathematicians while encouraging their performances. Excellence in mathematics was a means of upward mobility, through Lycée and polytechnic to Arcueil and the First Class.[31] Those who were not good at it deplored or wondered at its ascendency. Bernardin de Saint Pierre, a writer capable, in Biot's opinion, of exemplary taste, style, and sensibility, could not understand why his essay into natural philosophy, his *Etudes de la nature* (1784), had not succeeded among scientists. The book was scarcely calculated to win over mathematicians. It offered among much else a suggestion to replace lightning rods by laurel trees and a proof that the earth

---

28. Young, *Quarterly review*, 2 (Nov 1809), 343, quoted in Frankel, *Centaurus*, 18 (1974), 241–243, and Young, *Lectures on natural philosophy* (1807), 2, 670, quoted in Leslie, *Diss.* (1842), 732.

29. Leslie, *Diss.* (1842), 730, 740, 729, resp.

30. Babbage to John Herschel, 21 Feb 1816, quoted in Miller, *BJHS*, 16 (1983), 17; Prevost, diary entry, 9 June 1825, quoted in Weiss, *Prevost* (1988), 89.

31. Dhombres, *AIHS*, 36 (1986), 265, 267, 275–276; Dhombres and Dhombres, *Naissance* (1989), 168–195; cf. Picon, *Rev. his. sci.*, 42 (1989), 157–158.

resembles a football, not, as all geometers then knew, a pumpkin; and also the unfriendly observation that "the authority of great names only too often serves as a defense of error."[32] It is said that in his frustration Bernardin asked Napoleon, who prided himself on his own modest attainments in mathematics, why the savants paid no attention to his work. "Do you know the differential calculus, M. Bernardin?" "No." "Well, go learn it, and you will be able to answer your question yourself."[33]

The grandest slam at the "arrogant sterility" and "insolent tyranny" of mathematics during the Empire was given by the poet Lamartine. "There was a universal league of mathematical studies against thought and poetry. Figures alone were permitted, honored, protected, rewarded.... Since that time I have abhorred numbers, that negation of all thought; and there has remained with me against this exclusive, arrogant power of mathematics the same feeling, the same horror, which remains with the convict against the cold, hard, irons riveted to his limbs."[34] We have the same from Chateaubriand, who complained that the exercise of mathematics dries up the heart, and from Thomas Macaulay, who could "scarcely bear to write upon mathematics or mathematicians. Oh for words to express my abomination of the science."[35]

We proceed to examples of the arrogance, achievement, and limitations of Laplacian physics.

---

32. Biot, *Mélanges*, 2 (1858), 6–10; Levallois, *Vie des sciences*, 3 (1986), 287; Crosland, *Society* (1967), 19; Outram, *Hist. sci.*, 16 (1978), 158.

33. Quoted by Dhombres, *AIHS*, 36 (1986), 286. On Napoleon and mathematics, Fischer, *Napoleon* (1988), 39, 43, 172, and Dhombres and Dhombres, *Naissance* (1989), 665.

34. Lamartine, *Des destinées de la poésie* (1834), quoted in Herivel, *BJHS*, 3 (1966), 117. Cf. Home, in Métivier et al., *Poisson* (1981), 153, and Etienne Geoffroy's complaint that the mathematician Joseph Fourier attempted "to exercise [in Egypt] the same intellectual domination that Lagrange and Laplace are customarily accorded in Paris" (letter of 1800, in Gillispie, APS, *Proc.*, 33 (1989), 457).

35. Chateaubriand, *Génie du christianisme* (1807), 1, 21, quoted by Stafford, *Stud. 18th cent. cult.*, 11 (1982), 270, and Dhombres and Dhombres, *Naissance* (1989), 428–431; Macauley, lament of 1818 when an undergraduate at Cambridge, quoted by Gascoigne, *Cambridge* (1989), 272.

2. A MATHEMATICAL FIGLEAF

Clairaut and Newton

As the cases of electricity and magnetism illustrate, a primary objective of astronomizing physics was to find the precise dependence of the forces between interacting elements on the distance between them. The business faced the grave difficulty that, insofar as the merit of any proposed law of force was its fit with observation or experiment, it could always be altered a little as circumstances seemed to require. How build solidly on so shaky a basis? Cartesians repeatedly objected that Newton's methods would lead to physics by finagling. The objection might appear far-fetched: but in 1745 precisely the sort of fiddling that the Cartesians had anticipated was practiced on the law of gravity itself. In that year, Alexis-Claude Clairaut, Jean d'Alembert, and Leonhard Euler each independently made the non-discovery that the law of inverse squares did not give the observed rate of advance of the moon's perigee.

Clairaut thereupon proposed to add an inverse-cube force to Newton's square. That did doctrine less violence than the repulsive suggestion made by his confrère at the Paris Academy, Pierre Bouguer, that different parts of planets might attract according to different laws, Newton's being a fortuitous average. There was a row in the Academy over these proposals. Its remaining Cartesian fellow travellers had the satisfaction of having their worries confirmed while its few true Newtonians, led noisily by Buffon, protested that the law must be left as promulgated, simple, esthetic, and universal. The apparent exception in the lunar motion had to be sought in secondary causes. The row subsided when the moon-struck mathematicians carried their approximations further.[36]

Clairaut's amendment of Newton's rule may have indicated a disinterest in the true cause of gravity (he had no opinion, he once said, about the nature of attraction, "which lies beyond my power"), but it did not imply withdrawal from the project of finding

36. Clairaut, MAS, 1745, 577–578, and Buffon, ibid., 493, 497–500, 551–552; Clairaut, MAS, 1748, 421, according to which he had found the problem in his calculation in 1747; Clairaut, Théorie de la lune (1752).

correct force laws for physics.[37] An account of his treatment of the refraction of light and the rise of liquids in capillary tubes will show the practical problem of deducing the laws of short-range forces. It will also provide a benchmark for evaluating the approach to the same subjects that Laplace took half a century later.

In the *Principia*, Newton examined the motion of a small body traveling between two infinite parallel planes and urged or impelled toward (or away from) one of them by a constant force perpendicular to both. (The force vanishes everywhere except between the planes, and no other force acts on the body.) The motion between the planes will be a piece of a parabola, as in Galileo's analysis of the trajectories of cannon balls; and the parabolic piece will be tangent at each end to the straight lines along which the body proceeded, and will proceed, when free from the force. In figure 3.2.1, Aa and Bb are the planes, GH and IK the initial and final lines of motion, and HI the parabola. Newton showed that the sine of incidence of the body on one plane stands in a constant ratio to the sine of emergence from the other irrespective of the angle at which the body strikes the first plane. The proof runs in the demanding synthetic style by making use of the geometry of the parabola.

Produce the tangents at H and I until they intersect at L; with LI as radius, draw the circle QNP. Two properties of the parabola are required: that L bisects HM and that HM^2/MI equals the latus rectum of the curve, which depends on the strength of the force. Now NM is the same for all angles of incidence since it equals the fixed separation of the planes, IR. (HM is bisected at L, whence MO = OR and ON = OI; equals added to equals give MN = IR.) From a well-known proposition about secants and circles,

$$MI \cdot MN = MQ \cdot MP = (ML - LQ)(ML + LQ) = ML^2 - LQ^2 = ML^2 - LI^2.$$

This is the relationship desired. Divide both sides by ML^2:

$$(MI/ML^2) \cdot MN = 1 - (LI/ML)^2, \text{ or}$$

$$(LI/ML)^2 = 1 - (\text{latus rectum}) \cdot (\text{plate separation}).$$

The right side, and hence LI/ML, are therefore given for the problem. But in triangle LMI, $LMI = i$ is the angle of incidence and $LIM = \pi - r$ is the supplement of the angle of emergence. By the

---

37. Clairaut, *MAS*, 1739, 263 (quote).

law of sines,

$$\frac{\sin i}{\sin r} = \frac{LI}{ML} = \text{constant}.$$

Newton generalized the problem by supposing the space between the planes to be filled with a succession of layers of infinitesimal thickness each characterized by a slightly different, constant acceleration. The motion in each will be parabolic and the preceding derivation therefore holds for any continuous law of force that urges or impels the body between the planes. Newton concluded coyly that "these attractions bear a great resemblance to the reflections and refractions of light...discovered by Snel."[38]

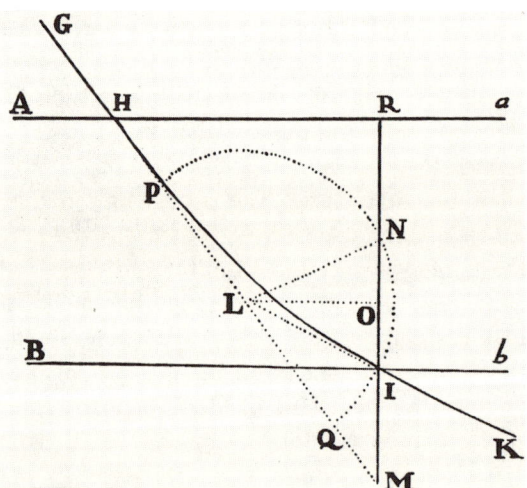

FIG. 3.2.1 Newton's diagram of the path of light through a refractive medium. GHIK is the path, $\angle$LMI the angle of incidence, $\angle$MIK the angle of refraction. Newton, *Principia*, Bk. I, Prop. XCIV, Cajori ed., 226.

Newton's approach does not set up the problem of deducing either the law of force at an optically active surface or the resultant trajectory of light particles. Nor does he connect the index of refraction analytically with the magnitude of the suppositious optical force. The treatment in the *Opticks* moves even further from these

---

38. *Principia*, Bk.I, Props. XCIV, XCV; quote from Cajori ed., 229.

problems. There Newton shows that Snel's law follows from the assumption that the "motion or moving thing" retains its velocity parallel to the planes when passing between them. "And this demonstration being general, without determining what light is, or by what kind of force it is refracted, or assuming anything farther than that the refracting body acts on the rays in lines perpendicular to its surface; I take it to be a very convincing argument of the full truth of this proposition [Snel's law]."[39]

Clairaut handled the mathematics so as to reintroduce the physical problem that Newton had suppressed in one of his fits of easy instrumentalism. For his rhetorical purpose, which was to ease the entry of Newton's methods into French mathematical physics, Clairaut went back to the derivation of the *Principia*, which showed, he said, how usefully and yet how harmlessly the concept of attraction could be applied to ordinary problems in optics. "He said so little about the matter that I think I should begin with an explanation very similar to his, which I put forth the more willingly because it shows that I do not reject the belief that attraction can be replaced by its mechanical cause."[40] Clairaut's sop to his colleagues' sensibilities was to replace the attraction of the surface of the body by a pressure within a shallow "atmosphere" surrounding it. In figure 3.2.2, the limit of the atmosphere runs through A, where the particle of light (here Clairaut did not mince words) begins to feel a pressure $X(x)$ toward the plane surface through B. Let the initial velocities at A be v_x down and v_y to the left; under $X(x)$ the particle describes the unknown curve AMB. The sine of the angle of emergance from any plane through the curve, at M say, is easily obtained from the differential calculus. We have from figure 3.2.2 $\sin\gamma = dy(dy^2+dx^2)^{-1/2}$. At any point, $dy/dt = v_y$ and $d^2x/dt^2 = X$; therefore $(dx/dt)^2 = v_x^2 + 2Q$, where $Q(b)$ is the integral of Xdx from $x = b$ to $x = 0$. Consequently,

$$(dx/dy)^2 = (v_x^2 + 2Q)/v_y^2, \text{ and}$$

---

39. *Opticks*, Bk. I, Part I, Prop. VI; quote from Cohen ed., 80–81. Cf. Sabra, *Theories of light* (1967), 302–308.
40. Clairaut, *MAS*, 1739, 263 (quote), 268–271.

154 LAPLACE'S SCHOOL

$$\sin\gamma = (1 + dx^2/dy^2)^{-1/2} = v_y(f^2 + 2Q)^{-1/2},$$

where $f^2 = v_x^2 + v_y^2$. But $\sin i = v_y/f$ and $r = \gamma(b)$, whence

$$\frac{\sin r}{\sin i} = (1 + 2Q/f^2)^{-1/2}.$$

As Clairaut observed, the whole argument goes through in precisely the same way if the pressure of the atmosphere is replaced by an attractive force acting just within the surface.

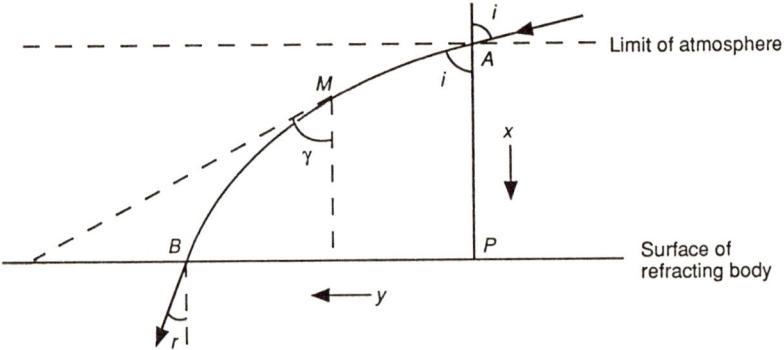

FIG. 3.2.2 Geometry of Clairaut's theory of refraction, 1739. The light path is AMB; i is the angle of incidence, r that of refraction.

The factor $(1+2Q(b)/f^2)$ is the square of the index of refraction μ. To derive μ from first principles required finding the force law $X(x)$ and its effective range b, which Clairaut supposed to be short but measurable. He regarded this problem, and the associated one of working out the curve described by light particles undergoing refraction, as a serious research goal; and he took a first step toward it by computing μ for the case that $X \sim x^{-n}$. He brought the same point of view to the analysis of other phenomena apparently regulated by short-distance focus, notably the rise of liquids in narrow tubes.

Clairaut again began where Newton—or, in this case, one of Newton's collaborators, James Jurin—left off. Jurin's experiments had made plausible the proposition that the force that caused water to rise in capillary tubes was exerted by a ring around the tube just above the water.[41] Clairaut liked the experiments but not the theory: a proper account, he said, would sum "all the attractions of every particle of the glass on all the particles of the water," with due regard for the different directions and magnitudes of the forces.[42] A pleasant and fertile field thus opened to the analyst. ABCDEFGH in figure 3.2.3 designates the walls of the capillary tube, YIZ the meniscus, and MNP the level of the water outside the tube; IKLM are imaginary, infinitely narrow canals of unit cross section. Clairaut deduced an expression for the height $Ii = h$ of the capillary column from the principle that the canal must be in equilibrium under the static forces acting on it: the pressure at $L(p_L)$ equals the pressure at $K(p_K)$.

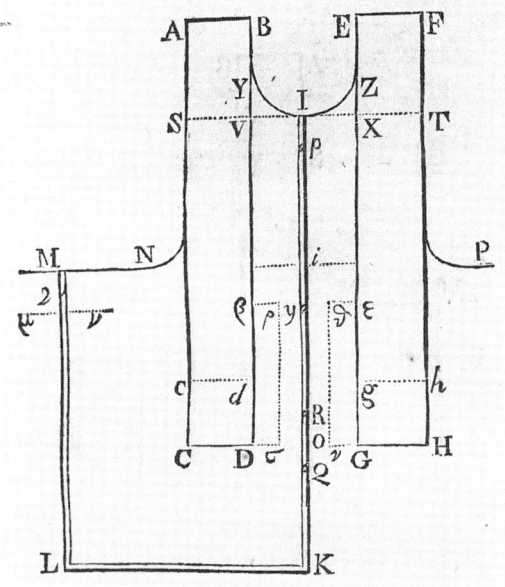

FIG. 3.2.3 Clairaut's diagram for an analysis of the forces implicated in capillarity, 1743. Clairaut, *Figure* (1808^2), *passim*.

---

41. On Jurin, Heilbron, *Physics at the Royal Society* (1983), 67–72.
42. Clairaut, *Figure de la terre* (1743), §lix; edn. of 1808, 113 (quote). The analysis that follows is paraphrased from ibid., §lv–lx (105–122 of 1808 edn).

156 LAPLACE'S SCHOOL

Let the forces between two particles of water, and between one of water and one of glass, follow the same "law" and differ only in intensity. Call the first $kF(x)$ and the second $hF(x)$, and assume that F is of short range, "sensible only at a very small distance." Now $p_L = g\text{ML} + \int kF(x)dx$, where g is the acceleration of gravity and the integral arises from the pull of the water on particles close to the surface MN. (The particle v at distance x beneath MN feels a downward force if $x < R$, the effective range of the attraction. The limits on the integral are 0 and R, or, what is the same, 0 and ∞, since $F(x) \sim 0$ for $x > R$.) The pressure at K is gIK less the attractions A_I on the particles around I and A_O on those around O. Clairaut obtained expressions for these attractions by taking all the material within the tube and the tube prolonged beneath CH into account: by this method, characteristic for his approach, he could extend his integrals to infinity although the forces act sensibly only over distances of the order of the interior radius (= b) of narrow tubes.

A particle at R within a short distance x of O will be drawn upwards by the tube and downwards by the water contained within an extension of the tube below CH; a particle at Q will have a similar experience; hence $A_O = -(h-k)\int F(x,b)dx$, where b indicates that the attraction depends upon the radius of the tube. At P a short distance x below I, we may suppose the tube made of water, since, by symmetry, the force at P exerted by the glass cancels. That allows us to write the downward pull of the water as $k\int F(x)dx$, without dependence on b. This is countered by the upward pull of the tube (which has been assimilated to water), $k\int F(x,b)dx$, and of the meniscus YZXV, say $\int G(x,b,h,k)dx$. In all,

$$A_I = \int F(x)dx - k\int F(x,b)dx - \int G(x,b,h,k)dx.$$

The condition of equilibrium gives for the height h = Ii,

$$h = ((2h-k)/g)\int_0^\infty F(x,b)dx + (1/g)\int_0^\infty G(x,b,h,k)dx.$$

From all this Clairaut extracted only that analysis can express the results of experiment. "Without pursing the calculation further to evaluate [the integrals] according to the different functions of the distance that might be taken to express the law of attraction, one easily sees that there is an infinite number of such laws for which the preceding equation for...[Ii] will give an appreciable height when the diameter of the tube is very small, and, on the contrary, a height nearly null when the tube is a little large. It is obvious that a law of attraction exists that will make Ii inversely proportional to the diameter, in accordance with experiment."[43] It is also obvious that Clairaut thought it desirable and feasible to search for the law. His instrumentalism, as indicated by his willingness to tinker with the sacred theory of gravitation, did not extend to introducing a force of which the most useful mathematical property was that its law need not (and in practice, could not) be found. That brilliant cynical advance in mathematical physics was the work of Laplace.

The Laplace transformation

Laplace attacked the theories of refraction and capillarity in the hope of finding the laws that had eluded Clairaut. He pondered for long, he wrote, before discovering the solution, or, as he might better have said, before learning how to avoid the problem. The solution, and the basis of Laplace's molecular physics: if the forces are sensible not over very short distances, as in Clairaut's formulation, but only over insensible distances, then the form of the force function does not matter at all.[44] Laplace could always manage the calculations so that the force occurred only in integrals that ran from zero to infinity. No parameter representing the range—the thickness of an attracting layer, or a fraction of the diameter of a capillary tube, or the power of the distance law—survived the calculations. And this very fact, which would have been a blemish in a realist theory, was a virtue to Laplace. It enabled him to complete his calculations.

An easy introduction to this aspect of Laplace's physics may be obtained from his elucidation of refraction. In figure 3.2.4, a light particle with speed f enters the refracting region at A. It then suffers the accelerating force, which bends its trajectory as shown. Laplace calculated the ratio of $\sin r$ to $\sin i$ much as Clairaut had done:

---

43. Ibid., 121 (§ix).
44. Laplace, *Oeuvres* (1878), 4, 349–351, and 14, 217–218 (texts of 1805–6); cf. Bickerman, *Centaurus*, 19 (1975), 184–185.

$$(dx/dt)_s^2 = 2\int_0^s X(x)dx + v_x^2,$$

$$(dx/dt)_{2s}^2 = 2\int_0^s X(x)dx + (dx/dt)_s^2 = 4\int_0^s X(x)dx + v_x^2.$$

(The integration occurs in two steps because the force reaches its maximum at $x = s$ and goes to zero above $x = 0$ and below $x = 2s$.) Because of the short range, s may be taken as ∞ in the integral, to which Laplace assigned the value K. Hence

$$\frac{\sin r}{\sin i} = \frac{v_y/(v_y^2 + (dx/dt)_s^2)^{1/2}}{v_y/f} = (1+4K)^{-1/2},$$

which recovers Clairaut's result. But whereas Clairaut regarded the result as provisional, Laplace accepted it as final. For him, the experimental value of the index of refraction determined the value of K; for Clairaut, the appropriate choice of X would allow a derivation of the value of the index. Like the forces it invoked, Laplace's system could not penetrate sensibly beneath the surface of things.[45] This fig-leaf physics had the enthusiastic support of the mathematicians in Laplace's entourage. Malus may be allowed to speak for them all: "[short forces] have the advantage that their *overall* influence appears alone in the calculated results, so that the law of the phenomena is independent of the function of the distance according to which they exercise their action."[46]

Superficial need not mean useless. That much at least emerged from Laplace's elaborate reworking of Clairaut's theory of capillarity. The new approach obtains the net uplift on the central canal within the capillary tube from a consideration only of the force acting between water particles. The glass walls of the tube are too far from the canal to affect it directly; they participate by attracting the contiguous layer of water, thereby holding the column against the downward pull of the fluid just beneath the meniscus. The resulting curvature of surface provides the capillary pull, as follows.[47]

45. Laplace, Oeuvres (1878), 4, 235–239, 248, 275.
46. Malus, Mem. divers savants, 2 (1811), 338.
47. Laplace, Oeuvres (1878), 4, 353, 356, 358–365. Cf. the paraphrase in Dhombres, Rev. hist. sci., 42 (1989), 53, 56–61.

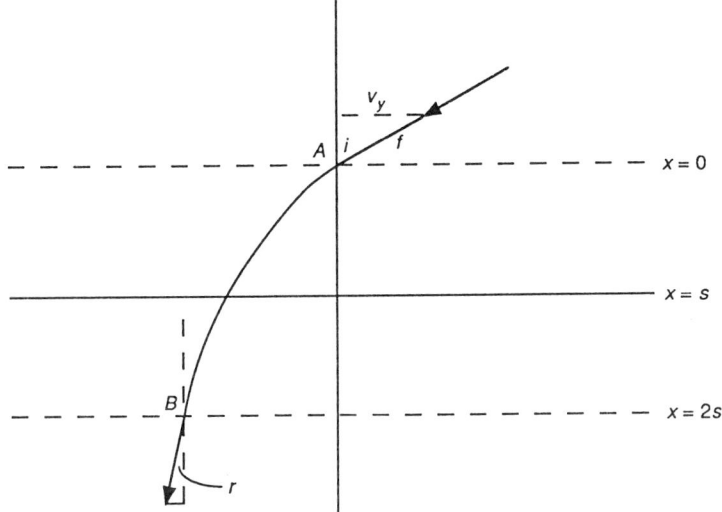

FIG. 3.2.4 Geometry of Laplace's account of refraction. The refractive force acts detectably only within a distance s of the surface $x=s$.

In figure 3.2.5, the shaded area around S is an element of the surface dA of a watery sphere of radius u. The volume element $d^3V = dAdu$ attracts the particle of water at P in the direction PD with a force $d^3F = d^3VX(p)\cos\omega$, where $p = $ PS and X is the force sensible only at insensible distances. If P is a point on a canal of unit cross section along the line QP, the element of attraction on the canal is

$$d^4F = (u^2\sin\theta)X(p)\cos\omega \cdot d\phi d\theta du dr.$$

The integration can be accomplished easily by ingenious substitutions employed by Laplace. The law of cosines applied to ΔPOS gives:

$$u^2 = r^2 + p^2 - 2rp\cos\omega,$$
$$p^2 = r^2 + u^2 - 2ru\cos\theta. \quad (3.2.1)$$

To integrate r, with u and θ fixed, differentiate the lower equation (3.2.1), obtaining $u\cos\theta = r - p\,\partial p/\partial r$; with which, after obvious substitutions, the upper equation (3.2.1) becomes

160 LAPLACE'S SCHOOL

$$u^2 = 2r^2 + u^2 - 2ru\cos\theta - 2rp\cos\omega,$$

or $\cos\omega = \partial p/\partial r$. We then have

$$d^2F = \int_0^{2\pi} d\phi \int_b^\infty X(p)(\partial p/\partial r)dr \cdot u^2 \sin\theta \, d\theta \, du.$$

The limits of the second integral indicate that the canal begins a distance b above O and runs far enough for X to become insensible.

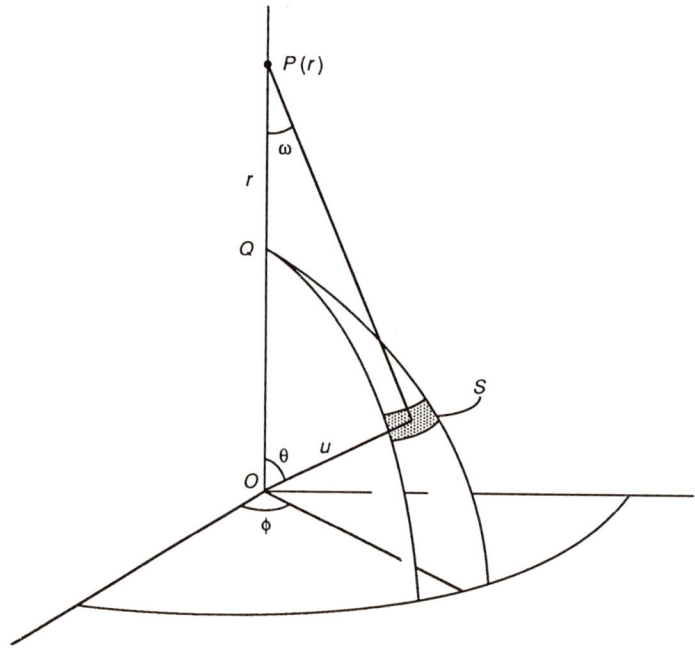

FIG. 3.2.5 Geometry of Laplace's calculation of capillary force. The force on a particle of fluid at P arises from the sum of attractions of bits of the fluid of volume $s\,du$.

Call the integral $Y(p(u,\theta))$. To integrate over θ, with u fixed and $r=b$, differentiate the lower equation (3.2.1) to obtain $\sin\theta = (p/bu)(\partial p/\partial\theta)$. Then

$$F = 2\frac{\pi}{b}\int_0^a\int_0^\pi Y(p(\theta,u))(\partial p/\partial\theta)d\theta \cdot u\,du$$

$$= 2\frac{\pi}{b}\int_0^a\int_{b-u}^{b+u} Y(p)dp \cdot udu,$$

where a is the radius of the watery sphere. Let the integral of Y be Z:

$$F = \frac{2\pi}{b}\int_0^a [Z(b+u) - Z(b-u)]udu.$$

The first term, $Z(b + u)$, vanishes because of the shortness of the range of X. There remains, with the substitution $w = b - u$,

$$F = \frac{2\pi}{b}\int_b^{b-a} Z(w)(b-w)dw.$$

In the case of interest, the canal touches the sphere, so that $b = a$ and $Z(w) = 0$ for values of $w \geq b$. Then

$$F = \int_0^\infty(-2\pi Z(w))dw - \frac{1}{b}\int_0^\infty(-2\pi)Z(w)wdw,$$

$$\equiv K - H/b. \qquad (3.2.2)$$

Equation (3.2.2) is the goal of the exercise.

As Laplace observed, K is much larger than H. (Z contributes to the integrals only when w is very small; in H, the contribution is diminished by multiplication by values of w close to zero.) K represents the cohesive force of water; it is analogous to the quantity K in the theory of refraction.[48] H represents the force of capillarity, which vanishes for a flat surface ($b = \infty$). The fundamental experimental finding that the height of capillary rise goes inversely as the diameter of the tube is an immediate consequence of equation (3.2.2). Figure 3.2.6a indicates the theory as developed so far: K represents the force that a cylindrical column of liquid terminating at the plane CD would exercise on the column EF, H being the amount to be subtracted to reduce the plane to the spherical surface AEB. In figure 3.2.6b, the downward force on the column is $K + H$, because the curvature removes the material ABCD that, in the case

---

48. Laplace, *Oeuvres* (1878), 4, 362–363.

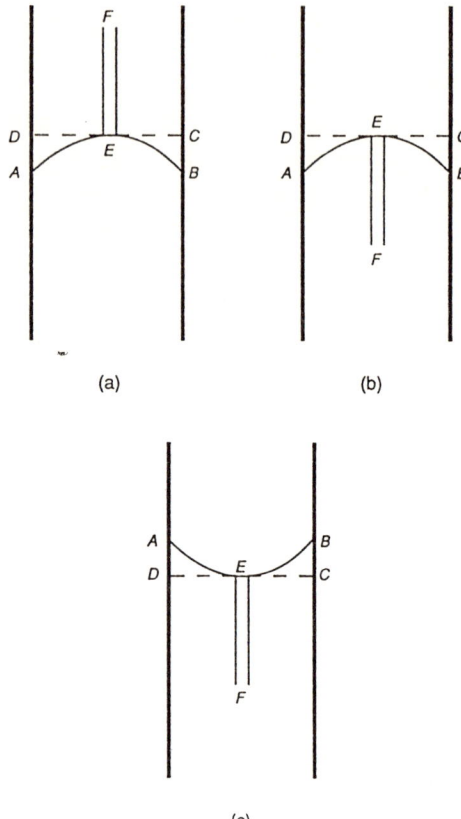

FIG. 3.2.6 Diagram to elucidate Laplace's theory of capillary action: (a), on a canal external to the meniscus convex to the air; (b), on a canal within the liquid with meniscus convex to the air; and (c), on a canal within the liquid with meniscus concave to the air.

of a plane surface, would pull up on EF. In the case of interest, figure 3.2.6c, the attraction from ABCD opposes that of the cylinder below CD, and the force upon EF drops to $K - H$.

Consider the equilibrium of the canal in the style of Clairaut. We have, for the left branch (figure 3.2.3), $p_L = K + g\rho ML$, and, for the right, $p_K = K - H/b + g\rho lK$, or,

$$\text{Height} = li = H/bg\rho, \qquad (3.2.3)$$

where ρ, the density, has been introduced to extend the theory to liquids other than water. Recall that Clairaut could not specify a single force law that would provide this relationship. It was a flaw in his theory. Laplace obtained equation (3.2.3) as a consequence of every force law that has no effect at sensible distances. Thus he gave a second example, to go with optical refraction, of the successful subjugation of molecular physics to analysis. What the constancy of the ratio of the sines of the angles of incidence and refraction was to the one, the inverse proportionality between capillary rise and tube diameter was to the other.[49] As appears from figure 3.2.7, $1/b = \cos\phi/r$, r being the radius of the tube.

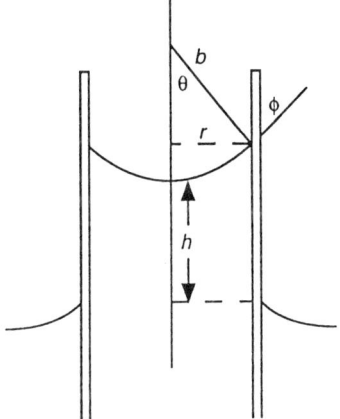

FIG. 3.2.7 Geometry relating the capillary rise h to the radius of the tube r, the radius of the meniscus b, and the angle of contact between the meniscus and the tube wall ϕ.

It remained to subject the analysis to experiment. Laplace called on several men of Arcueil to make the measurements. Gay-Lussac found the internal radius of the capillary tubes from their weight when filled with mercury, thus introducing "the precision of astronomical measurements" into the subject, according to Laplace.

---

49. Ibid., 4, 349; cf. ibid., 14, 233 (text of 1806). Laplace wrote the law in the form $h = H \cos\phi r \rho g$; ibid., 373–374.

Haüy and Louis-Jacques Thenard dropped three pipettes of known radii into water and oil of oranges. Their results appear below.

Table 3.2.1

	Water mm	Oil of oranges mm
First tube	13.50	6.8
Second tube	13.333	6.6667
Third tube	13.875	6.75

Laplace was delighted: "The smallness of the deviations...proves the exactness of the law of the elevation of fluids in inverse proportion to the diameter of the tubes." It also pointed to the desirability for more measurements, of equal precision, of other physical phenomena. "The need for similar experiments will be felt increasingly as physics perfects itself and enters the domain of analysis. The theoretical results can be obtained with great precision, and, by comparing them with very exact experiments, these theories can be brought to the highest degree of certainty of which the natural sciences are capable."[50]

Laplace's circle applauded his treatment of capillary rise as the cynosure of physical theory. It tied the material world together, according to Haüy, who wrote in this respect like a big-bang cosmologist: Laplace used the same formulas to account for capillarity as he did "to explain the greatest phenomena of the universe." Furthermore, added Biot, having found "with exactitude the true cause" of the ascent of fluids in narrow tubes, Laplace was poised to enrich chemistry with a proper theory of affinity. Only a truly superior mind could have mastered the relevant mathematics, astronomy, physics, and, eventually, chemistry. Haüy again: "It is to this superiority...that we owe this beautiful theory of the phenomena of capillary tubes."[51]

---

50. Laplace, as quoted by Crosland, *Gay-Lussac* (1978), 50, and in *Oeuvres* (1878), 4, 403–404, 495; cf. Dhombres, *Rev. hist. sci.*, 42 (1989), 74.

51. Quotes from Haüy, *Traité* (1806²), 1, 213, xxvii–xxviii, resp.; and from Biot, in Fischer, *Physique mécanique* (1806), 146 (quote), 148. Cf. Dhombres and Dhombres, *Naissance* (1989), 512–514, who seem to take the Laplacian school of capillarity at its own evaluation.

Critics will carp. We already know Young's views about Laplace's theories—"unnecessarily intricate...[and without] any novelty." John Playfair objected that Laplace had inverted cause and effect: not the concavity of the meniscus, but the attraction between the liquid and the glass, sustains the rise. Gauss objected that Laplace did not prove, but only assumed, an essential piece of the phenomena, the constancy of the angle of contact. "In my opinion this centerpiece of Laplace's theory in no way deserves to rank with the precise mathematical foundation of his other theories, but rather seems to have the character of the vague aperçus that physicists earlier entertained about the entire phenomenon."[52] Nonetheless, Laplace's treatment of capillarity dominated the subject during the 19th century, in part because H is closely related to the constant of surface tension. In figure 3.2.7, the radius of the tube is r, the upward tension at the glass surface is $\gamma\cos\phi$ per unit length, and the suspended column weighs $g\pi r^2 h$. Hence

$$\rho g \pi r^2 h = 2\pi r \gamma \cos\phi,$$

$$h = (2\gamma/\rho g)(\sin\theta/r) = 2\gamma/b\rho g,$$

whence $H = 2\gamma$. More fundamentally, the phenomena represented by K occur in gases as well as in liquids. J.D. van der Waals reproduced and built on Laplace's derivation of H and K in his famous thesis of 1873, which modified the perfect gas law to take K into account and obliterated the distinction between liquids and gases. Lord Rayleigh, while criticizing Laplace for leaving out molecules and for obscuring the meaning of K, nonetheless had recourse to it, under the guise of an "intrinsic pressure," in his own fundamental theory of capillarity and van der Waals' forces.[53] Profound physical discoveries eventuated from the implacable mathematizing of distance forces.

Naissance (1989), 512–514, who seem to take the Laplacian school of capillarity at its own evaluation.

52. Young, *Quart. rev.*, 11 (1814), 43; Playfair, *Natural philosophy* (1810), 1, 181–182; Gauss to Bessel, 27 Jan 1829, in Gauss, *Briefwechsel* (1880), 489.

53. Van der Waals, in Rowlinson, *Van der Waals* (1988), 139–142; Rayleigh, *Scientific papers* (1899), 3, 397–399, 421–423, 465–468 (texts of 1890, 1891).

3. SOUND AND FURY

Lagrange and Newton

Newton's derivation of his formula for the speed of sound in air has not earned him good marks from the cognoscenti. "A dark and convoluted argument," says Johann Bernouilli; "perhaps the most difficult and obscure place in the *Principia*" (d'Alembert); "contradictory..., vague, and indeterminate," fit for a physicist but not for a mathematician (Lagrange); and, to put the same point with politesse, "one of the most remarkable features of [Newton's] genius" (Laplace).[54] Nonetheless, an entire 18th-century-full of mathematicians could do no better. After Lagrange had cleaned up Newton's mess, he had only confirmed the original result: the speed of sound is the square root of the ratio of the average pressure to the average density ($c = \sqrt{p_0/\rho_0}$) of the medium through which the sound waves pass.[55]

The continental mathematicians gave Newton a bad rap. His demonstration suffered from its old-fashioned form and limped under the assumption of a special case; but it was also pioneering, ingenious, and generalizable. The argument is easy enough to follow with an occasional translation into the language of analysis. In Newton's physical picture, elastic particles, EF, FG, etc. (figure 3.3.1), of equal, microscopic extent, vibrate one after the other, each going through the same notion, *assumed* to be simply periodic, a time Δt behind the particle immediately before it.[56] The sum of all the physical lines EF, FG defines the extent of the "pulse," λ; hence, if τ designates the period of the harmonic motion, $\Delta t/\tau = EF/\lambda$. After a time t, the points E, F, G have been displaced by the wave to ϵ, ϕ, γ. Note that their original position was not necessarily that of equilibrium; indeed, as the next step shows, Newton started to follow the motion when EF stood at its maximum displacement

---

54. Bernoulli, *Recherches physiques et géométriques sur...la propagation de la lumière* (1736), 38, and d'Alembert, *Traité de l'équilibre et mouvement des fluides* (1744), 219, quoted by Lagrange, Acc. scienze, Turin, *Misc.*, 1 (1759), ii; Lagrange, ibid., ii, vii, 4; Laplace, *Oeuvres* (1878), 14, 297 (text of 1816).

55. Lagrange, Acc. scienze, Turin, *Misc.*, 1 (1759), 91.

56. The following paragraphs paraphrase *Principia*, Bk II, props. 47 and 49; ed. Cajori, 375–381.

(and hence underwent its maximum acceleration) from its quiescence before the sound wave struck.[57]

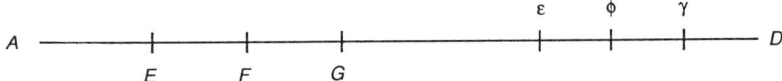

FIG. 3.3.1   Diagram for Newton's theory of the propagation of sound: the elastic "particles" in the line AD are displaced by a sound wave so that EF goes to $\epsilon\phi$, FG to $\phi\gamma$, etc.

This step takes us from figure 3.3.1 to figure 3.3.2, in which PI represents t; the equal arcs HI and IK, Δt; the circumference $2\pi a$, τ; and the lengths PL, PM, and PN, the displacements $E\epsilon$, $F\phi$, and $G\gamma$, respectively. The motion dilates EG into $\epsilon\gamma = EG + G\gamma - E\epsilon$ = EG − LN. From the similar triangles KHQ and IOM, LN/KH = IM/a, whence the expanded volume becomes EG − IM · KH/a = EG(1 − IM/$\bar{\lambda}$). The substitution of $\bar{\lambda} = \lambda/2\pi$ for EG/KH comes from $\Delta t/\tau$ = HI/$2\pi a$ = EF/λ and KH = 2HI, EG = 2EF. Assume in accordance with Boyle's law that the elastic force exerted by a "particle" of air is inversely proportional to its compression. Therefore the elastic force or pressure in $\epsilon\gamma$ is

$$p = \kappa m / EG(1 - IM/\bar{\lambda}). \tag{3.3.1}$$

Here $k = p_0/\rho_0$ converts density to pressure and $m = \rho V = \rho_0 V_0$ is the mass of air in $\epsilon\gamma$ or EG.

The difference between the pressures acting on opposite faces of $\epsilon\gamma$ may be expressed according to equation (3.3.1) as

$$\Delta p = \frac{km}{EG}(\frac{1}{1-HL/\bar{\lambda}} - \frac{1}{1 - KN/\bar{\lambda}}) = \frac{km}{\bar{\lambda}EG}(HL - KN). \tag{3.3.2}$$

Small quantities representing the excursions of the particles have been dropped from the denominator.

---

57. This point bothered Roger Cotes, the editor of the second edition of the *Principia*, who tried to make Newton place the origin of the motion at the initial quiescent position. Newton refused, and, as will appear, carried forward the calculation consistently. Cotes to Newton, 23 June and 30 Jul 1711, in Newton, *Correspondence* (1975), 167–171, 183–185.

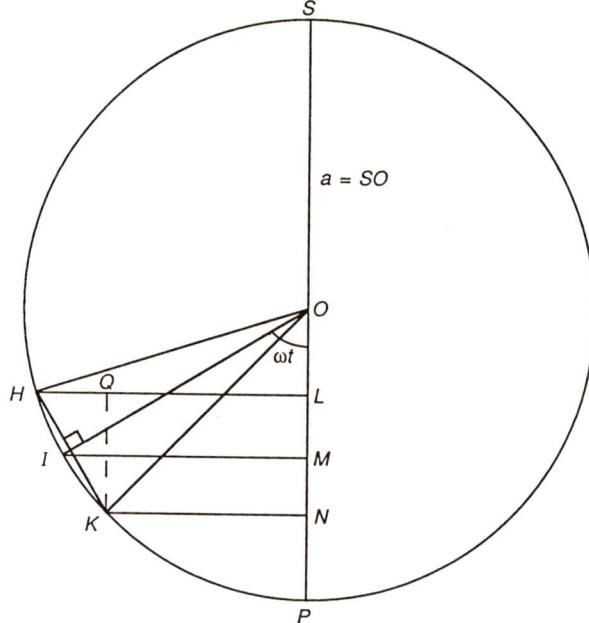

FIG. 3.3.2 Newton's representation of simple harmonic motion: the angular distance from OP represents time; the circumference of the circle, the period; and the distance along the diameter from P, the displacements of the elastic particles.

Consider now the situation when ϵ coincides with E (the leading edge of the particle EF has returned to its original position). K coincides with P and, from equation (3.2), $\Delta p = (km/\lambda EG)HK$. (When $K \to P$, $HL \to HK$, $KN \to 0$.) But $HK/EG = 2\pi a/\lambda = \bar{a}/\lambda$. Hence the accelerative force on $\epsilon\gamma$, $\Delta p/m$, is $ka/\bar{\lambda}^2$. We have another value for this acceleration from the supposition of simple harmonic motion, namely $a\omega^2 \cos\omega t = a\omega^2$ at $t = 0$, where $\omega = 2\pi/\tau$. Equate the two:

$$\frac{ka}{\bar{\lambda}^2} = (\frac{2\pi}{\tau})^2 a, \quad (\frac{\lambda}{\tau})^2 = k.$$

Now λ/τ is none other than c and $k = p_0/\rho_0$. Q.E.D.

Lagrange criticized Newton for putting a "paradox," for purporting to demonstrate that, irrespective of the law of force obeyed by the vibrating particles, the period of their motions would be the same. (The paradox arose from assuming the special case of simple harmonic motion and supposing its consequence to hold in general.) Lagrange's more general investigation dispelled the paradox by arriving at the very expression for c that Newton had deduced from his special assumption.[58] Here is the more general (and elegant) investigation.[59] In figure 3.3.3, the particles that occupied the volumes $\sigma\,dx$ (σ is the cross section of the air mass perpendicular to AB) left and right of x at $t = 0$ have moved into the volumes σ EFGH and σ GHIJ, respectively, at time t. The net pressure acting

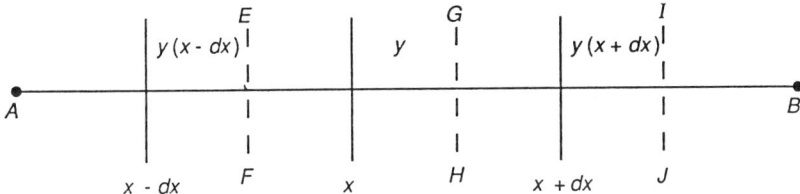

FIG. 3.3.3 Diagram for Lagrange's theory of the propagation of sound: the pulse displaces the air at the face $x-dx$ a distance of $y(x-dx)$ to the face EF; that at face x a distance y to the face GH; and that at the face $x+dx$ a distance $y(x+dx)$ to the face IJ.

on the face GH is easily obtained from the assumption, made natural by Boyle's law, that the pressure anywhere in a gas is proportional to its density there. Density is inversely proportional to volume. Hence the calculation amounts to finding the volumes $V_1 = \sigma(\text{EFGH})$ and $V_2 = \sigma(\text{GHIJ})$.

From the figure, V_1 and V_2 are $\sigma(1+y'-y''dx/2)dx$ and $\sigma(1+y'+y''dx/2)dx$, respectively, where the primes indicate differentiation with respect to x. Thus the net force on GH, f, may be written (with m = mass of air originally in σdx):

58. Lagrange, Acc. scienze, Turin, *Misc.*, 1 (1759), 6–7, and 2 (1760/1), 13, 15–16.
59. Ibid., 41–42.

$$f = \sigma(p_1 - p_2) = \sigma \frac{p_0}{\rho_0}(\rho_1 - \rho_2) = \sigma \frac{p_0}{\rho_0} \cdot m\left(\frac{1}{V_1} - \frac{1}{V_2}\right)$$

$$= \sigma\left(\frac{p_0}{\rho_0}\right)m\left(\frac{y''}{\sigma(1+y')^2}\right) = \frac{p_0}{\rho_0}my''.$$

From Newton's second law, $f = m\,\partial^2 y/\partial t^2$, and

$$\frac{\partial^2 y}{\partial t^2} = \frac{p_0}{\rho_0}\frac{\partial^2 y}{\partial x^2}.$$

Lagrange observed that the solutions to this equation, $y = \phi(t - x/c)$, show that $c = (p_0/\rho_0)^{1/2}$ is a velocity, which must be that of the propagation of sound.

But was it? Despite finagling with the numbers, Newton could not bring his theoretical value of c (some 1000 ft/sec) up to the best measured value (1142 ft/sec). His desperate means of accounting for the discrepancy of over ten percent was to stock the atmosphere with solid particles of water vapor through which sound might move instantaneously.[60] Lagrange did not worry much about the shortfall; the match was good enough for a mathematician if not for a physicist. "It amounts to only a tenth....In any case, it should not be surprising that the theory differs a little from measurement where absolute quantities are concerned, since we know that complicated experiments can never furnish the simple data free from extraneous circumstances that pure analysis requires of them." But if necessary, analysis could always save the numbers. The analyzer needed only to take some other relationship between pressure and density than that prescribed by Boyle's law. Lagrange proposed $p = \rho^n$, which would increase the value of the speed of sound by $n^{1/2}$. To bring theory into agreement with experiment, n had to equal about 1.3, which, as Lagrange said, would have required departures from Boyle's law so large that he could scarcely credit that any good physicist could have missed them. "I only give the hypothesis that the elasticity is proportional to $\rho^{1.3}$ as a trifling conjecture."[61]

60. Heilbron, *Physics at the Royal Society* (1983), 79–80.
61. Lagrange, Acc. scienze, Turin, *Misc.*, 1 (1759), 91–92, and 2 (1760/1), 152–153, resp.

Laplace's approach

Few better conjectures, if any, than Newton's and Lagrange's for saving the speed of sound came forward before 1800. Nor were there useful direct advances in theory; the young Turks around Laplace who worked out the promising ideas of their mentor thought that only Lagrange's account was worth considering.[62] But theoretical advances in other directions made the problem newly tractable around 1800. To a head filled with the caloric theory of gases, a way to augment the velocity of sound theoretically was not far to seek. Biot later represented the successful approach as a matter of common sense. Anyone should have been able to deduce that a sound wave must heat the medium through which it passes; otherwise it would condense saturated vapors through which it tried to pass. To save the sound in saturated vapor, compression would have to release heat. That presented no trouble to the informed savant of 1800. Several examples of the adiabatic heating (or cooling) of gases were then circulating in the journals of science.[63]

Laplace first indicated this line of thought in 1803, in two notes supporting the caloric model of gases as described in the *Statique chimique* of his friend Berthollet. Laplace reasoned from the assumption that the pressure in a gas is proportional to the number of molecules (n, say) pressing immediately against unit surface of the containing vessel and to the density of free caloric (w). Let the vessel be a box with floor area A and height h, and let it contain N molecules. Then $n = (N/Ah)(Ad)$, where d is the closest distance between molecules, and $w = Q(V,T)/V$, where Q is the total free caloric in the box, V is the volume Ah, and T is the temperature. Laplace asked for the amount of caloric set free in an isothermal compression to half the original volume. Since by hypothesis, $p \propto (nw)/A = NQd/V^2$, and since in isothermal processes $pV = \text{const.}$, we have $Qd/V = \text{const.}$

In his first note, Laplace mistakenly stated that the number of molecules acting immediately on the surface A doubles during the

---

62. Poisson, Ecole poly., *Jl*, 7:14 (1808), 325.
63. Biot, *MSA*, 2 (1809), 96–98; *JP*, 47 (1798), 186 (adiabatic cooling); Desormes and Clément, *JP*, 89 (1819), 324 (earlier examples of adiabatic heating); Kuhn, *Isis*, 49 (1958), 133–135.

compression: it follows that d remains constant ($N = nVdA$, N and A constant), and therefore also the caloric density Q/V: in diminishing the volume by half isothermically, half the free caloric must be removed. Now a sound wave passing through a gas produces volume changes, and the consequent play of caloric affects the velocity. Indeed, "it produces the excess of this velocity over that given by the ordinary theory, as I have assured myself by calculation."[64]

The insight was deep, the calculation mistaken. In his second note, Laplace recognized that d varies as $V^{1/3}$, so that in an isothermal compression to half volume the free heat sinks to $Q/2^{2/3}$ or over $0.6Q$ rather than the $0.5Q$ of the earlier calculation. "The gas gives off less than half of the heat it originally contained, which agrees with experiment and with the observed velocity of sound."[65] That even in this second state of enlightenment Laplace was far from a derivation of this elusive quantity appears from the calculation done by Biot under his direction and published in 1802.

Biot's unnecessarily elaborate mathematics may be simplified without loss as follows.[66] To augment c, one need only replace p_0 in the usual formula with a larger quantity, which has its origin in an increase in temperature occasioned by the caloric freed by the sound wave during its passage. Biot begins with a slight rewriting of Lagrange. The increase of the volume $\sigma\,dx$ in time dt (see figure 3.3.3) is $\sigma\,dx(1 + \partial y/\partial x)$; whence the specific increase in volume, $\Delta V/V_0 = dx\,\sigma(1 + \partial y/\partial x - 1)/dx\,\sigma)$, is $\partial y/\partial x$. But $\Delta V/V = -\Delta \rho/\rho$, which gives, to the second order of small quantities, $\partial \rho/\partial x = -\rho_0 \partial^2 y/\partial x^2$. The net force acting on the volume at time t is $-\sigma(\partial p/\partial x)dx$, which must equal $\rho_0 \sigma dx \partial^2 y/\partial t^2$. Hence,

$$\frac{\partial^2 y}{\partial t^2} = -\frac{1}{\rho_0}\frac{\partial p}{\partial x} = -\frac{1}{\rho_0}\frac{p_0}{\rho_0}\frac{\partial}{\partial x}(-\rho_0 \frac{\partial y}{\partial x}) = \frac{p_0}{\rho_0}\frac{\partial^2 y}{\partial x^2},$$

where the relation between pressure and volume is that stipulated in Boyle's law. Biot repeated this calculation with the hypothesis

$$p = p_0(1 + s)(1 + ks) \sim p_0(1 + (1 + k)s), \qquad (3.3.3)$$

---

64. Laplace, *Oeuvres* (1878), *14*, 329–331, on 331; cf. Arnold, *AHES*, *28* (1983), 275–276.
65. Laplace, *Oeuvres* (1878), *14*, 331–332.
66. Biot, *JP*, *55* (1802), 173, 178–181.

where s is written for $-\partial y/\partial x$ and $\rho = \rho_0(1+s)$ as before. The factor k takes account of the increase in pressure, above that owing to compression, induced by the adiabatic heating. Proceeding as before, Biot found

$$\frac{\partial^2 y}{\partial t^2} = \frac{p_0}{\rho_0}(1+k)\frac{\partial^2 y}{\partial x^2}. \qquad (3.3.4)$$

We have arrived at another Laplacian k. Unlike the quantities in the theories of refraction and capillarity, however, the k in equation (3.3.4) was incalculable only in practice. Biot did what he could to estimate its value. According to old measurements by Guillaume Amontons, air heated by 80° Réaumur increases its pressure by a third; according to Gay-Lussac's then recent and immaculate results, in the same temperature range the volume increases by 35 percent.

In equation (3.3.3), the first factor on the right refers to changes in pressure deriving from changes in volume, the second to augmentations of pressure owing to caloric effects. If the measurements of Amontons and Gay-Lussac had been made at constant volume, we might gather from (3.3.3), as Biot did, that

$$\frac{\Delta p}{p_0} = 0.33 = k\frac{\Delta \rho}{\rho_0} = 0.35k,$$

and thus that $k = 0.95$. Biot took the uncorrected theoretical value for c to be 915 ft/s and the best measured value (that done by the Paris Academy in 1783) to be 1038 ft/s. But $915(1.95)^{1/2} = 1278$, not 1038. He had overshot the mark. Working backwards from the answer, he got $k = 0.29$; using Amontons' result that the pressure increased by a third in 80°R, he set the increment/degree at 1/240; and thence he gathered that the increase in temperature, $\Delta T = (0.29)(240°)$, produced in the condensations of a sound wave was 69°. Was that implausible? "[Not] if you take into account that we deal with very small quantities of air in contact with highly conductive walls and that our thermometers have a considerable mass in comparison with the air in which they are immersed....How easy it is to reconcile in this way calculation and experiment in one of the most important theories of mathematical physics [*physique mathématique*]!"[67]

67. Biot, *JP*, 55 (1802), 182.

The reconciliation did not impress Poisson, who did not mention his co-disciple Biot in the lengthy memoir he devoted to the problems of sound in 1808. Lagrange had understood it all, according to Poisson, for he had seized the cardinal point about the development of heat. Was that the full and sufficient cause for the annoying discrepancy between theory and measurement? "It seems to me that such a conclusion follows if no objection can be made to the [mathematical] analysis from which the velocity of sound is deduced, and if at the same time all the physical circumstances that can affect the result have been taken into account in the calculation."[68] No doubt. But Poisson had to satisfy himself with squeezing the data in the manner of Biot.

To obtain the suppositious rise in temperature needed to give k its observed value, Poisson likened the condensation caused by a sound wave to a heating at constant pressure to the required temperature T_2 followed by an isothermal compression to the initial density ρ_1. His only experimental datum was Gay-Lussac's measurement of the dilation of gases at constant pressure, $V = V^*(1 + \epsilon T_c)$, where the asterisk indicates a value at the freezing point of water and T_c expresses centigrade degrees. Let the measurement of the speed of sound be done at $T_1°$C, where the air's density, ρ_0, is $\rho^* V^*/V_1 = \rho^*/(1 + \epsilon T_1)$. Compressed by a sound wave, the air at T_1 acquires a density $\rho_1 = \rho_0(1 + s)$; after heating at constant pressure to T_2, a density $\rho_2 = \rho^*(1 + s)/(1 + \epsilon T_2)$; and, after isothermal compression back to T_1, $\rho_3 = \rho_1$. Boyle's law holds along the isothermal, whence the final pressure p_3 may be derived from the initial:

$$p_3 = p_1(p_2/p_3) = p_0(1 + s)\frac{1 + \epsilon T_2}{1 + \epsilon T_1} = p_0(1 + s)(1 + \frac{\epsilon \Delta T}{1 + \epsilon T_1}). \quad (3.3.5)$$

In order to bring this expression into a form parallel to Biot's, Poisson took ΔT to be proportional to s, say ms, which gave for k the value $m\epsilon/(1 + \epsilon T_1)$.

Before putting in some numbers, it will be useful to change notation. As Poisson observed in a later treatment of the problem,[69]

---

68. Poisson, Ecole poly., *Jl*, 7:14 (1808), 325–326.
69. Poisson, *ACP*, 23 (1825), 15.

Gay-Lussac's law can be combined with Boyle's to give

$$p = \text{const.}\rho\,(1+\epsilon T_c).$$

This may be expressed as

$$p = \text{const.}\,\rho\,(1/\epsilon + T_c) = \text{const.}\,\rho\,(T^* + T_c) = \text{const.}\,\rho T\,,$$

where $T^* = 1/\epsilon$ is the temperature of the freezing point of water above "absolute zero." Poisson's physical assumptions now appear as $\rho_1 = \rho_0(1+s)$, $\rho_2 = (T_1/T_2)\rho_1$,

$$p_3 = \rho_0(1+s)(T_2/T_1) = \rho_0(1+s)(1+ks)\,, \qquad (3.3.6)$$

where $k = m/T_1$.

Poisson's uncorrected theoretical value for $c(T)$ at T^* was 279.4 m/s. At 6°C, where the hardy academicians of 1738 made their measurements, $c(6) = 282.4$ m/s; here the density of air has been computed from Gay-Lussac's law of expansion with $\epsilon = 0.00375$. The academicians obtained 337.2 m/s, whence the empirical value of $k = 0.425$. We have then that $m = kT_1 = (0.425)(1/\epsilon + 6) = 116°C$. Hence a dilation or compression of 1/116 would produce a rise of temperature of one degree. "Admitting this result, which cannot be confirmed by any direct experiment, the difference in velocity [of sound], which Newton was the first to notice, can be made to disappear."[70]

Enter γ

The outcome again was a brave fizzle: as in the cases of optical refraction and capillarity, the essential quantity could not be deduced from the theory. Contrary to the situations controlled by the short-range forces, however, further articulation of the underlying theory allowed a reformulation that made possible a most attractive interpretation of the constant k. The factor $ms/T_1 = \Delta T/T_1$ is the correction owing to the heat of compression, ΔQ. If we assume that the heating takes place so quickly that the gas retains its heat, then $\Delta Q = C_v \Delta T$, where C_v is the specific heat at the compressed volume. Laplace's group pictured the difference

---

70. Poisson, Ecole poly, Jl., 7:14 (1808), 360–364. The value 1/116 became standard; Kuhn, Isis, 44 (1958), 137; supra, §2.2.

whereas C_p measures the increase of the sum of free and latent caloric, the latter being the amount of heat required to enlarge volume at constant pressure. Therefore, $C_p - C_v$ indicates this latent heat, which would become free in a sudden compression that raises temperature by ΔT. But this same ΔT is supposed to be the rise of temperature in the thought expansion from ρ_1, T_1 to ρ_2, T_2, at constant pressure p_1. Therefore,

$$\frac{\Delta T}{T} = \frac{\Delta Q}{T_1 C_v} = \frac{(C_p - C_v)}{T_1 C_v} \cdot \frac{p_1 \Delta V}{R} = \frac{C_p - C_v}{C_v} \cdot \frac{\Delta V}{V_1} = (\frac{C_p}{C_v} - 1)s = ks.$$

According to equation (3.3.4), the correction factor is the square root of $1 + k = C_p/C_v$. Laplace announced this result in 1816, without any derivation and without mentioning k.[71]

It remained to evaluate C_p/C_v. Again, no measurements strictly to the point existed. Again, one made do with what was available, in this case numbers then recently obtained by Delaroche and Bérard. They began with two samples of gas at the same temperature $T_1 = T_2$ and occupying the same volume $V_1 = V_2$ but under different pressures p_1 and p_2. (Evidently the samples contained amounts of gas in the ratio of the pressures.) The experiment consisted of measuring the heats ΔQ_2 and ΔQ_1 given off during cooling at constant pressure to a common final temperature. Laplace supposed, "with many physicists," that the heat in a sample of air kept under a constant pressure is proportional to its volume; and he claimed that, on this supposition,

$$\gamma = \frac{C_p}{C_v} = \frac{p_2/p_1 - 1}{\Delta Q_2/\Delta Q_1 - 1}. \quad (3.3.7)$$

Delaroche and Bérard set $p_2/p_1 = 1.36$ and measured $\Delta Q_2/\Delta Q_1$ as 1.24. Therefore $\gamma = 1.5$, which brought the speed of sound up to 345.9 m/s, a little beyond the 337.2 m/s of the academicians of 1738. With greater plausibility than Poisson or Biot, Laplace could assign the remaining discrepancy to errors in the measurements:

71. Laplace, Oeuvres (1878), 14, 298; cf. Sebastiani, Physis, 23 (1981), 401–402. Fox, Caloric theory (1971), 159–161, gives a later argument of Laplace's containing the evaluation of γ but no reference to Poisson's formula, from which the reconstruction offered in the text proceeds.

"the smallness of the difference establishes irrefutably that the excess of the observed value of the speed of sound to that calculated by the Newtonian formula is due to the latent heat developed by the compression of the air."[72]

Does equation (3.3.7) follow from the assumptions? Recent historians who have tried to deduce it have had to feign implausible hypotheses to succeed.[73] No doubt, some improper assumption seems required, since Laplace worked with the misinformation that C_p depends strongly on the temperature. In fact, however, no special hypotheses are required to gain equation (3.3.7). From Laplace's basic assumption, the heats in the two gases at the start are $k_1 V_1$ and $k_2 V_2$, and, at the end, $k_1 V_1'$ and $k_2 V_2'$, where the k's depend only on the pressures and the primes indicate final states. Hence, $\Delta Q_2/\Delta Q_1 = k_2 \Delta V_2/k_1 \Delta V_1 = k_2/k_1$. (The equality of the volume differences follows from $V_1 = V_2$, $T_1 = T_2$, $T_1' = T_2'$.) We need only derive the k's.

In its expansion at p_1, the first sample loses heat $C_{p_1}(T_1'-T_1) = k_1(V_1'-V_1)$. But $V_1'-V_1 = R/(T_1'-T_1)p_1$. Hence $k_1 = p_1 C_{p_1}/R$. To get k_2, imagine a sample of gas at p_2 equal in quantity to that held at p_1, T_1, and occupying the same volume V_1. The temperature of this sample, T_3, is $(p_2/p_1)T_1$. Now cool it at constant volume to the pressure p_1 and temperature T_1. The heat given up is

$$k_2 V_2 - k_1 V_1 = V_1(k_2 - k_1) = C_{V_1}(T_3 - T_1) = C_{V_1}(p_2/p_1 - 1);$$

$$\frac{k_2}{k_1} = 1 - \frac{C_{V_1}(RT_1/V_1)(p_2/p_1 - 1)}{p_1 C_{p_1}} = 1 + \frac{C_{V_1}}{C_{p_1}}(p_2/p_1 - 1).$$

There is therefore no trouble to obtain

---

72. Laplace, *Oeuvres* (1878), 14, 299. Delaroche and Bérard's measurements at the higher pressure were poor; the support for his theory that Laplace squeezed from their numbers was spurious. Cf. Kuhn, *Isis*, 44 (1958), 137, and Sebastiani, *Physis*, 23 (1981), 404–405.

73. Fox, *Caloric theory* (1971), 162–163, rightly criticizes Finn, *Isis*, 55 (1964), 13–18, and, in *DSB*, 15, 357–358, expresses reservations about his own demonstration, also rightly.

$$\frac{p_2/p_1 - 1}{\Delta Q_2/\Delta Q_1 - 1} = \frac{p_2/p_1 - 1}{k_2/k_1 - 1} = \frac{C_{p_1}}{C_{V_1}} = \gamma. \qquad \text{Q.E.D.}$$

Although neither the theory nor the experiment was secure nor their agreement outstanding, they confirmed Laplace in his belief that the ratio of the specific heats was the quantity he needed. Two ways were then open to complete this corner of Laplacian physics. One was to improve the numbers with experiments undertaken ad hoc. Gay-Lussac eventually obliged, in 1822, with excellent values for γ, which, to Laplace's great surprise, seemed to be the same for all values of pressure and temperature employed in the experiments. The best value, $\gamma = 1.3748$, brought the speed of sound up to 337.1 m/s at 16°C, the comfortable temperature at which a new set of academicians, activated by Laplace, had remeasured it. The measurement gave 340.9 m/s. Laplace ascribed the discrepancy of less than one percent to water vapor in the atmosphere.[74]

Again the figleaf

The other route to setting the speed of sound, and thereby the caloric theory firmly within Laplacian physics, was to derive the factor C_p/C_V from a model built on short-distance forces. Here Laplace followed two strategies. The first derived the laws of Boyle and Gay-Lussac, on which the theory of 1816 depended, from the model. The second introduced the model directly into the mechanics of the propagation of waves. The model: each gas molecule carries an envelope of caloric; hence each suffers repulsions between its envelope and its neighbors', and attractions between itself and the environing envelopes and molecules; but in practice, the caloric-caloric repulsion greatly outweighs the other forces present. Each molecule constantly loses part of its envelope owing to bombardment by radiant heat from other molecules; each molecule constantly captures itinerant caloric; in the steady state, the gain balances the loss. We are again ready to calculate.[75]

Let the gas of density ρ be contained in a cylindrical vessel of radius a and height h, and let every molecule have a dose of caloric

---

74. Laplace, *Oeuvres* (1878), 13, 303–304 (text of 1822); *ACP*, 19 (1822), 436–437 (announcement of measurements by Gay-Lussac and Weber.)

75. Laplace described the model in several places between 1821 and 1823, e.g., *Oeuvres* (1878), 13, 277–282, and 14, 305–307.

c. We ask for the force dF exerted on a narrow column of gas beginning just under the lid and running to the bottom of the cylinder. As usual, the force is proportional to the "masses" or "charges" of interacting elements, in this case to $\rho c dV$ and $\rho c dV'$, where the dV's are the elementary volumes, and to a short-range force $\phi(f)$, where f is the distance between the centers of dV and dV'. Thus the elementary force is $H\rho^2 c^2 dV dV' \phi(f)$, H being a constant. Take for dV a portion of gas with area $rdrd\theta$ and thickness dz (figure 3.3.4), for dV' an element a distance s below the lid with cross section dA and depth ds. Then the downward force in the vertical direction exercised on the column by all the caloric within the cylinder is

$$df = H\rho^2 c^2 dA \int_0^{2\pi} d\theta \int_0^h ds \int_0^s dz \int_0^a \{\phi(f) \cdot (s-z)/f\} r dr.$$

Because of the short range, a and h may be taken as infinite and the integrations performed precisely as in the theory of capillarity. We have therefore for $p = p_z = dF/dA$, the pressure against the lid,

$$p = 2\pi H \rho^2 c^2 K, \qquad (3.3.8)$$

where K is the uncalculable constant that represents the result of the integrations.[76]

Now let our molecules suffer from bombardment by radiant caloric, which knocks off bits of their envelopes. The loss for each molecule will be proportional to the density of caloric in the gas, ρc, and to the quantity of caloric in its envelope, c, whence loss equals $\alpha \rho c^2$. Let the gain G be proportional to the absolute temperature—an ad hoc assumption needed to reach the intended conclusion—and so equal to βT. In the steady state,

$$\rho c^2 = (\beta/\alpha) T . \qquad (3.3.9)$$

The combination of equations (3.3.8) and (3.3.9) gives

$$p = \text{const.} \, \rho T ,$$

which, as Laplace emphasized, includes the laws of Boyle and of

---

76. Laplace, ibid (1878), 13, 280, reaches equation (3.3.8) after a rehash of his treatment of attractions between spherical masses. Although irrelevant in detail, it indicates again the close tie between his physics and his astronomy.

180 LAPLACE'S SCHOOL

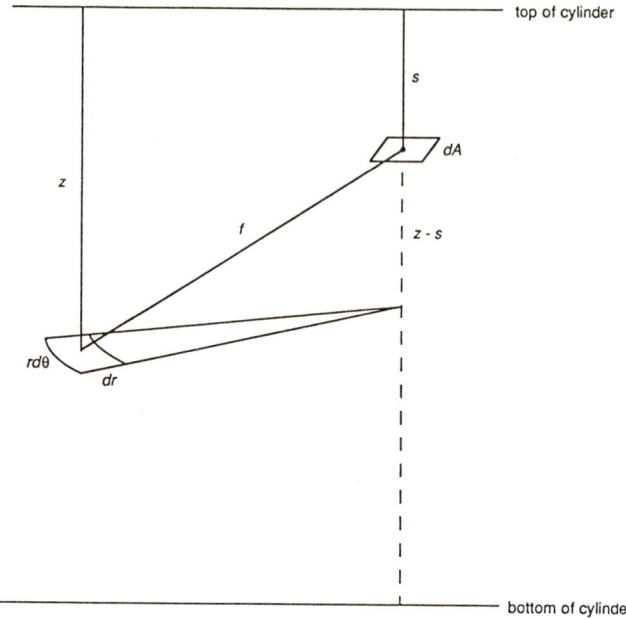

FIG. 3.3.4 Diagram for Laplace's calculation of the force exerted by the caloric envelopes around the molecules of a gas contained in a cylinder on a thin column of gas along the cylinder's axis: dA is the cross-section of the cylinder, and $rd\theta drdz$ an element of volume of the gas outside the cylinder.

Gay-Lussac. He thus made it plausible that the Standard Model could yield the basic laws of gases, although nothing authorized taking the constant, which contained H and other quantities dependent on molecular species, to be the same for all gases.

We proceed to the disappointing climax. In figure 3.3.5, an air molecule A of mass m situated on the axis of a cylinder suffers a repulsive force from the caloric around B. The vertical component of the repulsion is $df = mc_A \cdot \rho_B c_B \cdot H \phi(f)(s/f) \cdot d\theta z dz ds$, where ρ is the density of the air and c the caloric per molecule, and θ measures the azimuth around the cylindrial axis. The total force exerted on the molecule at A by all the caloric in all the air in the cylinder is

$$F_A = 2\pi H c_A m \int_{-\infty}^{\infty} ds \int_0^{\infty} (c\rho)_B (zs/f) \phi(f) dz$$

$$= -2\pi H c_A m \int_{-\infty}^{\infty} (c\rho)_B(s)\psi(s)ds,$$

where $\psi(s)$ is the result of integrating over z ($c\rho$ assumed to be a function only of s). Now develop $(c\rho)_B$ in the neighborhood of A, that is, around the point x on the cylinder's axis. To first order, $(c\rho)_B = (c\rho)_A + s\partial(c\rho)_A/\partial x$. With this substitution,

$$F_A = -2\pi H(c^2\rho)_A m \int_{-\infty}^{\infty} s\psi(s)ds - 4\pi Hc_A m \partial(c\rho)_A/\partial x \int_0^{\infty} s^2\psi(s)ds$$

$$= -4\pi HKcm\, \partial(c\rho)/\partial x. \qquad (3.3.10)$$

Equation (3.3.10) follows from its predecessor by dropping the subscript A; by noticing that the first integral, being an odd function of s, vanishes, and that the second, being even, can be doubled and run from zero to infinity; and by invoking the all-purpose solution, "set the integral equal to K."

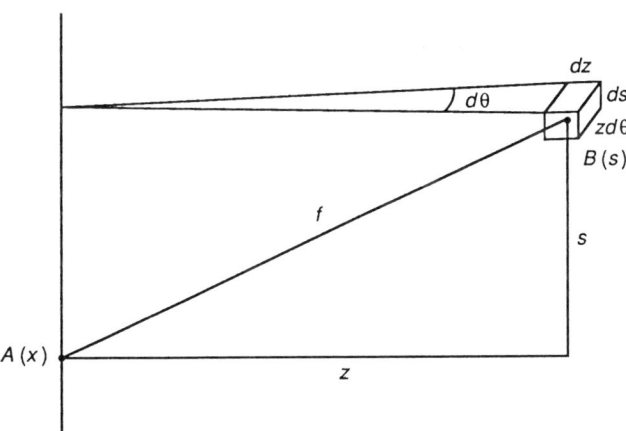

FIG. 3.3.5 Diagram for finding the force exerted by an enclosed gas on a gaseous molecule A within the enclosure. The element of volume a distance f from A is $zd\theta \cdot dz \cdot ds$.

We require another standard piece of sleight-of-hand. Let $\rho \partial c/\partial x = -c\partial \rho/\partial x$. Just let it.[77] Then the force on A becomes

77. Ibid., 5, 134. Cf. Finn, *Isis*, 55 (1964), 16.

$-4\pi HKc^2(1-\epsilon)\partial\rho/\partial x$. Let the displacement of A under F_A from its equilibrium position be $q(x,t)$. The analysis behind Lagrange's and Biot's derivations of the speed of sound may be repeated without a hitch to give

$$F_A = m\,\partial^2 q/\partial t^2 = 4\pi HKc^2 p(1-\epsilon)\partial^2 q/\partial x^2. \qquad (3.3.11)$$

Recall the result of the previous argument about pressures, equation (3.3.8). Putting it into equation (3.3.11),

$$\frac{\partial^2 q}{\partial t^2} = \frac{2p}{\rho}(1-\epsilon)\frac{\partial^2 q}{\partial x^2}.$$

The correction factor now is the square root of $2(1-\epsilon)$. Laplace had to show that it is nothing other than γ.

Let the heat Q in any element of gas be constant during the passing of a wave: the heat that appears in it, and that raises its temperature during compression, is gained from that latent portion of Q. Laplace makes Q a function of ρ and $p = \text{const.}\,\rho^2 c^2$. Then, from

$$\frac{dp}{d\rho} = \frac{[\partial Q/\partial \rho]_p}{[\partial Q/\partial p]_\rho} = -\frac{[\partial Q/\partial T \cdot \partial T/\partial \rho]_p}{[\partial Q/\partial T \cdot \partial T/\partial p]_\rho} = \frac{p}{\rho}\cdot\frac{C_p}{C_V}. \qquad (3.3.12)$$

Another expression for $dp/d\rho$ may be obtained from $p = \text{const.}\,\rho^2 c^2$:

$$\frac{1}{2}\frac{dp}{d\rho} = \frac{p}{\rho} + \frac{p}{c}\frac{dc}{d\rho},$$

or, by substitution in equation (3.3.12),

$$2\left(\frac{p}{\rho} - \epsilon\frac{p}{\rho}\right) = \frac{p}{\rho}\frac{C_p}{C_V}, \qquad (3.3.13)$$

where ϵ has dropped in from its definition $\epsilon = -(\rho/c)(dc/d\rho)$. Equation (3.3.13), which Laplace obtained around 1823, was the final object of his quest; it made $2(1-\epsilon) = C_p/C_V$, as advertised. "It follows that the speed of sound is obtained by multiplying the Newtonian formula by the square root of the specific heats, which is the theorem I gave without proof [seven years ago], in the *Annales de physique et de chimie* for 1816."[78]

---

78. Ibid., 5, 136–137 (quote).

This rigmarole brought Poisson back to the problem of adiabatic processes. He observed that the measurements of Clément and Desormes, which had been published three years before Laplace's last attempt to develop the theory of caloric based on short-distance forces, made possible a better calculation than he had been able to make with the numbers available in 1807, and that quite ordinary reasoning could produce Laplace's factor γ. In their prototypical experiment,[79] Clément and Desormes started with air at atmospheric pressure and temperature (p_1, T_1). They then withdrew some air, reducing the pressure at constant temperature $(p_2, T_2 = T_1)$, briefly reestablished connection with the atmosphere $(p_3 = p_1, T_3)$, and then let what remained in the experimental vessel cool to the original condition $(p_4, T_4 = T_1)$. The compression in the third step, $s = (\rho_3 - \rho_2)/\rho_2$, can be expressed in terms of the pressures measured in the experiment: since $\rho_3 = \rho_4$ and $\rho_2/\rho_4 = p_2/p_4$, $s = (p_4 - p_2)/p_2$. Similarly, the fractional rise in temperature during the compression $u = (T_3 - T_1)/T_1 = (p_1 - p_4)/p_4$. From their numbers, Poisson made $s = 0.01355$ and $u = 0.00473$. Now, according to equations (3.3.5) and (3.3.6), which contain Poisson's theory of 1807, $ks = \Delta T/T_1 = u$, whence $k = u/s = 0.3492$ and $c = (p_0/\rho_0)^{1/2}(1.3492)^{1/2} = 331.97$ m/s, something short of the Bureau des Longitudes' 340.89 m/s. After correction for the difference in temperature at which the measurements were made, Poisson came out 6.23 m/s short, "which can be attributed to errors of observation."[80]

The same line of thought leads easily to Laplace's correction γ. Take a volume V_1 of air at temperature $T_1 - \theta$, where θ is a small quantity, and heat it at constant pressure to $T_2 = T_1$; the undertaking will require an amount of heat $C_p\theta$ and occasion a dilation s. Now compress the gas to V_1; by hypothesis, the temperature increases by $\Delta T = ksT_1$ (that is the definition of k). If it now cools down to its original temperature at constant volume, it will give up heat $C_V(\theta + \Delta T) = C_V(\theta + ksT_1)$. The heat must have come from what was supplied during the initial expansion:

79. Desormes and Clément, JP, 89 (1819), 330–332.
80. Poisson, ACP, 23 (1825), 7–10 (quote). Cf. Sebastiani, Physis, 23 (1981), 405–410.

$$C_p \theta = C_V(\theta + ksT_1).$$

Now in the initial expansion, the pressure remained constant, hence $s = \Delta\rho/\rho = \theta/T_1$. We therefore have from the preceding equation,

$$1 + k = C_p/C_V = \gamma.$$

Poisson proceeded to something quite new in this tangled story: a single equation that summarized the application of the laws of Boyle and Gay-Lussac to adiabatic processes. Rewriting the expression $\Delta T = ksT_1$ in differential form, he had

$$\frac{dT}{T} = (\gamma - 1)\frac{d\rho}{\rho};$$

which, on the assumption that γ is constant, produced by integration several rules now familiar, including $p = \text{const.}\rho^\gamma$.[81] None of this followed from, or required, the physics of short-distance forces.

---

81. Poisson, ACP, 23 (1825), 13–15.

4 VARIETIES OF MERCANTILIST MATHEMATICS

One of the main benefits that Colbert promised himself in creating the Paris Academy of Sciences was a decent map of France. Colbert needed to know the extent of the kingdom for long-term civil estimates and for short-term military maneuvers.[1] Colbert's academicians, particularly Picard, liked the cartographic project for itself and because it provided the means for finding a reliable value for the radius of the earth, which "would be very subservient to several designs of his in astronomy."[2] The union thus established between cartography mainly for the state and geodesy mainly for the geometer prospered during the Enlightenment, especially in France but also in Austria-Hungary, the Papal States, and Great Britain. The first section of this chapter describes the progress of French cartography; the second, the efforts to ascribe a shape to the earth.

1. CARTOGRAPHIC CONTROL

A French business

The first register of the Paris Academy's mathematical transactions records that Colbert "wished that people would work at making maps of France more reliable than any yet made and that the

---

1. Lagarde, in Picolet, *Picard* (1987), 248–249, 251.
2. Francis Vernon to Henry Oldenburg, 19 Jan 1669/70, in Oldenburg, *Correspondence*, ed. Hall, 6 (1965), 433.

Company would prescribe the method to be used by those to be employed in the project." The same message appears high on the list of activities for the then new Academy recorded in the notebooks of its leading member, Christiaan Huygens, with the notable addition, and precedence, of a more academic objective: the Academy was "to measure the size of the earth [and] to advise about methods to make maps more exact than earlier ones."[3] The double project quickly came into being under Picard, who, however, understood the political importance of placing the map ahead of the measure. After outlining the elaborate procedure he proposed to follow, he told the academicians that an arc of a meridian through Paris, when rigorously measured astronomically and trigonometrically, would not only provide the basis for the most accurate map ever made; "also, we would have the benefit of being able to determine the size of the earth." Picard's political and scientific good sense appears again in his proposal to give the size in a unit that "would remain for posterity and not depend on our particular [practices]." The unit would be the length of a pendulum beating seconds in Paris. "So that the measure of the size of the earth first found by the difference in the heights of the pole, and with respect to the heavens, would be associated with the diurnal motion as a fundamental and convenient [quantity] available to every nation."[4]

With his 58-inch quadrant equipped with lenses and micrometer threads, Picard could measure the angles between distant objects to an accuracy 30 or 40 times greater than the inventor of the trigonometric technique, Willebrord Snel, could manage. And that although the setting of the telescope by hand without mechanical help and the displacement of the mechanical from the optical center of the instrument introduced substantial sources of error.[5] With his 10-foot sector, he found the difference in latitude $\delta\lambda$ between a church tower in Amiens, some 115 km north of Paris, and a

3. "Procès-verbaux de mathématiques," quoted by Levallois, *Vie des sciences*, 3 (1986), 262; Huygens, quoted by Taton, in Picolet, *Picard* (1987), 208.

4. Picard, report to the Academy, 31 Jul 1669, quoted by Taton, in Picolet, *Picard* (1987), 218–219, and Levallois, ibid., 263. Cf. infra, §5.2.

5. Snel seems to have been content with triangles that closed (i.e., summed to 180°) with an error of minutes; Picard's probably closed to 15 or 20 seconds ("probably" because his field notes have not survived and he adjusted his published results to sum to 180°). Westphal, *Zs. f. Instr.*, 4 (1884), 154–156, and 5 (1885), 260; Levallois, in Picolet, *Picard* (1987), 229–234; supra, §2.1.

pavillion in the village of Malvoisine 40 km to the south. Great care was required in making the measurements, which suffered from the flimsy mounting of the instrument and the difference in the expansion of its iron body and brass limb; an error of a minute of arc in the sky translates into almost 2 km on the ground. Picard gave the zenith distance of his reference star to five seconds or less: and from the distance he calculated the arc between the stations to be a little under 1°23', and the length of a single degree to be around 57,060 toise (1 toise = 1 ts $\doteq$ 2 meters). He effected the conversion from anglular to linear measure by laying out a pair of measuring rods one after another along a string to constitute a baseline, the ends of which he incorporated into his triangulation. The great length of this baseline, some 12 km, measured with great care and expense, contributed importantly to Picard's success. It was almost forty times longer than Snel's and five times longer than the previous maximum, a line of 1064 ts set out in 1645 by the Jesuits Francesco Maria Grimaldi and Giambattista Riccioli in a faulty determination of the distance between Modena and Bologna.[6]

Among Picard's associates in laying out bases and finding the arc was Gian-Domenico (Jean-Dominique) Cassini, brought from Bologna in 1669 to assist in the completion of the Paris Observatory. Picard and Cassini also oversaw a project to map the environs of Paris as a step toward fulfilling Colbert's commission. Cassini, who knew how to raise money for large projects, understood that Parisian astronomy would prosper along with French cartography; as he put the point for public consumption, "the chief goal the Academy set itself in its astronomical observations has always been their application to the advancement of geography and navigation."[7] In furtherance of this union, the Academy dispatched missions at the King's expense during the 1670s to find the longitudes of distant places by comparing their observations of eclipses of Jupiter's moons with the times of disappearance given in tables prepared by Cassini for the meridians of Bologna and Paris.

    6. Picard, *Mesure* (1671), 138–139, 147–148, 171–173, 187–189; Westphal, *Zs. f. Instr.*, 4 (1884), 189–191, and 5 (1885), 262; Smith, *Plane to spheroid* (1986), 68, 71–76.
    7. Cassini, *MAS*, 1666–1695, *8* (1730), 39; Picard, "Rapport," 21 Jul 1669, in Taton, in Picolet, *Picard* (1987), 218.

The most notable of these expeditions were those of Picard to Denmark, in 1671, to discover the longitude of Tycho Brahe's palace of Uraniborg, and of Jean Richer to the Caribbean, in 1672–73, to make measurements important for French interests in the sugar islands. It was on this voyage that Richer learned that a pendulum clock that beat seconds in Paris was slow in Cayenne; a tropical lethargy that, in the opinion of Newton and others, subverted the ancient assumption that the earth is a perfect sphere and greatly complicated the task of measurers of the globe.[8] Jesuit missionaries added further data points; and Cassini kept track of it all, on a polar plot 24 feet in diameter traced on the stone floor of one of the towers of the Paris Observatory. A map made from this hard copy was published in 1696.[9]

The world proved easier to map than France. After Picard and Cassini had done some surveying around Paris and Provence, during the course of which Picard perfected the application of the spirit level to the telescope, the Academy received the charge "to draw up a map of France with the greatest accuracy possible." That was in 1679. The first step, completed five years later, was a sketch of the kingdom, based on astronomically determined coordinates, which showed that ordinary maps had misplaced the western coastline of France into the Atlantic Ocean by as much as 150 km; in all, the academicians reduced French territory by no less than a fifth. That might help to account for the reluctance of Colbert's successor to support the project that Picard had proposed in 1681: the extension of the arc from Amiens to Dunkirk in the North and from Malvoisine to Perpignan in the South, to serve as the backbone for a series of triangles to cover all of France. No doubt also Picard's death in 1682, the political and religious difficulties of the end of the century, and the wars of the Spanish succession made it hard for ministers to perceive that the precise and expensive triangulation of an invisible line 950 km long from the English Channel to the Gulf of Lyon was necessary for grasping the real estate of France. Not until 1718, six years after Cassini's death, did his son Jacques

8. Cassini, *MAS*, 1666–1699, 8 (1730), 50; Olmsted, *Isis*, 34 (1942), 120, 126; Gallois, *Ann. géogr.*, 18 (1909), 289–291.

9. Cassini, *MAS*, 1666–1699, 8 (1730), 51; Gallois, *Ann. géogr.*, 18 (1909), 307–310; Wolf, *Hist. Obs. Paris* (1902), 62–63; Brown, *Cassini* (1941), 40–42, 62.

and his nephew Giacomo Maraldi complete Picard's meridian from sea to sea.[10]

Jacques Cassini and Maraldi measured out a baseline along the beach near Perpignan with toise sticks aligned by telescope. To make possible the inclusion of the baseline in the system of triangles, they set up tree trunks as markers on either end. Taking the baseline that Picard had measured near Amiens as their given, they calculated through the series of triangles what the base on the beach at Perpignan should be. The calculated and directly measured lengths of the distance between the tree trunks—some 7246 ts— agreed to within 3 ts before correction and precisely afterwards. A similar result, to within a single toise, was achieved with a base laid out across the dunes near Dunkirk. That was a singular coincidence, since many of Jacques Cassini's angles differed by 10 minutes, and a few by whole degrees, from later measurements made from his stations. His instruments, which included a large zenith sector made by Graham and a quadrant with a relatively small error of center, were at least as good as Picard's. Nonetheless, having determined the length of the triangulated arc from Dunkirk to Perpignan to be 8°31'12", Cassini gave the weighted average value of a degree over the distance as 57,061 ts, astonishingly, suspiciously, close to Picard's.[11] Cartography and territory seemed secure.

As a sidelight, Cassini calculated that a degree of the arc north of Paris was slightly smaller than one taken to the south. The difference between consecutive degrees, amounting to some 31 ts, fell well within likely errors of measurement. Nevertheless, Jacques Cassini deemed it significant. Cassini *père* had reached a similar conclusion from measurements he had made toward extending Picard's arc. Cassini *fils* had no interest in overlooking what he thought were facts that confirmed his late father's opinions; on the contrary, he might well have calculated that by emphasizing them he could make his own reputation separate from his father's. If confirmed, the variation in the length of degrees along a meridian would indicate that the earth is not a perfect sphere. That agreed

10. Gallois, *Ann. geóogr.*, 18 (1909), 291–298; Jacques Cassini, *Grandeur* (1720), 1–5; Konvitz, *Cartography* (1987), 5–8; Lagarde, in Picolet, *Picard* (1987), 253–259.
11. Jacques Cassini, *Grandeur* (1720), 37, 46, 188, 192–194, 246–247; Delambre, *Grandeur* (1912), 4–5, 10–15, 18, 22, 14; Westphal, *Zs. f. Instr.*, 4 (1884), 156–157, and 5 (1885), 262.

with deductions from Newton's principles and also with the principles of Descartes as developed by Huygens. But Newton and Huygens required the degrees to shorten toward the equator, and Cassini had found them to grow.[12] The implied challenge inspired much geodetic work, which usually piggy-backed on projects, like military or civilian cartography or the metric survey, which held greater interest for those who paid the bills than wrangles about the shape of the earth.[13]

On June 1, 1733, Jacques Cassini and his cousin Giovanni Domenico Maraldi left Paris at right angles to its meridian. They were on their way to St Malo, to begin to trace a parallel of latitude. By the end of 1735, they had triangulated all the way to Strasbourg, and the younger generation, represented by Jacques Cassini's son, César-François Cassini de Thury (Cassini III), had brought the line in the West to Brest. They had been commissioned by Philibert Orry, Comptrolleur-Général of France, to make a national map based upon a network of primary geodetic triangles covering the entire country. Most of the later work fell to Cassini III and Maraldi, who devoted themselves particularly to examining the coast from Cherbourg to Nice; they found it easier to survey than the wooded mountains of the hinterland, freer from suspicious peasants, and, above all, of greater interest, because of its strategic importance, to the state that supported them. By 1740, the enterprize had measured 400 primary triangles and a good many bases to a claimed accuracy (in cases where a base was measured twice) to around one part in 20,000. Four years later it had doubled its triangles and completed its assignment. The first Cassini national map, which finished the project started by Colbert, Picard, and Cassini I, was published in 18 sheets in 1744.[14]

12. J.D. Cassini, *MAS*, 1713, 190–191; Jacques Cassini, *Grandeur* (1720), 9; Levallois in Lacombe and Costabel, *Figure* (1988), 45–52; Costabel, ibid., 98–103.

13. Infra, §4.2 for the wrangles. It may be worth recording that English mathematicians favored an oblong over a flat spheroid ("that being better fitt for motion, and more likely to preserve the Axe in its proper place") before they had studied the *Principia*, and judged the argument that the earth's spin should make the equator higher "but a Conjecture...without any Observations that so it is." John Wallis to Edmund Halley, 4 Mar 1687, in Oldroyd, in Hunter, *Hooke* (1989), 211.

14. Konvitz, *Cartography* (1987), 10–16, and *Cartographica*, 19 (1982), 1–2; Westphal, *Zs. f. Inst.*, 5 (1885), 269. Cf. Vannereau, in *Cart*. (1978), 229–230.

With this achievement, Cassini III supposed that he had established the basis for fuller mappings by lesser men. He reckoned without the Army. In 1746, toward the end of the War of the Austrian Succession, Cassini (that is, Cassini de Thury, Cassini III) extended the family's triangles into Flanders at the request of the commanders of the invading force. The military engineers on the ground could not map it with the accuracy that the Cassinis had attained. Cassini did his mapping just behind the front lines. "I had no fear of being taken by the enemy," he said, "for I expected to be treated like M. de Maupertuis, who, having accompanied the King of Prussia into war, was taken prisoner to Vienna, where the Emperor and Empress treated him with a good will that made his captivity sweet and glorious." Cassini escaped captivity but not glory. In 1747 Louis XV came to review his troops in Flanders. He compared Cassini's maps with the terrain and was astonished at the agreement. "I wish the map of my kingdom drawn up in the same way," he told Cassini, "and I charge you with the task."[15]

Cassini set up an extensive operation centered on the Observatory, trained surveyors, raised capital (beginning in 1756, when, at the outbreak of the Seven Years War, the treasury withdrew its support), and remapped France at a scale of 1:86,400 (1 ligne to 100 toises). He insisted that his men take all the angles in their triangles (which closed usually to under a minute of arc), use quadrants rather than planchettes, and go back over previous work when the original targets could be found. By 1783, they had found the distances of 400 villages and 3000 mills, bridges, and isolated houses from the meridian and the perpendicular through the Paris Observatory.[16] On Cassini de Thury's death in 1784, the business was continued by his son and long-time co-adjutor, Jean-Dominique Cassini (Cassini IV), who also succeeded to the family fief of the Paris Observatory. The field work, which required some 40,000 triangles, had been finished by the outbreak of the Revolution, but not all the engraving. Eventually the map appeared, on 180 sheets. Although its latest historian judges it empty of scientific or technical novelty and useless, because of its scale, to civil engineers, it had

---

15. Cassini de Thury, "Description," in his *Relation* (1775), 128 (quote), 129–132, 147–148 (quote).

16. Cassini de Thury, *Description* (1783), 9–12.

sufficient value for strategic planning and national pride that the Dépôt de la guerre confiscated it in 1793. The revolutionary government hired Cassini's engravers, who by then had finished 165 of the 180 sheets, to complete the map, and discharged Cassini (Cassini IV), whose opinions were not revolutionary, from the Observatory.[17]

The reliability of the second Cassini map declined with distance from the primary grids through Paris. The triangles in those grids closed with an accuracy of perhaps 10 seconds of arc; elsewhere, to 20 seconds. Baselines were measured by placing sticks end to end, much as in the manner of Picard. The angles in the primary triangles were taken by sector and telescope; the field measurements, by plane table and compass.[18] Although Cassini III had bettered the positioning of the telescope on the sectors and the reading of the limb, he neglected to strengthen the mountings of his instruments, whose wobblings probably nullified the effects of his improvements.[19] Some serious error crept in especially along the borders in the North and in Flanders and the Rhineland. But they were as nothing compared with the errors around Munich, where, in 1761/2, Cassini III measured a baseline and triangles to link up with a net run west from Vienna by Joseph Xaver Liesganig, S.J., the surveyor-general of the Holy Roman Empress. Later French surveyors found that the angles in one of Cassini's Bavarian triangles summed to $190°$.[20]

The remarkable account that Cassini wrote about his travels to the East makes his purposes as muddy as his measurements. The title states the purposes as determining the size of a degree of latitude, mapping Germany, and finding the precise coordinates of the chief towns where astronomers had made observations. The preface gives determining the degree the first importance; the first pages of the work give military interest in cartography as the primary incentive for the project and finding the coordinates of German

17. Konvitz, *Cartogr.*, *19* (1982), 6, 12–14, 24–25, 30, and *Cartography* (1987), 17–22; Gillispie, *Polity* (1980), 115–117. Guillaume, *Proc. verb.* (1894), *2*, 218–226, 476–487, 601–602.
18. Schmidt, *Kartenauf.* (1973), *1*, 10–13; Westphal, *Zs. f. Instr.*, *5* (1885), 274.
19. Westphal, *Zs. f. Instr.*, *4* (1884), 191–193.
20. Schmidt, *Kartenauf.* (1973), *1*, 13, 45; Westphal, *Zs. f. Instr.*, *5* (1885), 274; Berthault, *Ing. géogr.* (1902), *1*, 49.

observatories as the second (the latitude of even so prominent a place as Munich was known only to within eight minutes of arc, "while in France we dispute only about seconds"); later on, determining the degree reappears, but as the third objective of the trip.[21] Like some modern savants, Cassini adapted his message to his audience, and, like many writers, forgot for whom he was writing as he proceeded.

Cassini left Paris on May 2, 1761 and made good speed following his own map to Strasbourg. He then triangulated his way across Wittenberg, "where I was astonished to find roads as good as those in France," to arrive in Vienna 54 weeks after his departure. The emperor received him warmly, despite gloom cast by the poor weather, which threatened to hide the impending transit of Venus. Observing this rare event meant as much to fashionable women as to men, according to Cassini; "for in Vienna, as in all German courts, science is not foreign to the fair sex."[22] Heaven heard their prayers. "The clearness of the sky, and the brilliance of the stars attracted the gazes of the princes and princesses and I had the satisfaction of hearing from the mother of the Elector, and the princesses, the names of the principal stars, which they know as well as the names of their courtiers." The main princelings of Southern Germany had a similar and equally unexpected knowledge of astronomy. The duke of Wittenberg claimed to be a disciple of Euler's; the Palatine Elector corresponded with French telescope makers; the Margrave of Baden-Domlac had all the latest maps, clocks, and instruments, "in a word, everything necessary to cultivate astronomy."[23]

The Prince Bishop of Passau proved the most "enlightened and magnificent" of all. While trying to extend his triangles from Bavaria into Austria, Cassini ran into heavily wooded hills impossible to see through or over. He spied one a little higher than the rest,

21. Cassini de Thury, *Relation* (1765), ii-v, xvii (quote), 3-8, 126. In *Relation* (1775), ix, Cassini puts the degree and the military map in first and second place; "what advantages could not be derived [from trigonometrically determined] points in wartime!" (xxvi).

22. Cassini de Thury, *Relation* (1765), vii-viii (quote), ix-xiv (quote), 26, 30-31; so great was the interest, there were not enough telescopes to satisfy it.

23. Ibid., xiv (quote), xx-xxiv (quote), 51. Cassini was much less impressed with the level of science in the Germanies in *Relation* (1775), xix.

but found that it too afforded no place for a sight. The hill belonged to the bishop. On learning Cassini's purpose (which one is not recorded), the bishop offered to fell 2000 trees, to open an unimpeded view. Cassini refused so great a sacrifice. The bishop thereupon built him an observatory on top of a tree 120 feet tall, large enough for the prince and his court to occupy while Cassini made his observations.[24]

Nothing could have been more accurate, in his opinion. The triangles linking baselines in the Palatinate and near Munich gave calculated values for the bases that differed from the measured ones by under one part in 7000. The 30 triangles linking Paris and Vienna made their separation 531,000 ts, which Cassini pointed out agreed precisely with expectation for a spherical earth with Picard's value for a degree of a great circle. (The arc of the great circle through Paris and Vienna subtends about 9°18' at the earth's center.) Cassini confirmed this value by fireworks of a kind he had introduced in measurements made in France in the 1730s. A flash of gun powder set off in Paris at time t_p was received in Vienna at time t_v after transmission through 38 intervening stations. Let the arc of the great circle connecting the cities measure $x°$; the difference in time between them is $x/15$ hours. (Paris and Vienna are very nearly on the same circle of latitude; it takes the sun 24 hours to run around this circle of 360°, which makes 15° an hour.) Let the total delay in transmission be y hours. Then $t_v = t_p + y + x/15$. Now reverse the process: $t'_p = t'_v - x/15 + y$, where t'_p and t'_v are the times of arrival and departure of the signal at Paris and from Vienna, and, by hypothesis, the delay in transmitting backwards equals that in transmitting forwards. Eliminating y, we have $2x/15 = (t_v - t_p) + (t'_v - t'_p)$, which provides x in terms of measurable quantities.[25]

Cassini identified several sources of error in his measurements. Perhaps the most serious was that many of the church towers he used as sights did not have panoramic views; he could not always

---

24. Cassini de Thury, *Relation* (1765), xxiv–xxv, 76–77. In *Relation* (1775), xx, Cassini says he rejected the offer because the cutting would take too long.

25. Cassini de Thury, *Relation* (1765), xiv, xxvi–xxxii, 25, 31, 53–67, 82, 90–94, 114–122; Cassini observes (ibid., 122–123) that his technique of signalling would be most useful in time of war.

measure all the angles in his triangles and consequently had no good check on them. Laying out baselines also challenged accuracy; the sticks change length with temperature and it is easy to lose count of the number of times they are set down. The common people watching Cassini's men place toise sticks carefully end to end 7344 times between Munich and Dachau thought him mad. To preserve him from their chafing and interference, his servant gave out that he was an old sinner come to seek penitence and a place to build a hermitage.[26] This bright idea did not help Cassini's posthumous reputation. The baron von Zach, a close student of Cassini's errors and innovations (Zach reintroduced the technique of determining differences in longitude by light flashes) and, as we know, a vigorous promoter of astronomy and geodesy in Germany, condemned the man who had raised the art of trigonometrical survey to national importance as immorally incompetent, "le vieux pécheur en géodésie."[27] Zach's promotional methods and occasional plagiarisms earned him similar compliments from the congenial company of earth measurers. "No one knows better than Zach how to make the best use of his often modest knowledge and to impose on people who know little or nothing" (*Laiern und Halbwissern*).[28]

Enter the Army

While the Cassinis mapped all France, the French Army was developing its skills by drawing up detailed topographical maps of fortifications, border areas, and likely battlefields. Most of the work was done by *ingénieurs géographes militaires*, who traced their formation as a special corps back to 1691; but not until the War of the Austrian Succession, when they had the challenge of surveying large tracts of German territory, improving German maps, and emulating Cassini III's fateful exercises in Flanders, did these geographical engineers see geodesists and trigonometers in action. After the

26. Ibid., 46–48, 100–102; Cassini, *Relation* (1775), viii, 2–6.
27. Zach, quoted in Westphal, *Zs. f. Instr.*, 5 (1885), 374; Jahn, *Gesch.* (1844), 2, 131–132; Friedrich, *Peterm. geogr. Mitt.*, 117:2 (1973), 149–150; Levallois, *Vie des sciences*, 3 (1986), 283 (La Caille's use of light flashes in 1738).
28. Olbers to Gauss, 19 Mar 1822, in Schilling, *Olbers* (1900), 2:2, 177; supra, §2.1. Gauss, who once worked with Zach, thought him arrogant, impudent, and ignorant; Gauss to Olbers, 10 June 1824, ibid., 312–313, and to Bessel, 15 Jan 1825, in Gauss, *Briefwechsel* (1880), 444.

war of the succession the Corps had the leisure to develop its technique, sometimes with the help of Cassini de Thury, who, in 1755, drew up instructions for surveying the coast around the mouth of the Orme. During the Seven Years' War, the Corps produced as many maps as they had done in all previous engagements, with greater accuracy and in more detail.

Peace again brought time to perfect technique, to triangulate more regions, and, where possible, to attach the detailed maps (often at 6 lignes to 100 toises) to the geodetic network under construction by the Cassini corporation. For example, an important survey along the Rhine in Alsace in 1775–76, undertaken to detect meanderings of the river that might render defenses useless, rested on reference points attached to Cassini's triangles and a new baseline of some 2,000 ts tied through a new trigonometric network to a base measured out in 1743 under the aegis of the Paris Academy.[29]

During the late 1770s and the 1780s, the geographical engineers triangulated border strips in Provence and Dauphiné, Alsace and Lorraine, Liège (with permission!), and the Pyrenees (in collaboration with Spain). They also labored along the coasts of Normandy and Brittany, and throughout Corsica, where no fewer than three trigonometers and thirty-one geometers were employed under the general direction of Jean Joseph Tranchot, who would later disseminate French methods in the Germanies.[30] The maps and plans linked to the triangulations were drawn at a scale six times larger than the second Cassini map and covered with topographical detail of military interest; severe penalties awaited anyone who disclosed information about them to unauthorized parties.[31]

One piece of information that pride could not suppress was the discovery that the military trigonometers had surpassed their model: in many places, especially in Flanders and around Lille, they found that Cassini's engineers had set grid points with "absolute imperfection" and misplaced villages by significant distances. As the prestige of the Corps went up, the Cassini engineers tried to assimilate

---

29. Berthaut, *Ing. géogr.* (1902), 1, 13, 20–32, 37, 41, 45–46, 50, 57, 93, 119–120.

30. Ibid., 71, 83, 90–91, 95–96, 99–100, 103–104, 106–108, 113–114; Huguenin, *Imago mundi*, 24 (1970), 127; Konvitz, *Cartography* (1987), 39. The border and Corsican maps received high praise from the officers and academicians who examined them.

31. Berthaut, *Ing. géogr.* (1902), 1, 65, 74, 77, 106.

themselves to it, by copying its military uniform and claiming military authority and perquisites. When, as in the Alps in the late 1770s and Brittany in the 1780s, both sets of trigonometers measured the same ground simultaneously, one group sometimes mistook the other's signal towers for its own. Cooperation was therefore indicated, particularly to Cassini, who had fewer resources than the military. He requested that his men be given points established by the geographical engineers. The military declined, for security reasons. The officer in charge of the triangulation in the Dauphiné put the point to his superiors with an unpleasantly modern ring. "The privilege given Cassini's engineers [to map France] should not include parts of the frontier, knowledge of which should be restricted....We note that the King of Sardinia has had the same survey done [on his side of the border]....He has locked up the result and permitted no publication because in his country an academic interest cannot gainsay a political and military one."[32]

The good and essential work of the Corps des ingénieurs géographes did not protect it from the politics of the Revolution. The regular engineers had already infringed on the independence of their more specialized brethren; in 1791 they convinced the Constituent Assembly to suppress the Corps altogether. It returned in 1793 when one of its former members, General Calon, who had become a Jacobin and a deputy, centralized the state's cartographical work in the Dépôt de la guerre and established a school there for surveyors, geometers, and engineers. On Calon's fall in 1797 the corps again disappeared as an independent entity—for two years. Resurrected once more, in 1799, as *ingénieurs artistes*, the geographical engineers (to use the name they regained in 1809) became a principal instrument for the dissemination of French exact science.[33] This promotion had two causes. For one, the conquests of the French Army delivered up many new territories for triangulation. For another, the resurrected corps adopted as its premier apparatus the repeating circle perfected by its old rivals, the Cassini

---

32. Ibid., 73, 99, 102 (quote), 109–110; Darçon to Dépôt de la guerre, [1777], ibid., 72.

33. Ibid., 121–122, 136–138, 144–145, 150–151, 157–160, 175–176, 184–185.

academic interest, for the joining of the Paris and Greenwich observatories.[34]

An export business

The first major power outside France to feel the need for improved military maps was the diffuse Austro-Hungarian Empire. In 1764 Field-Marshall Count Leopold Joseph von Daun, the first general to defeat Frederick the Great in battle, ordered geodetic operations throughout his Empress' dominions, "because [he said] during the Seven Years' War we had the sad experience of discovering the unfortunate consequences that a lack of good maps [may have] on the most important military events." His most experienced colleague, Field-Marshall Franz Moritz Lacy, had complained about being stuck in unmapped woods within Austrian territory, unable to move his army safely, and about being quartered where no adequate provisions existed. "It has often happened that the cavalry has been sent to villages without stables and the infantry to places with only wretched huts and no draft animals to haul forage and bread."[35]

Hands were available to correct the situation: Vienna had had a military engineering school since 1717 and its army had an engineering corps, which, in acknowledgment of French precedence, took the toise as its "Fortifikationsklafter," its fundamental unit of measurement and construction.[36] The joint operations of Cassini III and Liesganig between Strasbourg and Vienna in the early 1760s gave the Austrians an example of the method, if not of the precision, of large-scale trigonometric surveys; and between Daun's call and the end of the century, when the French engineer geographers irrupted into the Germanies, several indigenous geodetic mappings took place. Austrian engineers did all or parts of Bohemia, Moravia, and Silesia in 1768; Galicia, acquired by the first partition of Poland (by Liesganig, who had risen to the status of Geniebau- und Navigations-Direktor) beginning in 1773; and Bukovina, Austria, Slavonia, Croatia, and Hungary in the 1780s. In all, the output of

34. Supra, §2.1.
35. Dorn, quoted in Jordan, *Zs. f. Verm.*, 28 (1899), 53–54; Lacy to Dorn, 2 May 1764, in Paldus, *Annahmen* (1919), 5.
36. Dörflinger, in *Cart.* (1978), 192–193; Nischer-Falkenhof, *Imago mundi*, 2 (1964), 83–87.

the so-called Josephinische Landesaufrahme, which ran from 1764 to 1787, occupied 3589 sheets. The survey work, accomplished mostly with the planchette, compass, and chain, without a general triangulation and without connection to Liesganig's meridian through Vienna, varied in method and quality from region to region. Much was done by pacing out distances on foot or horse; "use a quiet animal," the directions to the surveyors ran, and bring a billiard ball, "to test whether the planchette is horizontal." The resulting maps, drawn at three times the scale of Cassini's and never printed, were held in the deepest secrecy. The whole had to be redone, beginning in 1806, with French methods.[37]

The best and best known of the works initiated by Daun's call was the mapping of the Spanish Netherlands between 1771 and 1777 by General Joseph Johann von Ferraris, a graduate of the imperial military academy. The project compiled 4,000 pages of inventory of everything in Belgium of military interest—mills, mines, minerals, fodder, coal, quarries, buildings, populations. Unfortunately for Belgium, its population density easily met the condition for supporting advancing armies: at least 35 people per square kilometer for an army of 60,000 men.[38] Despite the scale of his operations, however, Ferraris did not establish a primary triangular grid, but anchored a system of secondary triangles to some 30 points fixed by Cassini's men around Dunkirk. Ferraris discussed the project with Cassini, "known in Vienna by the astronomical and trigonometrical operations he ran from the banks of the Rhine to the capital of Austria," and apparently received the assurance that he could refer his positions confidently to the grid formed by the Paris meridian and the triangulated perpendicular to it. Ferraris accordingly tied secondary triangles taken graphically—with the planchette—to the Cassini points in Flanders and converted distances on the ground to latitude and longitude with respect to the meridian and parallel of Paris.[39]

37. Westphal, *Zs. f. Instr.*, 5 (1885), 374–376; Berleithner, in *Cart.* (1978), 132, 140–143; Paldus, *Annahmen* (1919), 6, 8, 14, 18 (quote), 28, 38, 39, 41, 46–49, 53–54. Cf. Clotten, *Zs. f. Verm.*, 10 (1881), 387.

38. Bruwier, in *Cart.* (1978), 60–65; Bernleithner, ibid., 134–135; Carter, ibid., 260, 264–265.

39. Comments on Bruwier, in *Cart.* (1978), 25–26; "Explication...de la nouvelle carte des Pays-Bas autrichiens, qui sera une suite de la nouvelle carte de France," in Lemoine-Isabeau, in *Cart.* (1978), 53–58 (quote); Koeman, ibid., 289–291.

The planchette could scarcely be more accurate than a few minutes of arc, say about one part in 1000 for instruments of the size Ferraris used, and the errors accumulated from one mapped section to the next. When the Ferraris map, engraved in 270 large sheets at 1:11,520 as a "carte de cabinet" (for government and military use) was published at the Cassini scale of 1:86,400, exacting astronomers depreciated it as "disjunct from observations, without direction, [and] off by 2 or 3 leagues [1 league = 2000 ts] from Cassini's observations, which, however, are not the finest [either]." Nonetheless, the French military much prized Ferrari's printed map when, in 1793, the Spanish Netherlands again became a theater of war.[40] The French Army seized all copies it could locate and took the printing plates, which it corrected and used to issue new editions for the use of its officers. Among the provisions of the Peace of Vienna was the restitution of the purloined plates.[41]

In the Germanies, several mappings at 1:86,400 were attempted between the Seven Years' War and the arrival of the French in Napoleonic times. At the beginning of the period, between 1764 and 1770, a former *ingénieur géographe* named Michel, who had transferred his expertise to the Electoral court in Munich, surveyed Bavaria. At the end, in 1795, Johann Gottlieb Freidrich von Bohnenberger, a former student of theology turned astronomer by Zach and, later, a professor at the University of Tübingen, and Carl Ludwig von Lecoq, a general in the Prussian Army, started triangulated surveys in Württemberg along the truce line from the Netherlands to the Ruhr established by the Peace of Basel. Bohnenberger's work was slow, academic, and exact; Lecoq's was fast (he had 24 Prussian officers at his command), practical (he omitted everything without military significance), and sloppy (his officers did not always measure all three angles in their triangles and, when they did, they sometimes achieved closing errors of 2 minutes of arc).[42] They had a 10-inch Ramsden sextant, with which they should have done better.[43]

40. Koeman, in *Cart.* (1978), 292–297; Bernleithner, ibid., 132–133; Oriani, diary note of 8 June 1786, in Tagliaferri and Tucci, *Incontri*, 4 (1789), 64 (quote).
41. Berthaut, *Ing. géogr.*, 1 (1902), 151, 153, 157, 178, 183; Lorette, in *Cart.* (1978), 77–81, 85–87, 110–117.
42. Schmidt, *Kartenauf.* (1973), 1, 13, 15, 20–22, and in *Cart.* (1978), 275–277; Berthaut, *Ing. géogr.* (1902), 1, 62; Pfitzer, *Zs. f. Verm.*, 42 (1913), 40.
43. Zach to David, 6 Jan 1797, in Zach, *Briefe* (1938), 139. Zach praised some of

The old sinner in geodesy, having indirectly infected the Germanies via Austria with the virus of exact cartography, turned his attention to states peripheral to the continental system. In 1775 his son and heir, Jean Dominique Cassini, on a visit in Florence, tried to interest the Grand Duke in a triangulated survey of Tuscany, to be conducted by the latest sprig of the Cassini tree in person. Several Tuscan administrations had flirted with the project since the mid 1750s, when the lamp of Florentine exact science, Leonardo Ximenes, inspired by the work of his fellow Jesuits Boscovich and Maire, first proposed it. Cassini IV received sufficient encouragement to draw up plans—he would need instruments from London and Paris, mathematicians as assistants, 10 horses, and 40 mules. The Grand Duke referred the plan to competent advisers, who recommended the project but saw no reason that it should not be directed by Italians; as Ximenes, then risen to grand-ducal mathematician, put the point, referring to surveys by the French that will be described momentarily, "your subjects are not Americans, Lapps, or Africans, to need foreign astronomers to draw up a map and to measure a degree of a meridian." Ximenes recommended not only a map of Cassini's style, but also an inventory like Ferraris'; the project grew too large for so small a state, and was postponed for some forty years.[44]

The cross-channel link

The Cassinis' second initiative on the periphery enjoyed immediate success. Early in October 1783, a month after the conclusion of the Treaty of Versailles between Great Britain and the United States, Cassini de Thury proposed to the Royal Society of London a joint exercise that he had drafted in 1775—to determine the differences in latitude and longitude between the national observatories in Paris and Greenwich. Cassini claimed that the uncertainty in separation in longitude, ΔL, amounted to 11 seconds of time; the corresponding quantity in latitude, $\Delta \lambda$, he estimated at 15 seconds of arc. Privately he insinuated that 12 of the 15 seconds derived

---

the measurements made with the sextant but condemned the chronometer, with which longitudes were taken, as poor and cheap ("it cost only 60 pounds").

44. Mori, *Arch. stor. ital.*, 35 (1905), 376–380, 391–395, 417 (report by Ximenes, 26 Dec 1777, quote); Barsanti and Rombai, *Ximenes* (1987), 39–43, 74, 91–92. Cf. Devic, *Histoire* (1851), 67–68.

from an error in the estimation of latitude of Greenwich committed by Maskelyne and Bradley, and he offered to help the errant English correct their mistakes, "should their [other] occupations prevent them from doing so themselves." The English on their side doubted that French instruments could perform to British standards.[45]

During these preliminary skirmishes, the president of the Royal Society, Joseph Banks, replied to Cassini that the Society would take care of operations on its side of the Channel and handed the matter to his lieutenant, Joseph Blagden, an old Paris hand. Blagden asked the opinions of Henry Cavendish, Alexandre Aubert, and Alexander Dalrymple, an Army surveyor eager to map the coasts of England on French principles; none of them could understand the allegation of the error in the estimate of the latitude of Greenwich by the royal astronomers of Britain, but all agreed that the project should go forward.[46] Blagden had doubts, however, about the powers of the Frenchmen with whom the Society would have to deal. "Frequent conversations with Mr Cassini the elder [Cassini de Thury] whilst I was in Paris convinced me that he is a man of very moderate intellect." Blagden rated Cassini IV a little higher as an observer, but still not an Englishman. "Yet the scheme must be executed...notwithstanding all the difficulties which I foresee will attend the choice of persons."[47]

In addition to promoting national honor and promising precision in ΔL and $\Delta \lambda$, the project had the merit of fitting perfectly into the development of British cartography. Before 1745 few good maps of any part of Britain existed, a fact made clearly and cruelly evident to the British Army during the Scottish rebellion that year. To ease its way in Scotland should its services again be required, the Army commissioned a military map and survey of the Highlands. William Roy, a draftsman who would join the Army engineers and rise with

---

45. Howse, *Maskelyne* (1989), 151–152; Aubert to Deluc, 18 Nov 1783, in Aubert, *N&R, 9* (1951), 86; Hornsby to Banks, 25 June 1784, in Forbes, *Vistas in astr., 28* (1985), 173; Cassini de Thury, in Maskelyne, *PT, 77* (1787), 152 (quote).

46. Banks to Blagden, 18 Oct 1783 (Blagden Papers, B. 19, Royal Society), and Blagden to Banks, 21 Oct 1783 (DTC Papers, 3, 141, Natural History Museum, London); Maskelyne, *PT, 77* (1787), 154–164, 169–170.

47. Blagden to Banks, 18 Oct 1783 (DTC Papers, 3, 132–133, Natural History Museum).

geodesy to a general officer and a fellow of the Royal Society, took the major part in the survey, which was completed in 1752. A quarter century later about half of Britain had been mapped at a scale of an inch to a mile; but, like the contemporary survey in Austria, the maps of various localities were constructed on different principles and with varying accuracy, and no triangulated grid existed to integrate or assemble them. Roy had tried to create such a backbone. Again like the Austrians, he seized the opportunity of the end of the Seven Years' War to push his project. Nothing was decided before the American Revolution absorbed military attention. Roy pushed again after the Peace of Versailles, just when the French proposed the linking of observatories.[48]

Roy began work in 1784, laying out baselines at Hounslow Heath west of London and Romney Marsh near the Dover Coast, and advertising completion of the project by 1786. He measured the base over and over again, with instruments made by Ramsden: a surveyor's chain, wooden poles, and glass rods. Much to Roy's irritation, Ramsden did not finish the famous theodolite for angular measure early enough to enable Roy to conclude the project before 1788; perhaps one cause of delay was dealing with Cassini IV, who came to London to try to entice Ramsden to France, or, failing that, to place French apprentices with him. Cassini managed only to capture a Hugenot lens maker and to commission a quadrant.[49]

The big theodolite came into action on the last day of July 1787 (figure 4.1.1). Roy took as many sights in the field as possible before moving the instrument to Dover in its four-horse carriage, provided by Banks "with his usual liberality." At Dover, in September, Roy, Blagden, Cassini, Legendre, and Méchain drew up final plans for the cross-channel link, "thereby to establish for ever, the triangular connection between the two countries." The very bad weather relented for a few clear nights; the great theodolite zeroed in on French lights at Cap Blanc Nez and Montlambert, while the French at Calais sighted on flares at Dover furnished by the Woolwich arsenal. The resultant values for ΔL and $\Delta \lambda$ would have been truer

48. Close, *Early years* (1969), 4, 7–8; Wallis, in *Cart.* (1978), 167, 170–174; Skelton, *Geogr. jl, 128* (1962), 417–421; Widmalm, in Frängsmyr, *Spirit* (1990), 184–186.

49. Roy, *PT, 77* (1787), 188–189; Devic, *Histoire* (1851), 111–115; Close, *Early years* (1969), 19–22; Konvitz, *Cartography* (1987), 27; Gillispie, *Polity* (1980), 122–123; Widmalm, in Frängsmyr, *Spirit* (1990), 193–194, 199–200.

204 / VARIETIES OF MERCANTILIST MATHEMATICS

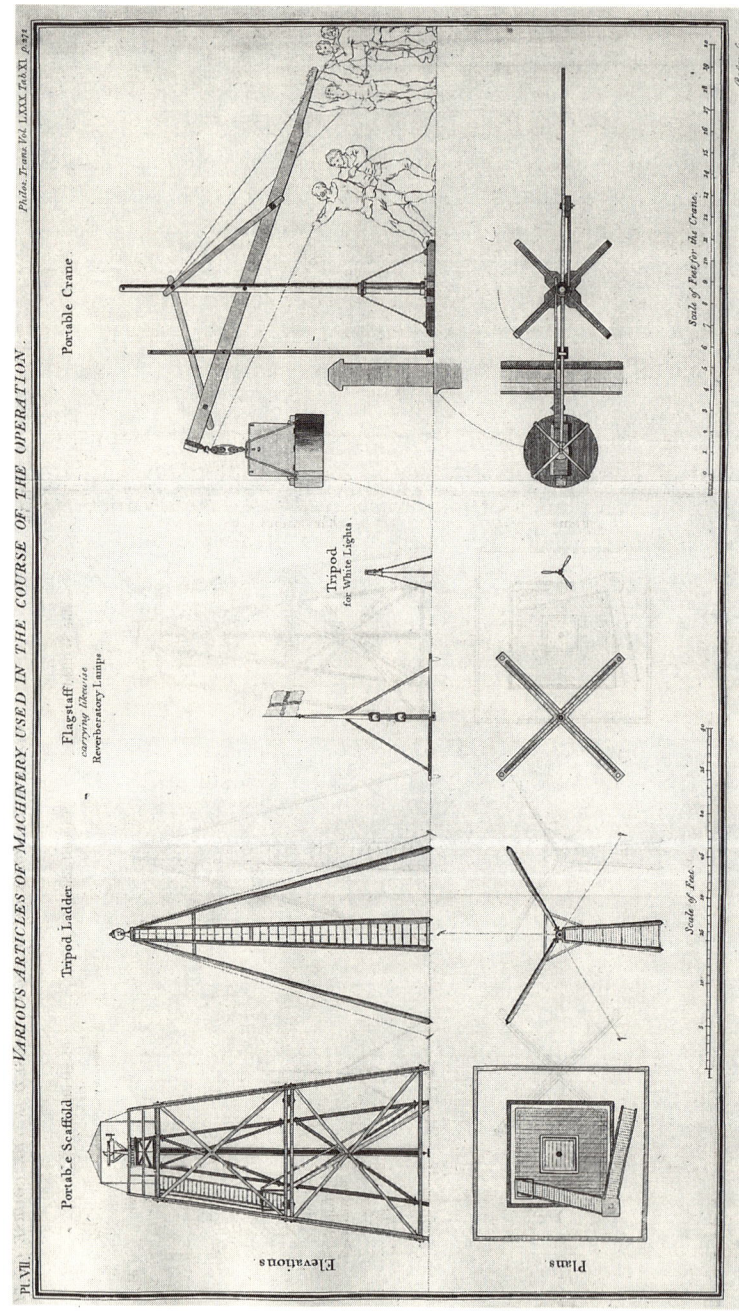

FIG. 4.1.1 Ramsden's theodolite in action. Note the crane (far right) to lift it onto the portable scaffold (far right). The reverberatory lamps and white lights marked the stations at right. Roy, *PT*, 80 (1790), 272.

had the English taken better account of the height of the station at Dover and had they not had to replace a level on the theodolite broken when a great wind blew the instrument off a scaffold.[50]

Roy's measurement of the Romney base differed from its value as calculated through the triangles from the Hounslow base by 4.5 inches in 28,532.92 feet, that is, by one part in 500,000. Using the triangles and the Hounslow base, and sighting on the lights from France, he made the Dunkirk base to be 6225.4 ts. The French got 6225.6 ts. With the Romney base, Dunkirk's came out three feet—half a toise—off.[51] As for the values sought at the outset, they agreed well with estimates Maskelyne had made while the work was in progress: $\Delta\lambda \simeq 2°38'26''$, making the error originally claimed by Cassini sink close to the 4".5 initially estimated by Maskelyne; and $\Delta L = 9\text{m}19.45\text{s}$, according to Roy's geodetic measurement, less than a second different from the value Bradley had found from observing Jupiter's moons and Maskelyne obtained during the course of the project by sending an assistant to Paris with a few good chronometers ticking Greenwich time. Maskelyne came closer, according to modern determinations, despite Roy's belief that "the trigonometrical operations...may be considered as infallible." The modern result may suggest that the squabble among Roy, Maskelyne, and Ramsden over who was the finest measurer of his time should be settled in Maskelyne's favor.[52]

The project to link the observatories of Greenwich and Paris had at least two slight but interesting, and one important outcome. On the slight side, Johann Georg Tralles used Roy's procedures and Ramsden's instruments on the first substantial triangulation in Switzerland, begun in 1788 and continued under the auspices of the Ökonomische Gesellschaft and the Canton of Bern. The Ramsden

50. Roy, *PT, 80* (1790), 111–114 (second quote), 115, 160 (first quote), 183, 265–266; Forbes, *Vistas in astr., 28* (1985), 178.

51. Roy, *PT, 80* (1790), 116, 122–126; Blagden to Banks, 30 Sep 1787 (DTC Papers, Nat. Hist. Mus., London); Westphal, *Zs. f. Instr., 5* (1885), 380–385, 420–424; supra, §2.1; Westphal (ibid., 424, 432) faults Roy for poor connecting triangles, inaccuracy in standardizing the measuring rods, and imperfect temperature compensation. Cf. Débarbat, *Échanges* (1990), 53–58.

52. Roy, *PT, 80* (1790), 225; Forbes, *Vistas in astr., 28* (1985), 174; Howse, *Maskelyne* (1989), 152–153; Widmalm, in Frängsmyr, *Spirit* (1990), 188, 195–198; Maskelyne, *PT, 77* (1787), 181–185; Roy, ibid., 214 (quote). The most elaborate previous estimate of ΔL was 9 min 35 sec; Wargentin, *PT, 67* (1777), 164.

instruments included a theodolite, ordered in 1794 and delivered in 1797, the only one to see service outside Britain. Tranchot thought it "magnificent" when he encountered it in Bern in 1812.[53] He was assisted by Ferdinand Hassler, who had worked with Zach, to whom he recommended himself by arriving at Gotha with an English sextant. This training helped Hassler to win the confidence of Thomas Jefferson and to set up, after many vicissitudes, a land survey in the best European style in the United States.[54]

The second slight spin-off from the measurements of 1787 came in 1803, when Napoleon took it into his head to view the invasion of England he was planning from the comfort of the French coast. How high above sea level would he have to stand to see the tide arrive in Dover? He addressed the question to the First Class of the Institute, which assigned it to Delambre. He calculated the requested height as a function of distance and atmospheric refraction; plugging in distances derived from the Roy-Cassini link, Delambre concluded that no suitable height existed near Calais or Dunkirk, but that Cap Blanc Nez would do nicely.[55] As it happened, Napoleon had no use for this information.

The most important consequence of the Paris-Greenwich project was the definitive establishment of the British Ordnance Survey. Sometime between 1785 and 1790 the Royal Society drew up a plan for a large trigonometric survey of Britain, at a scale of 2 inches to a mile; the Master of Ordnance, the Duke of Richmond, an enthusiastic cartographer, supported the general idea; and in 1791 the Ordnance Survey began, the first military organization in the world to enter wholeheartedly into trigonometric surveying.[56] The timing owed something to Ramsden's completion of a second large theodolite, commissioned by Richmond in 1790. It, or a copy, remained in service for seventy years. One of Richmond's first

53. Westphal, *Zs. f. Instr.*, 5 (1885), 373–374; Hammer, *Zs. f. Verm.*, 20 (1891), 447–448; Cajori, *Hassler* (1929), 16–25; Berthaut, *Ing. géogr.* (1902), 2, 327. Another project inspired by the Paris-Greenwich link was a survey of Jutland by Thomas Bugge in 1787 (Westphal, 429).

54. Greene, *American science* (1984), 35, 131; Cajori, *Hassler* (1929), 71–80, 118–127, 161–200.

55. Beaucour, Soc. sauvegarde chât. imp. Pont-de-Briques, *Bull. hist.*, 1 (1972), 198–204; Fischer, *Napoleon* (1988), 127–128.

56. Close, *Early years* (1969), 15, 25–27; Skelton, *Geogr. jl*, 128 (1962), 424–425; Wallis, in *Cart.* (1978), 169; Widmalm, *Kartan* (1990), 120.

objectives was to redo Roy's measurement of the Hounslow base, which experts faulted for not compensating for the flexure of the glass rods when supended or for the dilation of the chains with temperature. The result: Roy's errors and oversights did not amount to more than 5 or 6 feet in 27 miles. The new surveyors perpetuated their work in an appropriately military manner by replacing the pipes Roy had used to mark the ends of the base at Hounslow Heath with cannons sunk muzzle up. In a more delicate mood, they worried when their triangles failed to close within a quarter of an arc-second.[57] The Ordnance Survey continued around the British Isles beginning with the coastline; it reached Cornwall before the end of the century with the help of still another Ramsden theodolite, about half the size, and almost as good, as the bigger ones. The methods of the survey were soon in use elsewhere in the British Empire, particularly in India, where work with Ramsden instruments began just after the turn of the century, and some distinguished geodecists, for example Henry Kater, received their training.[58]

Imperial measures

In competitive geodesy, Britain could not approach France. General Bonaparte's fascination with maps gave added urgency to the need to survey the territories he and his colleagues conquered. He brought a large surveying crew on his expedition to Egypt in 1798—7 topographical, 13 military, and 12 civil engineers, plus three generals—who used the plane table, graphometer, Borda circle, marine chronometer, and a few astronomically determined stations to produce a map of the country eventually printed at 1:100000.[59] In 1801, as First Consul, Napoleon ordered that maps on the Cassini scale of 1:86400 be prepared for the four new French departments on the left bank of the Rhine, for the Batavian Republic

57. Williams, Mudge, and Dalby, *PT, 85* (1795), 414–418, 424–428, 438–439, 451, 484–486, 499, 509, 511; Richeson, *Land measuring* (1966), 181–183; cf. Westphal, *Zs. f. Instr.*, 5 (1885), 425–426, and Zach to David, 7 June 1786, in Zach, *Briefe* (1938), 119.

58. Williams et al., *PT, 87* (1797), 432–433, 507–508; Westphal, *Zs. f. Instr.*, 5 (1885), 431–432; Widmalm, *Kartan* (1990), 140.

59. Gillipie, *APS, Proc. 133* (1989), 466–469; Berthaut, *Ing. géogr.* (1902), 1, 193–194.

fashioned from the Netherlands, and for several other properties.[60] As Biot rightly said, "it is the war that...gave geodetic operations the great extention and extreme perfection they have acquired in all the states of Europe."[61]

The Dutch survey began tentatively when the French appointed a physician with the resounding name of Cornelius Rudolphus Theodorus Krayenhoff as military commander of Amsterdam when they took the city in 1795. No good map then existed for dividing up the Netherlands into departments on the French model. Krayenhoff was ordered to prepare one. He measured out a base line with an ordinary surveyor's chain on the frozen Zuider Zee and took his angles with a standard sextant. By 1800 he had finished. He showed his results to J.H. van Swinden, one of Holland's most distinguished natural philosophers. Van Swinden severely criticized his countryman for using old crude methods when a new French technique was only too much in evidence. Krayenhoff took the point. "I saw with regret that such an imperfect execution would give an unfavorable impression of the state of science in Holland and would compromise the fame of our nation in the domain of mathematics, astronomy, and geography." He thereby echoed the call to emulation that Roy had sounded with English understatement ("The honour of the nation is concerned in having at least as good a map of this as there is of any other country") and Zach had tooted with Hungarian enthusiasm ("I received your letter, and read in it those magic words, that you implore me, for the glory of the fatherland, to send you a sextant for the advancement of our national geography").[62]

In 1801 Krayenhoff began anew. Using a large Borda circle made by Lenoir, he revealed errors in the extension of Cassini's grid into Belgium committed by a French astronomer, Perny de Villeneuve, and attached his triangles instead to Delambre's more southerly, but more correct, stations. Later, on a commission from the French minister of war, he triangulated through East Friesland to link up with a net in Hanover completed by a French geographical engineer,

60. Schmidt, *Kartenauf.* (1973), *1*, 50, 81; Berthaut, *Ing. géogr.* (1902), *1*, 305–431, *2*, 283–390.
61. Biot, *Mélanges* (1858), *1*, 75–76, text of 1818.
62. Haasbroek, *Investigation* (1972), 12–15 (quote); Roy, *PT*, *80* (1790), 262–263; Zach to Schedius, 18 Feb 1801, in Zach, *Briefe* (1984), 75.

Epailly, in 1807. Epailly had proceeded in truly imperial style, with a crowd of officers and horses, and with no patience with peasants. "To keep up with the reconnaissance of triangles, the placing and raising of signals, and the observation of angles," he wrote his superiors, "it was necessary despite the outcries and complaints of the peasants to cut down their trees mercilessly, to knock holes in their church steeples...to enjoy everywhere the right of conquest and to ignore altogether refusal and resistance." In attempting to emulate the French, and to satisfy the cultural chauvinism of van Swinden, Krayenhoff allowed himself liberties with the data if not with the peasants. Perhaps van Swinden helped in tidying up. Krayenhoff sent him the raw data before publication with an invitation to make it sum to a conclusion of Parisian clarity. "I...submit these operations without withholding the smallest detail or arbitrary arrangement respectfully to your judgment as a university professor in order to make a statement as sharp as those given about the observations of the French astronomers." It appears that Krayenhoff's closing errors, claimed to be under one second, had a standard deviation of about 4 seconds. He also made errors in computation undetected by the university professor.[63]

In contrast to the Dutch survey directed by a Quisling doctor who learned on the job, the triangulation of the new departments on the left bank of the Rhine was conducted by one of the most experienced of French geographical engineers, Tranchot of the Corsican survey, whom Napoleon chose for the job over Delambre. Since the triumph in Corsica, Tranchot had worked on the resurvey of the Paris meridian run during the French Revolution. He had the technical but not the administrative experience required. Although Napoleon supported the work spiritually, and prescribed a fine for anyone who intereferred with the survey, the Army, feeling the pinch of a decade of constant war, could not keep up with Tranchot's requirements of men and material. The survey, begun enthusiastically in 1801 as a continuation of Cassini's map beyond Dunkirk, stopped at least twice for lack of funds. Nonetheless, Tranchot's staff of 20 or so engineers got enough exercise with the

63. Epailly, as quoted by Galle, in Gauss, *Werke, 11*:2, Abh. 1, 27–28; Haasbroek, *Investigation* (1972), 15–16, 86 (quote, slightly reworked from the original translation), 116, 150, 184, 215; Schmidt, *Kartenauf.* (1973), *1*, 29–34.

Borda circle. They ordinarily made three series of observations of every angle, and sometimes as many as six, each series requiring a dozen or more repetitions. Sometimes the sightings were made from scaffolds built for the purpose, and torn down immediately by local citizens for firewood, construction lumber, and indemnity for cutting trees to make the scaffolds; but more often from the old standby, the church steeple. In an instructive reversal of the usual military connection, the administration of the cathedral of Cologne argued successfully that the Prussian government should pay for the repair of lightning damage to one of the towers because it was essential to astronomical and geodetic work. The meticulous repetitions from scaffolds and steeples produced a set of Rhineland maps probably superior to anything the French had for metropolitan France, "the most beautiful topographical work of the time."[64]

Owing to the vacillating material support, Tranchot had completed the primary triangles, but not the full survey, by its discontinuance in 1814. The 1625 square leagues (about 26000 km^2) he had surveyed represented 150 man-years of work. Three more years of work remained. Napoleon grew very impatient, but to no avail; Tranchot would not finish by Waterloo. The peace treaty of 1815 awarded all the maps of Germany Tranchot had completed to Prussia as war spoil. The French copied them before delivery to the German governor of Paris, Friedrich Carl Ferdinand von Müffling, who happened to know something about map making. Müffling had learned surveying on the job, under Lecoq, in 1796. Thus armed, he became Zach's lieutenant in a survey of Thuringia begun in 1802, on commission from the King of Prussia, who wanted military maps of newly acquired territory. The survey had a Borda circle, Zach's favorite sextants and chronometers, and a clutch of mathematicians. In measuring the base, Zach arranged the rods, which he calibrated from a copy of a French standard (the toise de Pérou), so that one served as a nonius for the next. He marked the ends of the base most appropriately with sunken cannon barrels. Frequent measurement of temperature permitted appropriate compensation.[65] All this

64. Berthaut, *Ing. géogr.* (1902), 2, 188, 283–284, 288–294, 297–299 (quote, Berthaut's opinion); Dhombres and Dhombres, *Naissance* (1989), 232; Schmidt, *Kartenauf.* (1973), 1, 51–53, 59–61, 75–76, 84–88, 121–124, and in *Cart.* (1978), 278–279, 282. Tranchot's largest error in angular measurement was 3 sec of arc; his mean error for all his triangles, 0.72 sec. (Schmidt, 1, 97).

65. Berthaut, *Ing. géogr.* (1902), 2, 269–270, 283–285, 299, 413–421; Schmidt,

exactness made its impression on von Müffling. "Experience has taught us that with careful, twenty-fold multiplication [of observations with a Borda circle], the three angles of a triangle can close to within a second." When he came to write instructions for finishing Tranchot's work, he required that his men discard all triangles that did not close to under three seconds.[66]

Other notable German developments stimulated by French examples were the measurement of a base near Damstadt in 1808 by a Hessian civil servant, C.L.P. Eckhardt, and a mathematician, L.S. Schleiermacher, who aimed to link the observatories of Seeberg and Mannheim. They followed methods similar to those Zach pioneered in Thuringia. The Darmstadt base was tied to Tranchot's triangles in 1809. Eckhardt joined the French project. "I acknowedge gratefully [he wrote in 1835] that I thereby first had the opportunity to be initiated into the mysteries of the finer art of observation with transportable instruments."[67] Similar experience came to J.G. Soldner, who continued from 1808 to 1818 the triangulation of Bavaria begun by the French in 1801 with links to Cassini's old Paris-Vienna line and continued, with the collaboration of the Bavarian cartographic service and several Borda circles, until war forced a parting in 1805.[68]

C.F. Gauss worked with Zach in Thuringia before observing the gentle operations of Epailly in Hanover in 1805. Gauss came to French-style geodesy from a compulsion to extract the most likely value of a quantity from several measurements of it. In 1794, at the age of 17, he made Zach a present of a reduction of some very old data—Ulugh Beg's, from the 15th century—by the method of least squares, which he, Gauss, had just invented. Gauss first published the method a few years later in correcting Zach's efforts to obtain information about the shape of the earth from the many

---

Kartenauf. (1973), 1, 37–38, 64–68, 194–196; Zach to David, 15 Aug 1805, in Zach, *Briefe* (1938), 185–186; Westphal, *Zs. f. Instr.*, 5 (1885), 378–379.

66. Ibid., 212–226; Pfitzer, *Zs. f. Verm.*, 42 (1913), 43 (quote). Müffling had cooperated with the French earlier; Berthaut, *Ing. géogr.* (1902), 2, 109, 111; Schmidt, *Kartenauf.* (1973), 1, 207–210.

67. Hammer, *Zs. f. Verm.*, 34 (1905), 285; Geist, ibid., 35 (1906), 169–175; Schmidt, *Kartenauf.* (1973), 1, 78; Eckhardt, *Astr. Nach.*, 12 (1835), 130.

68. Berthaut, *Ing. géogr.* (1902), 2, 300–311, 326–328; Schmidt, *Kartenauf.* (1973), 1, 47–48; Amman, *Zs. f. Verm.*, 37 (1908), 154–155; Widmalm, *Kartan* (1990), 131, 138–139.

measurements that had accumulated by the end of the 18th century. In 1799, with support from Lecoq, whom he had helped with the mathematics of exact geodesy, Gauss asked Zach for permission to come to study astronomy in Gotha. He received a mixed message in reply: advice to give up astronomical observing (Gauss had notoriously poor eyesight) and a sextant to practice making observations.[69]

In 1803, Gauss heard that the King of Prussia had engaged Zach to survey Thuringia. He renewed his request, excited by the news that Zach would exploit the latest methods. "Zach took on the assignment only on the condition that the measurement would be done in such a way that, at least, it would rival the latest French and English surveys."[70] Gauss participated in the work, from 1803 to 1805, and learned to measure baselines, take angles, and determine differences in longitude by shooting off gunpowder. In 1805 Gauss met with Epailly and examined the French instruments used on the survey of Hanover; and he gained access to the data von Müffling had obtained while working with Lecoq. The information provided the basis of many evenings' amusement. As Gauss wrote later, "I can always get something interesting from many trigonometrical measurements, since their daily reduction always gives me some entertainment.... To God, it is all the same whether we have determined the place of a church tower to a foot or of a star to the second."[71]

Civilian scholars like Soldner in Bavaria, Eckhardt and Schleiermacher in Hesse, Bohnenberger in Würtemberg, and Gauss in Hanover, as well as military officers like Müffling, had at least reached parity with the French by the end of the Napoleonic era. In 1817 Laplace conceded that his countrymen would find something worth copying in the practices of the Germans.[72] Munich became a center

69. Galle, in Gauss, *Werke* (1863), 11:2, Abh.1, 3–4, 20–23; Gauss, ibid., 3, 138, 4, 140, 8, 136–137.

70. Gauss to Olbers, 22 Apr and 5 May 1803 (quote), in Schilling, *Olbers* (1900), 2:1, 149–151 (quote).

71. Lecoq to Gauss, 13 Apr, 17 May, and 27 Sep 1799, responding to Gauss' "Lieblingswunsch" to learn practical astronomy from Zach, in Gerody, Ak. Wiss., Göttingen, *Nachrichten*, 1959:4, 46, 53 (Gauss to Lecoq, 24 Apr 1799), 57, 58; Galle, in Gauss, *Werke* (1863), 11:2, Abh.1, 24, 26, 28; Gauss to Bessel, 15 Nov 1822, in Gauss, *Werke*, 9, 357, and *Briefwechsel* (1880), 412 (quotes); Gauss to Olbers, 22 Feb 1804, in Schilling, *Olbers* (1900), 2:1, 177–179.

72. Pfitzer, *Zs. f. Verm.*, 42 (1913), 1; Hammer, ibid., 34 (1905), 288–289, 292 (re Laplace); cf. Widmalm, *Kartan* (1990), 142–144.

of exact instrumentation by supplying the equipment of astronomy and geodesy. A disciple of Gauss', H.C. Schumacher, who became professor of physics in Copenhagen in 1815 and head of a trigonometrical survey of Denmark, used instruments furnished by Georg von Reichenbach, who had worked in London with Ramsden's dividing engine and returned to better it in Munich. After using a Reichenbach repeater in 1809, Zach declared it "a marvel, surpass[ing] in accuracy, convenience, and even in beauty the best I've known made in Paris or London."[73] Gauss prized Reichenbach's instruments in his survey of Hanover, which improved materially on Epailly's measurement. And when Gauss needed an instrument maker to manufacture his own important contribution to geodetic apparatus, the "heliotrope," he turned to a German workshop.[74]

The heliotrope, which furnished a reflection of the sun in place of a signal tower or church steeple as a foresight, was, in von Müffling's words, a god-send. Among other benefits, it made possible a survey through thick woods in Hanover that Epailly had regarded as impassable, since no signals could be seen through the trees. Gauss calculated carefully the direction from a low-lying station to a relative height, hewed down a line of trees, placed a heliotrope at the end of the alley, and observed without difficulty, at distances of up to nine miles, the reflected image of the sun.[75] Gauss reckoned that he and his colleagues greatly excelled their models. "Delambre's and Krayenhoff's measurements, I think, can not be compared with ours for exactness."[76]

2. THE SHAPE OF SCIENCE AND THE EARTH

By putting his faith in his flimsy facts about the relative size of degrees north and south of Paris, Jacques Cassini ran against a

73. Zach, *Mémoires* (1811), 26 (quote); Otto Mayr, *DSB*, 11, 354–355.
74. Gauss, *Werke* (1863), 9, 408–411; Galle, in ibid., 11:2, Abh.1, 57. Like the great Ramsden theodolites, Reichenbach's were divided accurately enough to obviate the need for repetitions; Herbst, *NTM*, 28:1 (1991), 62–68.
75. Galle, ibid., 32, 59–60; Gauss to Bessel, 26 Dec 1821, 15 Nov 1922, in Gauss, *Briefwechsel* (1880), 394–395, 406; Schmidt, *Kartenauf.*, 1 (1973), 226 (re Müffling).
76. Gauss to Gerling, 14 Nov 1838, in Gauss, *Werke* (1863), 9, 393.

214 / VARIETIES OF MERCANTILIST MATHEMATICS

carefully argued consequence of the theory of universal gravity. Newton knew that gravity allowed the earth to have the profile of a pumpkin and he observed that the tropical lethargy of pendulum clocks, as discovered by Jean Richer, required it. And more than that: Newton deduced the excess of the equatorial over the polar diameter from the mechanics of gravitating bodies and from a fiction about canals running to the center of the earth.[77]

Newton's apple

In figure 4.2.1, the canals AC and CQ, filled with a fluid of unit density, reach to the surface at A on the equator and Q at the pole. In order that a particle of the fluid at C be in equilibrium, it must be subject to equal pressures from the two canals. Newton's calculation of these pressures was a tour-de-force. Consider first the gravitational acceleration f_Q existing at Q and caused by the oblate spheroid formed by rotation of the ellipse AQB around the axis CQ. Evidently, $f_Q = f(b) + f_{exc}$, where $f(b)$ is the acceleration of gravity at the surface of a uniform sphere of radius b = CQ and f_{exc} is the acceleration deriving from the excess material contained between the oblatum AQB and the sphere CQ. If ρ_0 denote the average density of the earth, we have

$$f(b) = (4/3)\pi b^3 \rho_0 G / b^2 = \mu \rho_0 b,$$

where G is the gravitational constant and μ is shorthand for $(4/3)\pi G$. An analytic paraphrase of Newton's derivation of f_{exc} might run as follows.[78] The acceleration at Q along QC arising from the element of volume $dv = dx \cdot xd\phi \cdot dz$ (ϕ is the azimuth around QC) is

$$df_{exc} = (G\rho_0/r^2)\cos\theta \cdot xdxd\phi dz.$$

To obtain f_{exc}, integrate ϕ from 0 to 2π, x from x_c (its value at the circle of radius b) to x_e (its value at the ellipse QA), and z from $-b$ to b, as indicated in figure 4.2.2:

---

77. *Principia*, Bk III, prop. XIX, 424–427 (ed. Cajori), 2, 592–598 (ed. Cohen).
78. Cf. Todhunter, *History* (1873), 1, 2–17, and Costabel, in Lacombe and Costabel, *Figure* (1988), 108–113.

$$f_{exc} = 2\pi\rho_0 G \int_{x_c}^{x_e} \int_{-b}^{b} \frac{b-z}{[x^2+(b-z)^2]^{3/2}} \cdot x\, dx\, dz ,$$

where $x_c^2 = b^2 - z^2$, $x_e^2 = (a^2/b^2)(b^2 - z^2)$. Let $a = b(1+\epsilon)$, where ϵ is small in comparison with unity. The integrals are easily performed to first order in ϵ, with the result that $f_{exc} = \mu\rho_0 b(4\epsilon/5)$. Hence the acceleration at Q, f_Q, is $\mu\rho_0 b(1+4\epsilon/5)$, equivalent to the acceleration produced by a sphere of radius b and density $\rho_b = \rho_0(1+4\epsilon/5)$.

A similar analysis shows that in a prolate spheroid—one formed by rotating the ellipse of figure 4.2.1 around its long axis AQ— $f_A = \mu\rho_0 a(1-4\epsilon/5)$. But we do not want f_A, which refers to a prolate spheroid, but $f_{A'}$, the acceleration at the end of the semi-major

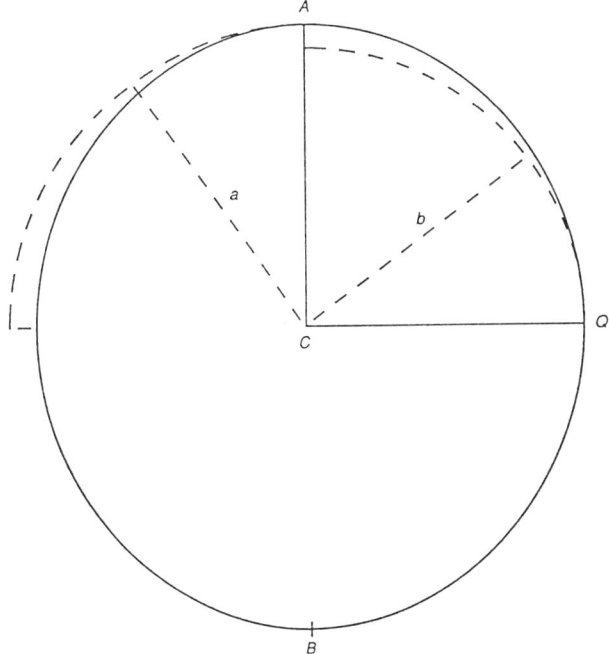

FIG. 4.2.1 Diagram for Newton's theory of the shape of the earth. C is the earth's center, AC and CQ liquid-filled canals in equilibrium under gravity and centrifugal force; the dotted lines are quadrants of circles circumscribing, and circumscribed by, the earth.

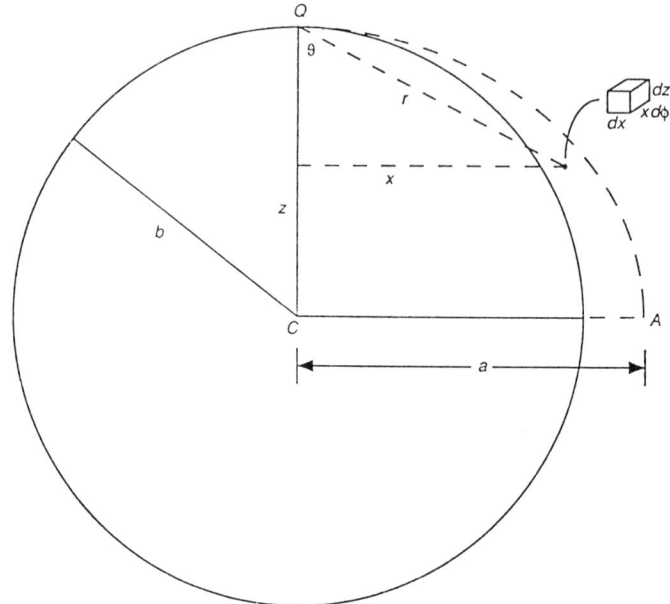

FIG. 4.2.2 Geometry for calculating the gravitational force of the "excess" material between the circumscribing sphere of radius a and the terrestrial ellipsoid. The element of volume a distance x from the earth's axis and a height z above the equation is $dxdz \cdot xd\phi$, where ϕ is the azimuthal angle around the axis CQ.

axis of an oblate spheroid. It is easy to see that $f_A' > f_A$: starting from a sphere of radius $a =$ CA, we obtain the oblatum by reducing one axis perpendicular to AC to CQ, that is, by removing the material earlier designated as "excess;" and we get the prolatum by a further removal, of an approximately equal excess, that reduces the axis perpendicular to AC and CQ to CQ. Hence f_A' lies about midway between f_A and $f(a)$. Newton accordingly took $f_A' = \mu \rho_0 a(1-2\epsilon/5)$: the acceleration at A on the oblate spheroid, the one of practical interest, is the same as that at the surface of a sphere of radius a and density $\rho_a = \rho_0(1-2\epsilon/5)$.

Let us return to the canals. In either of them, an infinitesimal layer of fluid of unit cross section a distance r from the center suffers a downward force $\mu \rho r dr$. The pressure at C arising from CQ is

$$p_{CA} = \int_0^b \mu\rho_0(1+4\epsilon/5)r\,dr = (\mu\rho_0 b^2/2)(1+4\epsilon/5).$$

In AC, each layer also experiences a centrifugal force, $\omega^2 r\,dr$, where ω is the earth's speed of rotation in angular measure (2π radians in 23 hours and 56 minutes). Hence, $p_{AC} = (\mu\rho_0 a^2)(1-2\epsilon/5 - \omega^2/\mu\rho_0)$. Equilibrium requires

$$a^2(1-2\epsilon/5-\omega^2/\mu\rho_0) = b^2(1+4\epsilon/5);$$

or, since $a^2/b^2 \sim 1+2\epsilon$, and $\omega^2/\mu\rho_0$, the ratio of the centrifugal to the gravitational force at the surface, is a small quantity, $4\epsilon/5 = \omega^2/\mu\rho_0$. From the value of the acceleration of gravity as measured at Paris, Newton calculated that $\omega/\mu\rho_0 = 1/289$. Hence, in Newton's rounded value, $\epsilon = 1/230$.[79] "And since the mean semidiameter of the earth, according to *Picard*'s mensuration, is 19615800 *Paris* feet, or 3923.16 miles (reckoning 5000 feet to a mile), the earth will be higher at the equator than at the poles by 85472 feet, or 17 [and] 1/10 miles."[80]

Neither the over-precision of theory, nor the under-precision of measurement, nor their conflict aroused much interest at first. Newton did not mention Jacques Cassini's challenge in the edition of the *Principia* (the third) that appeared late enough to take it into account. Cassini's mathematician colleagues in the Paris Academy had not yet—it was the 1720s—discovered the power or felt the prestige of computing via the gravitational theory. The main defense of the oblate earth was entrusted to John Theodore Desaguliers, paladin of Newtonian science and demonstrator of experiments to the Royal Society of London. Desaguliers rejected Cassini's conclusions as having no basis in fact (they supposed an accuracy three to six times finer than the measurements) or theory (they flew against Newton's gravity, in which "no Body had yet been able to shew any Flaw"). Moreover, they were easily refuted, thus. Take

---

79. Proceeding as ever by proportion, Newton worked with a fictitious sphere for which $\epsilon = 0.01$. In this case, $f_Q/f_A = (101/100)(126/125)(126/125.5) \sim 501/500$, and the ratio of the pressures, without reference to the centrifugal force, is $p'_{CA}/p'_{CQ} = af_A/bf_Q = 505/501$. To reduce 505 to 501, the centrifugal force must take off 4 parts in 500, that is, $\omega^2/\mu\rho_0 = 4/500 = (4/5)\epsilon$.

80. *Principia*, ed. Cajori, 427.

two metal hoops, fix them at right angles to one another, put an axis through their common center, and spin the whole. Make the axial hoop at rest slightly larger than the equatorial, as in Cassini's earth. As the speed of rotation increases, the figure becomes a sphere and at last an oblatum.[81] A pretty demonstration, though not to the point.

A check of the earth's shape lay open to academicians who liked to travel. Figure 4.2.3 is an ellipsoid of revolution about the y axis. The latitude of the place O is the angle λ made at the equatorial plane by the normal to the horizon; the definition is not obvious, but necessary to preserve λ as the height of the pole; according to d'Alembert, many early writers on geodesy perplexed their subject unnecessarily by the error of referring λ to the center of figure.[82] A more obvious definition made a degree along a meridian at the surface around O proportional to r, the radius of curvature of the meridian there. An easy calculation brings r in terms of the ellipticity $\epsilon = 1 - b/a$. In general, $r = (1+y'^2)^{3/2}/|y''|$, where the primes indicate differentiation with respect to x. In particular, in the oblatum of figure 4.2.3, $y^2 = b^2(1-x^2/a^2)$, and

$$r = (a^2/b)(1+(x^2/a^2)(b^2/a^2 - 1))^{3/2}. \qquad (4.2.1)$$

We may eliminate x in equation (4.2.1) in favor of λ from $\tan \theta = y'$ and $\theta = \pi/2 + \lambda$:

$$r = (b^2/a)(b^2\sin^2\lambda/a^2 + \cos^2\lambda)^{-3/2}. \qquad (4.2.2)$$

Note that $r(0) = b^2/a$, $r(\pi/2) = a^2/b$. For $\epsilon \ll 1$, equation (4.2.2) gives

$$r = a(1-2\epsilon + 3\epsilon\sin^2\lambda). \qquad (4.2.3)$$

Equation (4.2.3) lent itself to corroboration by measurement of the lengths of degrees at widely separated latitudes. A measurement in France would hardly do. The difference between a degree of the

---

81. Desaguliers, *PT*, 33 (1725), 344–345, and *Course* (1763²), 1, 391–393, 446–447.
82. D'Alembert, *Encycl.*, 6 (1756), 753.

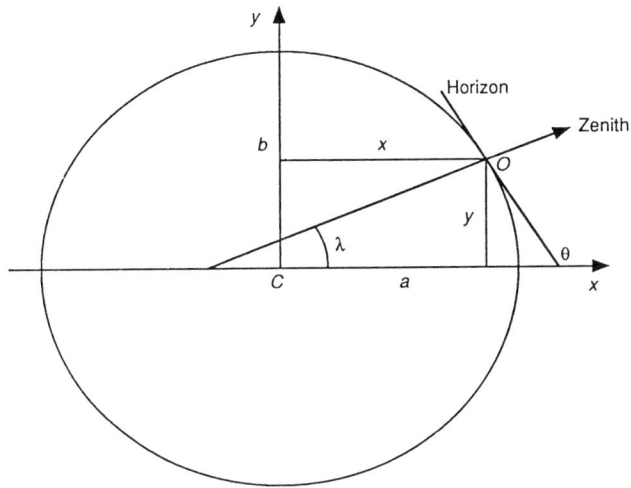

FIG. 4.2.3 Geometry of the Newtonian ellipsoid of revolution. The latitude λ of the place O is defined as the angle made by the perpendicular to the tangent at O with the major axis.

meridian around Dunkirk ($\lambda = 51°$) and one around Perpignan ($\lambda = 42.5°$) is, on Newton's model ($\epsilon = 1/230$),

$$\text{Du}° - \text{Pe}° = (2\pi/360)(r(51°) - r(42.5°))$$
$$= (3/230)(0.147)(57060) = 100 \text{ ts}.$$

Here the equatorial radius a comes from Picard's value for the length of a degree on the assumption of a spherical earth.

Now 100 ts corresponds to less than 7 seconds of arc in the determination of latitude, which, according to Louis Godin, a leading astronomer of the 1730s, was but a third of the likely error in observation.[83] Comparison of arcs around the equator and near the Arctic circle, however, was much more promising, if neither expense nor hardship figured in the equation:

$$\text{Ar}° - \text{Eq}° = (3/230)(\sin^2 66°)(57060) = 620 \text{ ts}.$$

83. Godin, MAS, 1733, 223–224.

The possibility that the business might yet be settled without leaving the soil of France was raised in 1724 by Giovanni Poleni, an applied mathematician at the University of Padua, and promoted briefly by Godin and Maupertuis at the Paris Academy in 1733. Poleni recommended measuring arcs along parallels of latitude rather than arcs along meridians of longitude.[84] The procedure had the advantage that, for the same triangulated distance, the latitude measurement would show a greater difference between the oblatum and the sphere containing it than the longitude measurement. The truth of this unlikely proposition may be gathered from figure 4.2.3; from $y' = \tan\theta$, $\theta = \pi/2 + \lambda$, and the equation of the ellipse, we have

$$x = a(b^2\tan^2\lambda/a^2+1)^{-1/2} \sim a\cos\lambda(1+\epsilon\sin^2\lambda). \quad (4.2.4)$$

At the latitude of Paris, the difference between a degree of the meridian on Newton's oblatum and one on Picard's sphere is

$$\Delta_{\text{long}} = (2\pi/360)(r(48.83°) - a) = -74 \text{ ts};$$

whereas the difference per degree of longitude is

$$\Delta_{\text{lat}} = (2\pi/360)a\,\epsilon\cos\lambda\sin^2\lambda = 93 \text{ ts}.$$

The latitudinal arc from St Malo to Nancy subtends about the same angle as the longitudinal arc from Dunkirk to Perpignan, although it includes only two-thirds the distance on the ground; and yet, over these equal arcs of 8.5°, the difference between Newton's earth and Picard's is about the same whether taken in latitude or longitude ($\Delta_{\text{lat}}/\Delta_{\text{long}} = 0.9$). A perhaps more persuasive consideration, which Godin developed, compared the Δ_{lat} between Newton's earth and Cassini's. On the first hypothesis, $\epsilon = 1/230$, $a = 3.276$ million ts, $a > b$; on the second, $\epsilon = 1/95$, $a = 3.255$ million ts, $a < b$; on the Paris parallel, $\Delta = 550$ ts/degree.[85]

The resumption of the mapping of France during 1733 gave an opportunity to put Poleni's proposal into practice. Jacques Cassini's analysis of the numbers he and his son obtained along the Paris parallel confirmed the family's prolate planet; the academicians'

---

84. For a thorough discussion of this episode, see Greenberg, *HSPS*, 13 (1983), 246–254.

85. Godin, *MAS*, 1733, 230–231.

computations went the other way, but not far enough to get Newton's oblate orb. The difficulty lay in the determination of longitude by eclipses of Jupiter's moons. The art did not give the arc to sufficient accuracy to discover the earth's profile.[86] Once again it appeared that the answer demanded uncomfortable travel. This time, however, sufficient interest had developed to procure the funds and find the men for the undertaking. In June 1733, while Jacques Cassini triangulated his way to Strasbourg, some of his younger colleagues, concerned about the Academy's reputation and ambitious for their own, proposed to measure a meridian at the equator. The following year, to lengthen the basis of comparison and to avoid reliance on Picard and Cassini, the Academy resolved to send an expedition to the Arctic Circle.[87]

Elusive ellipticity

The French expeditions of the 1730s are among the great adventures of civilized men. In the North, the surveyors, led by Maupertuis, braved freezing cold, glaciers, and mosquitoes; in the South, under the direction of Godin, Bouguer, and La Condamine, they faced fearsome heat, withering cold, high winds, huge mountains, and mosquitoes; in both, constant danger; and all to learn whether a degree of longitude contains a few more feet in Lapland than in Peru. Overall, the northerners had the better of it, despite the man-eating gnats. They had a frozen river on which to lay out a base line, a Graham sector to take latitudes, six stout soldiers and fifteen boats to carry their instruments, and a way of working that avoided fights; they returned to Paris in 1737, after an absence of sixteen months, with a measurement that, as Voltaire put it, "simultaneously flattened the poles and the Cassinis."[88]

---

86. Greenberg, *HSPS*, 13 (1983), 254–259.

87. Lafuente and Mazuecos, *Caballeros* (1987), 56–62; Cassini de Thury, *Relation* (1765), v: "our first measurement of the perpendicular to the Paris meridian was the cause and the epoch of all the new undertakings that the Academy made to measure the earth by going to the equator and the polar circle."

88. Brunet, *Maupertuis* (1929), 1, 33–58; Brunet, *Clairaut* (1952), 30–53; Nordmann, *Cahiers d'hist. mond.*, 10 (1966), 87–91; Smith, *Plane to spheroid* (1986), 174–176, 192; Voltaire, quoted in Saveney, *Rev. deux mondes*, 79 (1869:1), 28. Maupertuis' expedition lends itself to pictorial description: Martin, *Figure* (1987), 43–87; Taton, in Lacombe and Costabel, *Figure* (1988), 115–137.

In fact, although it included such eminent academicians as Clairaut, P.C. Lemonnier, and Anders Celsius, Maupertuis' expedition did not arrive at an accurate figure. Clairaut and Maupertuis had had little previous experience in astronomical observations, apart from a little practice at Thury in the autumn of 1735, under the guidance of the man whose measurements they were trying to discredit.[89] The best value derived by the northern expedition, 57,438 ts/degree, was out by over 200 ts, according to measurements made early in the 19th century by Jöns Svandberg, who imported a Borda circle from Paris for his purpose. The academicians may have erred by some 12 seconds in their latitude determinations, which made about 190 ts on a spherical earth of Picard's size; had the instrumental (or observational) error gone the other way, it would have swamped the effect Maupertuis sought (at $\lambda = 66°$, a degree of Newtonian meridian should be 125 ts bigger than one on a Picard sphere), and confirmed Cassini. The academicians further enlarged their result by failing to take into account the shrinkage of their toise sticks in the extreme cold. Combining his mistakenly large value for the Arctic degree with Picard's value for one around the latitude of Paris, Maupertuis deduced from equation (4.2.3) that $\epsilon = 1/121$, $a =$ Newton's (3.27 million ts).[90] And that amount of flatness, as he wrote James Bradley, "appears even more considerable than Sir Isaac Newton thought it." The result delighted Maupertuis, who wanted a decisive qualitative outcome, and did not concern himself overly with precision. As he had written Johann I Bernoulli just before setting off, "we will never have the earth's dimensions to the utmost exactness and I think that it will be enough for its inhabitants to know its size in general and that it is elongated or flattened and approximately the limits of its shape." His one-time instructor Cassini, caught up in technicalities, insisted instead on instrumental difficulties that vitiated the result. The British enjoyed the spectacle and awaited news from Peru.[91]

89. Maupertuis to Johann I Bernoulli, 12 Sep 1735, and reply, 13 Oct 1735, in Brown, *Science and comedy* (1976), 175–176. Cf. Terrall, *Isis, 83* (1912), 224–225.

90. Smith, *Plane to spheroid* (1986), 143, 184, 187–192; Levallois, *Vie des sciences, 3* (1986), 288. Cf. Lafuente and Mazuecos, *Caballeros* (1987), 70–81; Westphal, *Zs. f. Instr., 5* (1885), 273; Zach, in *Monatliche Correspondenz, 1* (1800), 113, 339.

91. Maupertuis to Bernoulli, 27 Mar 1736, in Brown, *Science and comedy* (1976), 178, and to Bradley, 27 Sep 1737, in Tweedie, *Stirling* (1922), 78 (quote); correspondence among Colin Maclaurin, Machin, and Stirling, 1738, ibid., 75, 89, 174. It is

The members of the expedition to Peru suffered from prolonged exposure to the Andean climate and to one another; they arrived in the New World late in 1735 and straggled back to the Old beginning in 1744. Among the many causes of delay were the weather, which during the winter sometimes permitted only 15 minutes of observations a month; the fatigue and discomfort of life in rude huts at the various stations; the hostility of the Indians, who deduced that men so engaged must be lunatics, criminals, or sorcerers; the unfriendliness of the Spanish viceroy; the difficulty of transporting delicate instruments up great mountains on mule back; the constant shaking from earthquakes; and antipathy among the French leaders of the expedition, Godin, Bouguer, and La Condamine.[92] There were rumors of another reason as well. "Mr Short [the telescope maker James Short] writes that an unlucky accident has happened to the French Mathematicians in Peru. It seems they were shewing some French gallantry to the natives' wives, who [the natives, apparently] have murdered their servants, destroyed their Instruments & burnt their papers, the Gentlemen escaping narrowly themselves."[93]

Nonetheless the final results of the Peruvian gallants agreed well among themselves and with later determinations: one baseline, measured twice by different subparties in 25 days of heavy work, came to 6272.74 and 6272.69 ts, or a precision of one part in 150,000; the second baseline, corrected for temperature expansion of the toise sticks, came to 5258.90 and 5258.95 ts; the degree of arc fell out as 56,750 according to Bouguer and La Condamine, and 56,768 according to Godin and Jorge Juan, a Spanish naval officer attached to the expedition. To acquire these numbers from the observations many tedious adjustments were required, for aberration, nutation, precession, and refraction, for example, and also for the effects of temperature on the instruments, the deviation of the sums of the angles (each of which had a likely error of 20 seconds

---

not clear where Maupertuis' error lay; Olbers to Gauss, 9 Feb 1821, in Schilling, *Olbers* (1900), 2:2, 66; Zach, *Mémoire* (1811), 15–16; Jahn, *Gesch.* (1844), 2, 157–158, 185. Cf. Terrall, *Isis, 83* (1912), 227–234.

92. Smith, *Plane to spheroid* (1986), 105, 109–111, 152, 166; Lafuente and Mazuecos, *Caballeros* (1987), 157–194, and Lafuente, in Lacombe and Costabel, *Figure* (1988), 139–150; Westphal, *Zs. f. Instr.*, 5 (1885), 270–271.

93. Maclaurin to Stirling, 6 Dec 1740, in Tweedie, *Stirling* (1922), 91.

or more) in the triangles from 180°, and the reduction of all measurements to sea level.[94] No doubt the calculations and extensive checking of instruments provided welcome employment on the windswept deserts when observations could not be made.

Combined with the result of the Northern expedition, the average Peruvian degree of 56,759 ts implied an oblate spheroid with equatorial radius a = 3,271,652 ts and an ellipticity ϵ = 1/310. (Combined with Picard's measure, ϵ = 1/315.) The slight difference between these values and Newton's (a = 3,276,433 ts, ϵ = 1/230) did not at first appear to be significant. French brains and bravado had demonstrated the truth that English lethargy had only guessed. Voltaire again: "Heros of physics, Argonauts of our time/who leaped the mountains, who crossed the seas...You have confirmed in uncomfortable places/What Newton knew without leaving his study."[95]

Picard had known nothing about aberration and nutation, although he had noticed changes in positions of his reference stars that derived from aberration. Maupertuis came after Bradley's discoveries, for which he corrected the Lapland observations and also Picard's. That reduced the length of the Paris degree from 57,060 ts to 56,925 ts, which, when combined with the Arctic degree, gave ϵ = 1/174; little different, according to d'Alembert's conception of error, from Newton's 1/230.[96] Still, the discrepancy amounted to a difference of six English miles in the length of the polar axis, which might be thought considerable. Therefore in 1740, before the results from the Peru expedition were known, the French redid Picard's arc and baseline; they thought it easier than repeating the Arctic expedition, whose result, in any case, they assumed to be more reliable than the measurements made by Picard and Cassini. Cassini himself (Jacques Cassini, Cassini II) recognized that he had been sloppy, especially in the mountainous southern reaches, where

---

94. Smith, *Plane to spheroid* (1986), 108, 142–143, 148, 151; Levallois, *Vie des sciences*, 3 (1986), 281; Marquet, in Lacombe and Costabel, *Figure* (1988), 193–196. According to Zach, *Mémoire* (1811), 13–14, the Peruvian measurers took the wrong signs in correcting for aberration and mutation, and should have reported an arc 3."1 longer than they did.

95. Voltaire, "Quatrième discours sur l'homme: de la modération," quoted by Levallois, *Vie des sciences*, 3 (1986), 276.

96. D'Alembert, *Encycl.*, 6 (1756), 754; Levallois, *Vie des sciences*, 3 (1986), 268, 293n4.

he had contented himself with deducing the third angle of some triangles from measurement of the other two, and so had no check on the accuracy of his proceedings. He arranged to have his son placed in charge of the measurement; but Cassini III was busy mapping France, and did little more than put his name on the title page of the final report. The work fell to a scrupulous observer, the abbé Nicolas-Louis de Lacaille. The report ascribed Cassini II's failings to his three-foot quadrant, which was awkward to employ in church towers and had no micrometers on its lenses. To correct these faults, the new expedition used a two-foot quadrant fitted with micrometers for the triangulations and a six-foot sector to take the height of stars.[97]

Lacaille began measuring in June 1739, redoing the old triangles where they sufficed—when they contained no angle less than thirty degrees and when all angles could be separately determined—and devising new ones when they did not. He spun beautiful new triangles across the Auvergne during the severe winter of 1739/40; he laid out his own baselines north and south; he and Cassini de Thury redid Picard's base at Amiens, and found it off by six toises, or one part in a thousand, an act of impiety that, according to Cassini, met strong opposition. They observed the thermometer regularly, but made no corrections for temperature, deeming them unnecessary. Finally, taking everything together, after two years' hard work, Lacaille confirmed that degrees of latitude in France behaved for the most part like degrees elsewhere and increased toward the pole.[98] Unhappily, the new value for a degree around Paris did not improve the fit: Lacaille and Cassini de Thury got 57,183 ts, which, on remeasuring Picard's base, they made 57,074 ts. The larger figure agreed fairly well with Lapland and Newton, and with the flattening of $\epsilon = 1/215$ obtained from combining the Arctic and the Equatorial degree, but the smaller and supposedly better figure, 57,074 ts, could not be reconciled with the others.[99]

---

97. Cassini de Thury, *MAS*, 1740, 279–281, 288; La Condamine, *Mesure* (1751), 104, 222, 229; Delambre, *Grandeur* (1912), 63–67. Cassini II and III used a quadrant little better than Cassini I's; Westphal, *Zs. f. Instr.*, 4 (1884), 159; Fauque, in Lacombe and Costabel, *Figure* (1988), 209–216.

98. Delambre, *Grandeur* (1912), 69–72; Cassini de Thury, *Relation* (1775), v; Gillispie, *Polity* (1980), 113–115; Westphal, *Zs. f. Instr.*, 5 (1885), 265–266.

99. D'Alembert, *Encycl.*, 6 (1756), 756.

Corrections for the effects of atmospheric refraction, the aberration of starlight, the nutation of the earth, the discrepancy between plane and spherical triangles, the thermal expansion of measuring rods, and the demonstrated or suspected differences among the toises of Lapland, Peru, Picard, and Lacaille, did not bring a degree of agreement among geodesists. The Academy named commissioners to reconcile the several measurements of the distance between Paris and Amiens. Bouguer informed a government minister that the commissioners had declared in public that they could reach no firm decision. "Is it not astonishing that we fancy that we know the lengths of a degree fairly well in various places on the globe, and that we do not have the same advantage for the environs of Paris? It is worthy of you, Sir, to provide for the definitive solution of this question." The resultant inquiry confirmed Lacaille's Paris degree of 57,074 ts.[100]

At this point the distressing possibility that the earth might not be an ellipsoid of revolution began to gain currency. Already in 1736 a resourceful *Biblischer Mathematicus* had seized upon the disagreement about the shape of the earth as a confirmation of scripture. "Hast thou surveyed unto the ends of the earth?" God asked Job. "Declare if thou knowest it all."[101] La Condamine acknowledged in 1746 that it might not be possible to know the earth's shape, if it has one; and the busy Jesuit mathematician Roger Boscovich took it for granted, when he started to promote geodetic surveys around 1750, that only God had an intuition of the true figure of our globe.[102] His vicar, Benedict XIV, commissioned Boscovich to run a meridian line from Rome to Rimini, and, while about it, to map the Papal States. It was a perfect opportunity for the skeptical geodesist. The Rome-Rimini arc had about the same average latitude as the Cassinis' line from Dunkirk to Perpignan; it would therefore allow a test whether at the same latitudes the length of a degree along all meridians is the same. Boscovich expected the test

100. Delambre, *Grandeur* (1912), 72–81, 143–145; Todhunter, *History* (1873), 1, 355–356; Levallois, in Lacombe and Costabel, *Figure* (1988), 67–73.
101. Schmidt, *Bibl. Mat.* (1736), 134; Job, 38:18. Schmidt set himself the task of naturalistic explanation of such puzzles as how Joshua's agents could have conducted a land survey (Joshua, 18:4–9) without modern equipment (Schmidt, 135–139).
102. La Condamine, *MAS*, 1746, 637; Lafuente and Pesset, *RHS*, 37 (1984), 249, 252–254, identify this pessimism with the beginning of experimental geodesy.

to fail, as it did; "prejudice for regularity and simplicity [he wrote] is a source of error that has only too often infected philosophy." He obtained the collaboration of a Jesuit colleague, Christopher Maire, a good astronomer with a taste for geography, "and, moreover, healthy enough to withstand the fatigue of travel."[103]

Boscovich and Maire started their work in rains the like of which had not been seen since the time of Noah. The Tiber flooded Rome twice. On the road, they narrowly escaped death by drowning, slipping, and sliding. Twice marooned in villages, they improved their time by preaching, "thinking we could fill the void in our geometrical occupations in no better way than by works of zeal." They were attacked by dogs and, worse, by peasants, who knew that no one could be idiot enough to do what Boscovich and Maire claimed to be doing; it followed that the Jesuits must be looking for buried treasure, guarded by hydraulic spirits. Reasoning thus, and needing nails, the peasants tore down the signal towers that Boscovich and Maire had built on the summits of the Apennines. The French translator of the report of the expedition, which Boscovich and Maire drew up in Latin, was astonished. "Who could have believed that there would be so strong a resemblance between the peasants of the Apennines and the Indians of the mountains of Quito?"[104]

Boscovich and Maire measured in the same way the French did, using instruments produced for them, after much delay, by a Father Rufo, who had built up the collection of mechanical models at the Gregorian College, but had never seen a zenith sector. Using old ship masts as rulers, they laid out baselines of about 12 kilometers along the Appian way and on the beach near Rimini; they set masts with small gaps in between, which they measured with dividers, so as to avoid inadvertent displacements; two measurements of the base differed from one another by one part in 1/72 ts in over 6037 ts, and by a 19th-century remeasurement by 28 cm in over 3000 meters. They anchored their angles more accurately than the French had done in Lapland; they determined the number of toises in their ship masts from a standard sent from Paris; and they ended with a

103. Boscovich, AS, Bologna, *Comm.*, 4 (1757), 353, 361; cf. ibid., 146–147, and La Condamine, *Mesure* (1751), 263–267; Boscovich and Maire, *Voyage* (1770), 1–2, 29 (quote), 31–32 (quote).

104. Boscovich and Maire, *Voyage*, (1770), 48–49 (last quote), 52, 57–63, 74–75 (first quote).

value of 56,979 ts for a degree along the meridian at a latitude of 42°.[105] The number did not agree well with the degree of 57,048 ts obtained by Lacaille and Cassini III at an average latitude of 43.5°. Boscovich took this discrepancy to confirm his expectations. "Look where you will," he wrote, "you will see nothing regular, nothing fixed or constant."[106]

Boscovich's result challenged not only previous assumptions but previous measurements. If it held up, d'Alembert wrote, "all efforts made so far to determine the figure of the earth will be a total loss." For his part, he hoped that the Jesuitical degree would fail and that a planned measurement in France would confirm 57,183 ts against the objectionable 57,074 ts. Everything depended on measurements at the very edge of attainable accuracy. "What observer will guarantee [his work] to two seconds? Those who are the most exact and the most sincere, would they dare assure an accuracy of 60 ts in a degree, since 60 ts does not suppose an error of 4 seconds in measuring the celestial arc, and none in geographical determinations?....Nothing therefore requires us to believe the meridians dissimilar."[107] Cassini III reached a different conclusion from the same premises: geodetic and astronomic measurements being uncertain, only Cassini's method of light signals gave mankind a chance to answer satisfactorily God's question to Job.[108]

Meanwhile, Boscovich snatched the initiative from the French academicians. He called attention to Bouguer's finding that a big mountain in Peru drew aside a plumb bob by a few seconds of arc, thus falsifying the vertical and the orientation of instruments in its neighborhood. How had the attraction of the Pyrenees affected the

105. Boscovich and Maire, *Voyage* (1770), 40–44, 262–340; Nikolic, *AIHS*, 14 (1961), 318–319; Wesphal, *Zs. f. Instr.*, 4 (1884), 161–162, and 5 (1885), 334. The standard sent from Paris was shorter by 8/75 ligne (0.012%) than the toise de Pérou, which differed by 1/20 ligne (0.006%) from the Lapland toise (Boscovich and Maire, *Voyage*, 40–41).

106. Boscovich and Maire, *Voyage* (1770), 32–33, 117, 489 (quote), 492.

107. D'Alembert, *Encycl.*, 6 (1756), 760 (quote), 761. Boscovich's measurement held up; Hahn, in Bossi and Tucchi, *Commemoration* (1988), 72, on Laplace's favorable opinion; Zach, *Mémoire* (1811), 23–24, on Zach's redetermination of the difference in latitude between Rome and Rimini (4."4 > Boscovich's value); Berthaut, *Ing. géogr.* (1902), 2, 359–361, on the French remeasurement of 1808–1809; Jahn, *Gesch.* (1844), 2, 184.

108. Cassini, *Relation* (1765), 130–133, and *Relation* (1775), v.

southern end of the Dunkirk-Perpignan line, and the pull of the Apennines the work in the Papal States? To obtain measures without mountains, Boscovich proposed to the Royal Society of London and the Austro-Hungarian Empress that they commission determinations of a degree across the plains in America and in Eastern Europe; and to the King of the Two Sicilies that, for comparison, he do the same, in the foothills of the Alps around Turin.[109]

The Royal Society commissioned Charles Mason and Jeremiah Dixon, already surveying in the New World, to measure an arc in Maryland; the Holy Roman Empress sent off Liesganig to the prairies of Hungary, where he measured out baselines using rafters from the Jesuit college in Vienna; and the King of the Sicilies asked the professor of physics in the University of Turin, Giambattista Beccaria, to try his hand at a degree in Piedmont. Boscovich had little confidence in the "cherished children" of the Academy, that is, the degrees taken in Lapland and Peru; he preferred to base geodesy on the comparison of different meridians measured at roughly the same middle latitudes, as in Maryland, France, Northern Italy, and Austria. He thought for a time to carry on the work himself, in California. The refusal of a travel permit by the Spanish authorities saved him from that barbarous place.[110]

The three projects stimulated by Boscovich only increased confusion. Mason and Dixon did not triangulate, but measured their degree directly on the ground; they had the advantage of a level terrain, and the best English instruments; they corrected for temperature and compared their measuring rods carefully with the "toise de Pérou;" "yet"—it is the opinion of the late 18th century—"the result was no better than the rest." The critic, von Zach, could not grant even that much to Beccaria or Liesganig. Beccaria had bobbled his arc by 900 ts ("an altogether intolerable error") and missed an angle by 13 minutes, according to measurements Zach took on the spot with a repeating circle made by Reichenbach; whereas Liesganig, according to his manuscripts, which Zach had

109. Boscovich and Maire, *Voyage* (1770), 36; Zach, *Monat. Corr.*, 4 (1801), 553–554; Nikolic, *AIHS*, 14 (1961), 315–316.
110. Chatelain to Boscovich, 25 Mar 1770 (The Bancroft Library, Univ. of California, Berkeley), and to Arnolfini, 17 Mar and 2 June 1767, in Arrighi, *Carteggio* (1965), 20–22, 28; Bernleithner, in *Cart.* (1978), 140–141.

acquired, had falsified his angles, taken one church tower for another, misplaced his endpoint by 4500 ts, and misidentified the star by which he fixed the latitudes of his end points.[111]

Zach's harshness toward Liesganig had a personal side that has a general interest. He had served as Liesganig's administrative assistant, but did no surveying. He wrote a friend thirty years later: "You will scarcely believe it, never once did I see a triangle, an angle, an azimuth, a reduction to the meridian or perpendicular, all that was reserved to the P.P.S.J. IHS," the fathers of the Society of Jesus. "I had no experience of this survey, the Jesuits kept it all secret from me....Liesganig was fundamentally a complete ignoramus." Elsewhere Zach complained of the unreasonable secrecy of the Austrian military authorities about geographical and statistical matters.[112] The secretiveness of the French and Piedmontese authorities about maps and mapping during peacetime has already been mentioned. Even the sea-girt British worried that their collaboration with the French in linking the observatories of Paris and Greenwich might compromise security.[113]

It is only fair to leave the last word to Liesganig, whose measurements were to some extent vindicated by Ludwig August von Fallon, who travelled the same route in 1806. Perhaps with his unwillingly sedentary assistant in mind, Liesganig spoke for all field workers impugned by armchair Zachs. "Whoever dreads steep and rugged paths and mountain crossings on foot, whoever shies away from storms, heat, and cold, and other continuous bodily discomforts," had no warrant to criticize; "to him," says Liesganig, "I wish a quiet life and a soft pillow."[114]

Liesganig, Beccaria, and Mason and Dixon's spokesman, Astronomer Royal Maskelyne, assigned all discrepancies in values for the earth's flattening to the unknown effects of mountains.[115] In order

111. Zach, *Monat. Corr.*, 8 (1803), 515–521 (first quote), and *Mémoire* (1811), 7–8, 25, 93 (second quote), 123, 136; Howard, *Zs. f. Verm.*, 17 (1888), 34–38; Westphal, *Zs. f. Instr.*, 5 (1885), 344.
112. Zach to Schedius, 23 May 1798, 26 Jan 1799, and 30 Apr 1801, in Zach, *Briefe* (1984), 45–46, 56, 80–81 (quote).
113. Lord Liverpool to Joseph Blagden, n.d. (Blagden Papers J.7, Royal Society).
114. Bernleithner, in *Cart.* (1978), 141–144; Liesganig, *Dimensio graduum* (1770), b4v (quote), 211–212.
115. *Jl des sçavans*, 25 (June 1767), 128–132; Maskelyne, *PT*, 58 (1768), 327–328. Maskelyne's predecessor at Greenwich, Bradley, had made the same point earlier; Bradley to Stirling, 2 Dec 1733, in Tweedie, *Stirling* (1922), 162.

to "convince those who will yield their assent to nothing but downright experiment," Maskelyne proposed to the Royal Society in 1772 to measure directly the attraction of a mountain on a plumb bob. Charles Mason, then back in England, was engaged to look for a suitable hill. He found one in Scotland, a nice symmetrical isolated height named Mt Schehallien.[116] During the summer of 1774, surveyors triangulated the neighborhood while Maskelyne measured the zenith distance of a certain star from temporary observatories just north and south of the base of the hill. Let the observed zenith distances be P and Q, the angle through which the hill pulls the plumb line off the vertical be δ, and the two zenith distances be α and β (figure 4.2.4). Then, since $\alpha = P + \delta$ and $Q = \beta + \delta$, $2\delta = Q - P - (\beta - \alpha)$, that is, the deviation δ is half the difference between the observed and the true differences in latitude. Maskelyne obtained $Q - P = 54.6$ seconds by play with his zenith sector; his surveyors found the arc NS, which, when reduced to an angle using the latest value for the radius of an approximately spherical earth, gave $\beta - \alpha = 42."94$, whence $\delta = 5."8$.[117] The exercise made plausible the irksome perturbation attributed to mountains by geodecists; it did not show, however, that taking mountainous attraction into account would bring all the determinations of the earth's flattening into accord.

Revolutionary geodesy

The ultimate datum on the figure of the earth collected during the 18th century—ultimate in both time and quality—derived from the labors of P.F.A. Méchain and J.B.J. Delambre. They had most excellent and well-wrought equipment—Borda circles, zenith sectors, temperature-compensated measuring rods—and the patience and fortitude of saints. Besides the bad weather, wretched terrain, and suspicious peasants that customarily afflicted arc runners, Méchain and Delambre faced the dangers of travel in a country racked by revolution and threatened by enemies on all its frontiers. Méchain took the southern beat, from Paris to Barcelona. His instruments perplexed the peasants, who prudently jailed him as a counterrevolutionary. Officials who understood numbers eventually pro-

---

116. Maskelyne, *PT*, 65 (1775), 495–499; Howse, *Maskelyne* (1989), 129–132.
117. Maskelyne, *PT*, 65 (1775), 500–542; Howse, *Maskelyne* (1989), 134–139.

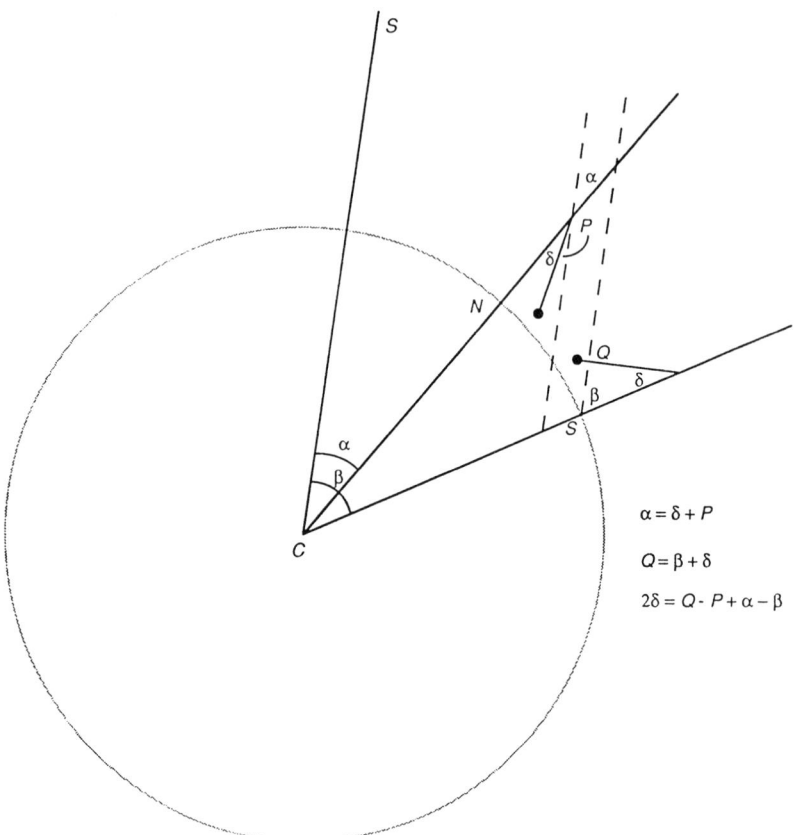

FIG. 4.2.4 Geometry of Maskelyne's measurement at Schehallien: the dotted lines, parallel to CS, are the true directions, α and β the true altitudes, and P and Q the apparent altitudes, of the star at the two observatories, and δ is the deviation of the plumb bob.

cured his release. He decided he would be safer in Spain, and so began his mapping there, with great efficiency until an injury compelled him to break off early in 1793. By the time he had recovered and completed the arc in Spain, with an immaculate determination of the latitude of the southern terminus at Montjouy, the Spanish government had closed the border to Frenchmen. Méchain passed his enforced leisure by observing stars in Barcelona, also with exquisite accuracy; but, to complete his misfortunes, the difference in latitude between Mountjouy and Barcelona, as deduced from the stars, came out 3.24 seconds larger than the difference Méchain had calculated from the 950 ts between the stations that he had obtained by earthly triangulations.[118] These miserable three seconds tormented Méchain for the rest of his life.

While Méchain stagnated in Barcelona, Delambre had one unpleasant adventure after another in the north of France. Anticlerical peasants had knocked down many of the church towers that had served his predecessors; when he hung lanterns as substitute sights, he was suspected of signalling to the enemy (France had been at war with most of Europe since April 1792). Armed citizens prevented him from carrying out his orders. His heroism may be gauged from his experience at Epinay, where, having been detained by local patriots, he tried to explain his mission in an impromptu lecture on geodesy. To no avail. He was packed off to a higher jurisdiction, in Saint Denis. The square teemed with volunteers waiting to go to the front, who did not have the prerequisites for the crash course in triangulation they forced him to give. "They did not receive me more favorably [than in Epinay]. Evening came on; you could not see. The audience was very large: the front rows heard without understanding; those behind heard less and saw nothing. They grew impatient and grumbled; some proposed one of those quick means [of dealing with suspect people] then so much in vogue." Were it not for a quick-witted official, who rushed Delambre into protective custody, the arc might have ended in Saint Denis.

The National Assembly found time while busy transforming France into a Republic, which it declared on September 22, 1792, to issue an order for Delambre's release. He returned to Paris to observe from the top of the Panthéon. That occupied him during January and February of 1793. It then took him two months to acquire a passport to return to the field. Although he had many explanations to make en route, especially as he approached the

---

118. Méchain and Delambre, *Base* (1806), *1*, 51–56; Delambre, *Histoire* (1827), 759–760.

frontier, his passport kept him out of jail. By the fall he was working around Dunkirk, repeating Lacaille's triangles whenever possible.[119]

While Delambre put his triangles together, the Revolution took apart the institutions on which his work depended. On August 8, 1793, it closed the Academy of Sciences, and, just before Christmas, it fired Delambre. He was then working in the Loire, where he had erected a tower 64 feet tall, which had the usual effect on rural sensibilities. News of his dismissal reached him in bitter cold weather, on the shaking tower, as he strove to serve the people. On his return to Paris, he was interrogated by a revolutionary committee. It studied his manuscripts without making much of them and pounced on a diploma from the Royal Society, which bore the arms of George III. Delambre found it prudent to retire from the service of his country.[120]

Shortly thereafter a powerful advocate of geodesy came to the rescue: that General Calon who had helped to improve the surveying technique of the Royal Geographical Engineers. Calon had become a deputy and a Jacobin; his previous service to the king inhibited him not at all from voting for the execution of Louis XVI, "without reprieve." His rampant republicanism may perhaps be associated with unfulfilled ambition and a sense of injury suffered in the late 1770s, when, as a veteran of 20 years, he had been ordered to do survey work of a kind usually assigned to his juniors.[121]

The suppression of his former corps presented Calon with a vacuum in which to fullfil his ambition. With the corps went the map department of the Dépôt de la guerre, the headquarters and storehouse of military cartography. It was replaced by an Agence temporaire des cartes with jurisdiction over all maps belonging to the state, including those held by the Commission du mouvement et de l'organisation des troupes, directed by Calon. Very soon all was confusion: the agency could not handle the flood of maps, and Calon, who became head of the Dépôt in April 1793, managed to

119. Méchain and Delambre, *Base* (1806), 1, 23-24, 41-44; Warmé, *Eloge* (1824), 15-16; Berthaut, *Carte* (1898), 1, 121-122; Méchain and Delambre, *Base* (1806), 1, 32-34.
120. Ibid., 1, 46-52.
121. Berthaut, *Ing. géogr.* (1902), 1, 43-46, 66, 102-103, 122-123.

recapture most of its archives. He then purged the place of suspect employees; repopulated it with the graduates of a school he established in astronomy, geography, engineering, history, languages, and engraving; and staffed his school and his service (which was under pressure to provide maps of the front and of captured territories to the Army) with all the unoccupied savants he could locate. At its height in the fall of 1796, Calon's empire included the former archives of marine cartography, a printing office, and the school, at an annual cost of 300,000 livres. His expansiveness caused inefficiencies and inspired jealousies, and by the spring of 1797 he and his empire had fallen.[122]

Among those Calon enticed back was Méchain, who had begun his career as a naval hydrographer. (Méchain had managed to get himself to Genoa in September 1794, after a narrow escape from corsairs.) Delambre also came to Calon. It was agreed, probably early in 1795, that they would triangulate the new departments and that, to prepare as firm a base as possible for the work, they would resume the measurement of the meridian. Then at the zenith of his influence, Calon persuaded the Committee on Public Safety to grant 220,000 livres for map-making, half of it to go to extending Cassini's Carte de France into the Rhineland and half to go toward finishing the meridian between the Loire and the Pyrenees.[123]

That occurred on May 13, 1795. Delambre left Paris on June 28 in the capacity of "astronomer of the Dépôt de la guerre" to resume his observations at Orléans. He had trouble finding intact steeples and the cost of erecting signals quickly depleted his cash. No one wished to accept the assignats (government IOUs) he carried; once again Calon tided him over. Also, three years of republican rule had not made peasants less suspicious and superstitious: they now tore down his signals on the theory that they attracted storms.

Méchain had his hardships, too, as he crawled north across the mountains toward his rendezvous in Rodez. In one place, an outcrop 600 ts high and 2 ts wide surrounded by precipices, he sat for days waiting for the clouds to lift long enough to work the Borda circle. He arrived within signalling distance of Rodez around

122. Ibid., *1*, 126–127, 130–132, 135–150.
123. Delambre, *Histoire* (1827), 756–762; Méchain and Delambre, *Base* (1806), *1*, 57–58; Berthaut, *Ing. géogr.* (1902), *1*, 162–164.

September 1, 1797, shortly after Delambre had completed his part of the work. There they stopped. Méchain would not accept Delambre's help and could not go on. "In this cruel situation," he wrote his collaborator, "I prefer to stay in this terrible exile, far from what I cherish in the world; I will sacrifice everything, give up everything, rather than return without finishing my part of the job." Delambre felt that he could not insist, and Méchain finished the arc the following September.[124]

After concluding his triangles at Rodez, Delambre joined Laplace at Melun to fix the ends of the baseline near Paris. Unfortunately, all the Borda rules needed were not ready until the following spring (1798). Their use proved extremely tedious. Following Boscovich's initiative, Borda's rulers did not touch; a platinum tongue advanced by a micrometer screw extended to cover gaps; and the tongue was calibrated to show not only distance but also temperature, by its expansion relative to the copper body of the ruler. Working this exact and exacting instrument, Delambre could make only about 90 lengths of the rule (about 180 ts) a day; it took 33 days to cover the entire distance of 6,075.90 ts. In the summer he redid the Perpignan baseline. That took six weeks. The length of the Perpignan base as calculated from the length of the Melun base and the triangles differed from the measurement on the ground by less than a foot. On October 22, 1798, everything required to calculate the length of the arc from Dunkirk to Barcelona was in hand. Méchain and Delambre returned to Paris, where an international committee had assembled to examine, approve, advertise, and propagate their results.[125]

Meanwhile, Delambre had uncovered an oddity: if the 45th degree of latitude bisects the measured arc, the value of the entire quadrant to which the arc belongs may be calculated to a very close approximation without knowing the ellipticity of the earth.[126] The truth of this unexpected proposition may be discerned as follows. The element of length along the ellipse (see figure 4.2.3) is

124. Ibid., 1, 165; Bigourdan, *Système* (1901), 135–145; Méchain and Delambre, *Base* (1806), 1, 66–80; Méchain to Delambre, 11 Nov 1797, ibid., 83.
125. Ibid., 1, 84–90; Delambre, *Rapport* (1810), 6; Berthaut, *Carte* (1898), 1, 123–124.
126. Delambre, *Grandeur* (1812), 302–303; Ten, in *Scientifiques et savants* (1990), 447–448.

$dx(1 + y'^2)^{1/2} = dx \sec\theta = -dx \csc\lambda$. From equation 4.2.4, $dx = -a(1 - \epsilon^2)\sin\lambda(1 - \epsilon^2\sin^2\lambda)^{-3/2}$. The length of an arc, $A_2 - A_1$, taken along the ground, is therefore the integral of $a(1 - \epsilon^2)(1 - \epsilon^2\sin^2\lambda)^{-3/2}$ from latitute λ_1 to latitude λ_2. To order ϵ^2 we have,

$$\Delta A = A_2 - A_1 = a\int_{\lambda_1}^{\lambda_2}\{1 - \epsilon^2(1 - \frac{3}{2}\sin^2\lambda)\}d\lambda$$

$$= a(\lambda_2 - \lambda_1)(1 - \epsilon^2/4) - (3a\epsilon^2/8)(\sin 2\lambda_2 - \sin 2\lambda_1)$$

$$= a\,\Delta\lambda\{1 - \epsilon^2/4(3\epsilon^2/8\Delta L)F(\lambda_2, \lambda_1)\}. \qquad (4.2.5)$$

If $\lambda_1 = 0$, $\lambda_2 = 90°$, then $F = 0$ and $\Delta A = 90a(1 - \epsilon^2/4) = Q$, the length of a quadrant. Substituting for a in the equation 4.2.5, we have

$$Q = \frac{\Delta A \cdot 90(1 - \epsilon^2/4)}{\Delta L\{1 - \epsilon^2/4 - (3\epsilon^2/8\Delta L)F\}} = \frac{90\Delta A}{\Delta L}\{1 + (3\epsilon^2/8\Delta L)F\},$$

to order ϵ^2. But $F = \sin 2\lambda_2 - \sin 2\lambda_1 = 2\sin(\lambda_2 - \lambda_1)\cos(\lambda_2 + \lambda_1)$. If λ_1 and λ_2 are symmetrically placed around 45°, then $\lambda_2 + \lambda_1 = 90°$ and $F = 0$ and the value of the quadrant does not depend on ϵ!

Since, as will appear, the meter introduced by revolutionary France was defined as a fraction of the quadrant, French geodecists could push their government to support extending the arc through Paris south of Barcelona to a station in the Balearics as far below the 45th parallel as Dunkirk is above it. In September 1802, Méchain approached Napoleon for support "for thus completing so beautiful a project, the greatest ever undertaken."[127] He died in Spain before he could realize the mission. Four years later Biot and Arago, backed financially and intellectually by Laplace and the Bureau des longitudes, undertook to extend Méchain's terminus to the island of Formentera. The expedition had its tedium and excitement. On the tedious side, Biot and Arago had to sit for three

127. Quoted ibid., 450.

months in the mountains of Majorca before they could detect by telescope, across the open sea, the flares that would link them to Méchain's grid. They could see with no trouble the town of Castillon de la Plana on the mainland, where Méchain had died, but the pinpoint of light, no brighter to them than a star of the fifth magnitude, eluded them. At last, on a clear night with clear eyes, success; "I would not know how to express the emotion we felt," Biot wrote in 1810, in his report of the trip. The natives who visited the observation hut also left him speechless. "When they entered the silent room...the contrast of civilization with barbarism, of the most sublime knowledge with the deepest ignorance, had something great and pitiful about it that touched the soul in a way I would not know how to express." As for excitement, Biot ran into some Ragusan pirates on his way home. He gave them what gold he had and a lecture on geodesy. They let him and his companions go. "I must say that they treated us very well, for pirates."[128]

The upshot of the measurement was that the meter as calculated over the whole arc from Dunkirk to Formentera differed by under 0.2 mm from the meter as calculated over the lesser arc from Dunkirk to Barcelona. Still unsatisfied, the Bureau des longitudes asked Biot to try his instruments in Britain, on an extension of the arc through Greenwich into the Shetland Islands. He arrived in 1817, on the invitation of the Ordnance Survey, which hoped to use the occasion to finish the match between the repeater and the theodolite. Unfortunately Biot, though appreciated by Mudge ("a very able man and a very diligent observer") fell out with Mudge's lieutenants, who held him responsible for French criticism of certain results of the survey, to be mentioned presently, and Biot took his angles on his own.[129] To the great glory of the British nation, however, the Shetlanders, unlike the Majorcans, understood his mission, and helped. They may have known more about the business than he did, for he located his headquarters on the Island of Unst on the meridian of Formentera, from which it differs by well over two

128. Biot, *Mélanges* (1858), *1*, 51–58, 60 (quote), 64–65 (quote), text of 1810; Ten, in Lacombe and Costabel, *Figure* (1988), 246–263. Arago was not so lucky; he travelled separately and spent some time in captivity; ibid., 75; Frenkel, *HSPS, 8* (1977), 43.
129. Close, *Early years* (1969), 66–68; the quote is from Mudge to Colby, 7 June 1817.

degrees. The error made for good rhetoric in the cause of a Europe united intellectually under French dominance. "The great arc [Biot wrote], starting from the Balearic Islands, crosses Spain, France, England, and Scotland to arrive at the rocks of Ancient Thule. When combined with the flattening of the earth deduced from pendulum measurements or the theory of the moon, will it not give for the fundamental unit, or the meter, the most complete and, if I dare say it, the most European determination that can be desired?"[130]

No ellipsoid of revolution

The data obtained by Méchain and Delambre further perplexed the would-be discoverer of the true shape of the earth. Were the earth an ellipsoid of revolution, the analyst could establish its flattening from two measured arcs and their mean latitudes, as follows. Let a and b be the unknown semi-axes. Then (figure 4.2.3 and equation 4.2.1) the radius of curvature r at a point P at the latitude λ is

$$r = (a^2/b)\left[1 + \frac{a^2 \text{ctn}^2 \lambda}{b^2 + a^2 \text{ctn}^2 \lambda}\left(\frac{b^2}{a^2} - 1\right)\right]. \quad (4.2.5)$$

(Other spheroids than ellipses would of course yield different functional relationships between r and λ.) Each measurement delivered a value of r (via the length of a meridional degree, $2\pi r/360$) and its associated λ; every pair of measurements defined an ellipse a, b, and a flattening $\epsilon = (a - b)/a$.

In 1783, when he first reviewed the measurements, Laplace deduced that $\epsilon = 1/250$, a value apparently confirmed by scattered determinations of the beats of pendulums, which gave Newton's number, 1/230. Six years later, making a new choice among the measured arcs, Laplace obtained $\epsilon = 1/279$, which did not agree with the best from new pendulum counts, which gave 1/359. Adding the result of Méchain and Delambre and ignoring that of Maupertuis (a disrespect later justified by the remeasurement by Svandberg), Laplace obtained the smaller flattening of 1/307.8, which agreed with a calculation by Legendre if not with the shape of the earth. Delambre almost reproduced Laplace's over-precise

---

130. Biot, *Mélanges* (1858), *1*, 65–66, 79 (quote)-83; Delambre to Blagden, 26 June 1808 (Blagden Papers, Royal Society).

number by computing ϵ from his survey with Méchain and the measurement in Peru: $\epsilon = 1/334$, recalculated by Delambre in 1817 as 1/308.64, and by Puissant as 1/309.67. The French had a strong political interest in establishing that all meridians have very nearly the same shape in order to save the rhetoric of the metric system.[131]

British analysts, unencumbered with the meter, doubted that any ellipsoid would do. After comparing eight sets of measurements, Roy declared unequivocally that the earth could not be a rotated ellipse. His successors, the first Ordnance surveyors, Williams, Mudge, and Dalby, concurred: "the measured degrees of latitude in different places are inconsistent with an elliptical meridian." Their redetermination of the distance between the Paris and Greenwich observatories could be accommodated on Newton's ellipsoid, $\epsilon = 1/230$; but that would disagree with the results of the expeditions to Lapland and Peru by an unacceptable 137 ts. Their determinations of the radius of curvature at several latitudes in England required a flattening of $\epsilon = 1/149$, as large as the suspect value of Maupertuis.[132]

Pushing his inquiries further, Mudge ran a meridian from near Bristol to the Isle of Wight. His equipment and technique were of the best: the Ramsden theodolite, a new Ramsden zenith sector, a base line laid out to two inches in over 26000 feet in accordance with Roy's procedures at Hounslow Heath. The result: at 52°2'20", a degree spans 60820 fathoms; at 51°35'18", 60864 fathoms; which is to say, that degrees grow longer towards the equator, exactly the opposite of their expected behavior on a Newtonian earth. How to explain such an embarrassment? Mudge could not cast the blame upon the mountains, for he had encountered only a few hills, so he called upon the oceans. "Meridianal operations carried on in insular countries are not so likely to afford just conclusions with regard to the different length of the degrees, as the same operations conducted in places very remote from deep seas."[133]

131. Levallois, *Vie des sciences*, 3 (1986), 288; Schmidt, *Kartenauf.* (1973), 1, 31; Méchain and Delambre, *Base* (1806), 3, 112; infra, §5.2. Laplace, *Oeuvres*, 2, 148 (text of 1799) tabulates r and λ for all the older measurements except Beccaria's: Lapland, Peru, Mason-Dixon, Boscovich-Maire, Liesganig, Méchain-Delambre.

132. Roy, *PT*, 77 (1787), 201, 207, 209-212; Williams, Mudge, and Dalby, *PT*, 85 (1795), 237-238, 527.

133. Mudge, *PT*, 93 (1803), 383-385, 425-426, 489, 492 (quote); Close, *Early years* (1969), 54.

Mudge's results were variously received. A Spanish mathematician, Joseph Rodriguez, who had accompanied Biot and Arago to Formentera, reasoned that because Ramsden's theodolite gave triangles that usually closed to under one second, Mudge must have blundered with the zenith sector to have ended with degrees that increased toward the equator. Working backwards from the French flattenings of 1/310 and 1/320, Rodriguez made out that Mudge had missed his zenith distances by the unaccountably large margin of four arc-seconds. That called forth a vigorous defense by Mudge's colleague Olinthus Gregory. What could be the motive of the "foreigner" Rodriguez? "[The] object *appears* to be no other than the depression of English (and perhaps other) ingenuity and exertion to the undue exaltation of the French character." Gregory's substantive defense was that Rodriguez had assumed arbitrarily that the earth has the shape of an ellipsoid; in fact, values of the flattening ran from 1/148 to 1/540; even Laplace had admitted, though circumspectly, that no single ellipsoid would fit the data. Seas, mountains, valleys, hills, special strata, all could draw a pendulum off the vertical by the four seconds that wrankled Rodriguez. Moreover, Mudge had been unfortunate in two of the stars on which he sighted; Herschel had shown them to be doubles, and not dependable points.[134] But Rodriguez made little impression on geodecists, who welcomed Mudge's bizarre finding as conclusive proof of the literally over-riding influence of local irregularities. Most knowledgeable people, including Delambre, accepted Mudge's work as reliable and the ellipsoid as illusory.[135]

During the second decade of the 19th century calculations converged on an average value for ϵ around what Laplace and Delambre made it. Von Müffling arrived at 1/315.6 from his mop-up of Tranchot's work between 1814 and 1817; Eckhardt, completing the survey he began in cooperation with Schleiermacher at Darmstadt in 1807, derived 1/309.97 with the help of the technique of least squares; the inventor of the technique moved from 1:187 to 1:334, to settle on 1:302.78 around 1820; and Laplace, exploiting the

134. Rodriguez, *PT, 102* (1812), 325–326, 336–338, and *PM, 41* (1813), 23–24, 30, 90, 95–97; Gregory, ibid., 178–179 (quote), 183–188. Cf. Miller, *BJHS, 16* (1983), 11.
135. Widmalm, *Kartan* (1990), 128; cf. Zach, *Mémoire* (1811), 20–22. According to Bessel, Mudge erred (though not enough to change the direction of the effect) in compensating for nutation; Jahn, *Gesch.* (1844), 2, 190–191.

technique as delivered by Legendre, showed how to combine measurements made by repeating circles and on two or more baselines. Other least squarers, Struve, Bessel, and Muncke, deduced 1/297.648, 1/300.705, and 1/307.7, respectively, the last, by Muncke, from 18 combinations of 7 sets of measurements. William Lambton, who logged the longest arc measured up to 1820, with almost 10° through India, settled on 1/310 after combining his results with arcs in France, England, and Sweden. Various assortments of measurements made around Geneva, Milan, and Padua made for a slightly flatter earth, at between 1/271.31 and 1/292.[136]

The consensus value, $\epsilon\tilde{=}1/300$, was recognized as an average. The earth appeared not to have a regular shape. Franz Xaver von Zach's brother Anton, who conducted a trigonometrical survey for the Austrians around Venice, declared in 1806 that the most that could be claimed for certain about the earth's profile was that the equatorial diameter exceeded the polar. It was he who recommended that the earth's shape be defined, not found, via the method of least squares. This definition, which agreed with one given by Clairaut, was adopted and pushed by Gauss. "On a small scale the earth is certainly no ellipsoid, but deviates in a wave-like way from the ellipsoid that represents the earth at large.... What we call the earth's surface in the geometrical sense is nothing other than a surface that everywhere cuts the direction of gravity at right angles and coincides with the surface of the sea." This wavy figure, the "geoid," has as its closest approximation an ellipsoid with a flattening of 1/298.3. That agrees almost exactly with James Ivory's weighted average of 1/300, and closely with the consensus figure of 1/310 concocted from the various heroic measures made with the techniques, and by the savants, of the ancien régime.[137] The story has the useful moral that not everything in the physical world, not even all well-defined mathematical abstractions of it, may be measurable.

136. Jahn, *Gesch.* (1844), 2, 165, 179–182; Galle, in Gauss, *Werke* (1863), 11:2, Abh.1, 35–36, 48–49; Lambton, *PT*, *108* (1818), 486, 488; Eckhardt, *Astr. Nach.*, *12* (1835), 134; Bru, in Lacombe and Costabel, *Figure* (1988), 226, 236–237.

137. Widmalm, *Kartan* (1990), 154–158; Levallois, *Vie des sciences*, *3* (1986), 284–285, 291, and in *Scientifiques et sociétés* (1990), 430–434; quotes from Gauss to Schumacher, 20 Dec 1823, in Galle, in Gauss, *Werke* (1863), *11*:2, Abh.1, 89, and Gauss, "Bestimmung," in *Werke*, *9*, 49, resp.; Gauss to Olbers, 29 Jan 1822, in Schilling, *Olbers* (1900), 2:2, 165. Leslie, *Diss.* (1842), 776–777.

5 THE MEASURE OF ENLIGHTENMENT

1. PLIGHT OF THE PEOPLE

Cain's legacy

God may have made the world according to weight and measure, but it was Cain who invented weights and measures, and thus—we have this from Flavius Josephus—"converted the innocent simplicity in which man had lived into a miserable existence dominated by fraud and deceit."[1] The existence of French men and women around 1790 was made miserable by, among other things, 700 or 800 differently named measures and untold units of the same name but different sizes. A "pinte" in Paris came to 0.93 liter; in Saint-Denis, to 1.46; in Seine-en-Montagne, to 1.99; in Précy-sous-Thil, to 3.33. The aune, a unit of length, was still more prolific: Paris had three, each for a different sort of cloth; Rouen had two; and France as a whole no fewer than seventeen, all in common use and all different, the smallest amounting to just under 300 lignes, royal measure, the largest to almost 600.[2]

France possessed nonuniform measures in law as well as by custom. Their multiplicity went with other relics of the feudal system, which maintained arbitrary rents and duties usually to the disadvantage of the peasant. A landlord wanted his bushels of grain or

---

1. Josephus, *Jewish antiquities*, cited by Kula, *Measures* (1986), 3.
2. Bouchard, *Prieur* (1946), 286; Talleyrand, *Proposition* (1790), in Miller, *Speeches* (1790) 60–63.

hogsheads of beer in the biggest measures in use in the neighborhood, and he preferred to sell according to the smallest. Nor were all seigneurs above enlarging the vessel in which they collected their rents; and since in many cases they possessed the only exemplars of their patrimonial bushel, no one could be certain that it did not grow in time. But one suspected. A frequent complaint in the *cahiers*, or notebooks of desiderata brought by representatives of the people to the meeting of the Estates General in 1789, was that "the nobles' measure waxes larger year by year." These same representatives castigated the oppressive confusion of customary measures as barbaric, ridiculous, obscurantist, gothic, and revolting, and demanded an end to them, and the establishment of a system of unchanging and verifiable weights and measures throughout the country, or at least throughout their region. Many urged that the King's measure, the royal foot, be made the law of the land.[3]

Sharpers and crooks whose practices were not sanctioned by ancient rights and wrongs and middlemen acting in analogy to money changers opposed the rationalization that menaced their livelihood. In 1747, shortly after returning to Paris with the vision acquired while measuring a piece of a meridian in Peru, La Condamine identified and condemned this special interest, which he proposed to abolish with the confusions that engendered it. His accusation echoed in France for decades and eventually bounced across the Channel. We read in the *cahiers* from Orléans that multitudinous measures "expose[d] people daily to swindlers" and in the records of Parliament that John Riggs Miller, one of its obscure and verbose Members, declared that they had but one purpose, "the perplexing of all dealings, and the benefitting knaves and cheats."[4]

Reformers laid down several requirements for a new system of weights and measures. It should not rest on an arbitrary unit, especially not on a king's foot; it must not offer enticements to cheaters; and it had to be easily reproduceable were its exemplars lost. Further, it had to be rational, so as to recommend itself to all nations, and become universal. The "toise de Pérou" had attained

3. Kula, *Measures* (1986), 164–226, esp. 191–196, 230, 236; Hyslop, *French nationalism* (1968), vii, 56. Cf. Garnier, in *Les mesures* (1984), 10.
4. La Condamine, *MAS*, 1747, 492–495, reported favorably in *HAS*, 1747, 82–88. Kula, *Measures* (1986), 209; Miller, *Speeches* (1790), 18.

some currency in France and in a few other countries, and the units used in Paris also had more than local authority. It would not do to impose them, however, as Talleyrand wrote Miller, since they had not been derived from nature or constructed "with the ceremony necessary to settle once and for all the opinion of all enlightened nations." Last and also first, the reformed system had to be simple, or, as Miller preferred to say, "on a level with the lowest and humblest capacity." It must not require "skill in calculation beyond what...the inferior orders of men commonly possess;" everyone should be able to confirm for himself the correctness of all transactions of interest to him, "the meanest intellect...on a par with the most dexterous."[5]

This last paragraph contains many buzz-words of the Enlightenment.[6] The replacement of the arbitrary and the capricious, of the feudal and historical, by the natural is the message of all the philosophes from Montesquieu to Condorcet. The natural coincides with the rational and the universal: when people cast aside customary belief and established abuse, they can reach agreements that all others, guided by their own reason, will accept. This reason is not the property of a few great intellects; everyone has the right to know, and to recognize, the truth. Any system that claims universal assent must be universally intelligible.

Mathematics and the rights of man

From the most remote times philosophers have taken number as the exemplar of the intelligible. Once one has grasped a principle in geometry, it was said, not even God Himself could understand it better, although, to be sure, He might know more theorems. Calculating people think for themselves; they despise the unintelligible, capricious, unfounded, authoritarian, and feudal as infringements on their thoughts and actions. In brief, mathematics is a science for free people. Or, to say the truth as the 18th century saw it, for free men. Newton's doctor, Sir John Arbuthnot, praised mathematics for giving "a manly vigour to the mind;" a tonic Newton took to such good effect that he mastered all "the noble and manly sciences" and became "the greatest man that ever liv'd."[7] All this will help to

5. Talleyrand, *Proposition*, in Miller, *Speeches* (1790), 68–69; Miller, ibid., 16, 35.
6. Crosland, *Stud. Voltaire eight. cent.*, 87 (1972), 297.
7. Quoted from, respectively, John Arbuthnot, *Works* (1751), 1, 9, 36, and James Jurin, *PT*, 34 (1727), dedicatory epistle to Martin Folkes.

construe the remark made by the representatives of the revolutionary government in 1799, on accepting the exemplars of the meter, liter, and kilogram. The metric measurers, they said, reaching for their highest compliment, had carried through their work "with the confidence of a male and republican spirit."[8]

Perhaps the most significant sign of burgeoning rudimentary numeracy during the 18th century was the multiplication of tables of numerical equivalence. A good survey of the tables in use in commerce has yet to be made; but there is no doubt that their number increased dramatically after the Seven Years' War. They came in several sorts: conversions of weights, measures, and moneys; total price of goods, tabulated by size and unit cost; tables of interest and annuities; agricultural yields; and so on. It did not require great prowess at mathematics to use these compilations; rather, a degree of numeracy comparable to the literacy of one who could read but not write. The spread of the metric system depended on this widespread rudimentary numeracy, and raised its level.

Commercial tables were required not only because of the number and uncertain equivalence of feudal weights and measures, but also, and perhaps primarily, because the arbitrary multiples and submultiples of the various units made computation burdensome and complex. Calculation of the price of a piece of cloth 2 yards 1 foot 4 inches square at 3 pence 2 farthings the square foot was a sufficient challenge. To change it into aunes, pieds, livres, and deniers, and to proceed to a problem in bushels and cubic king's feet, would have puzzled Archimedes. According to the Paris Academy, referring to the situation in 1790, people at ease with money computations could not handle weights and measures. "In the present state of affairs, a man who can calculate with sous and deniers cannot calculate with toises, pieds, pouces, and lignes, with livres, onces, gros and grains."[9]

The Paris Academy and many other scientific reformers supposed that by dividing the new standards and the revised coinage decimally they would eliminate the need for specialist computers.

---

8. P.L.C. Baudin, 4 messidor an 7 (1799), in Méchain and Delambre, *Base* (1806), 3, 650.

9. Borda, Lagrange, Lavoisier, Tillet, and Condorcet, *HAS*, 1788, 5 (text of 1790).

The decimal was not free from arbitrariness; but its simplicity and convenience could not be gainsaid, at least by practiced calculators, and, as the Academy observed, although not universal it is as natural as the human hand.[10] Only the hand of the learned had so far employed decimal arithmetic, and by no means universally, as Lavoisier pointed out in his *Elements of chemistry*, when urging his colleagues to state weights in decimal parts of whatever units they used. This natural arithmetic, "previously locked up in the domain of the sciences," was precisely what the reformers thought they sought. "Those who knew little will know everything; others will hurry to forget what they no longer need to know; all will accept as a true benefit a method of calculation that will save them time, study, and chances for error."[11]

Prieur de la Côte d'Or, the collaborator of Guyton de Morveau who had graduated to the all-powerful Committee of Public Safety (Comité du salut public), expected the decimal calculus to be the technical language of Utopia. "How happy we will be not to be forced to consult anyone about our prosperity, property, expenses, and drink, and to have nothing to do any more with people who often seek only to profit from our ignorance."[12] The connection between democracy and the decimal was made plain and explicit by Condorcet during the first year of the Revolution. Decimalization, he said, fit perfectly with the political program and mandate of the National Assembly. "It [the Assembly] wants to insure that in the future all citizens can be self-sufficient in all calculations related to their interests; without which they can be neither really equal in rights...nor really free."[13] Long after the promulgation of the metric system, Laplace advised Napoleon that its chief advantage as understood by its creators was not the destruction of feudal metrology but the division by tens.[14]

In order that the people be reminded of the benefit every time they used it, the new measures carried prefaces that designated

10. Ibid., 6; cf. Crosland, *Stud. Volt. eight. cent.*, 87 (1972), 299.
11. France, Comm. temp. des poids et mesures republicaines, *Instruction abrégée* (an ii), xvii, 171; Lavoisier, *Elements* (1790), 295-296.
12. Prieur, *Instruction* (an iii), 4-5; cf. Miller, *Speeches* (1790), 39, and supra, §2.2.
13. Condorcet, *Mémoires sur les monnoies* (Paris, 1790), 3-4, quoted in Champagne, *Role* (1979), 60.
14. Text of 7 May 1811, in Bigourdan, *Système métrique* (1901), 193.

decimal multiples or sub-multiples of the fundamental units. Initially the Academy reckoned that the beneficiaries would not be able to grasp the system, and so recommended common names, like "doigt" and "palme," rather than "deci-," "centi-," and "milli-," for submultiples of the meter. They erred in calculating revolutionary zeal. The Committee on Public Instruction (Comité d'instruction publique) preferred a clean sweep and adopted the systematic names. This occurred on August 1, 1793, on the eve of the suppression of the Academy.[15] One of the Academy's successors in metric matters explained that "it is almost impossible to reason correctly without a language aptly made."[16]

The language did not please the people. Classicists objected that the prefixes (enriched by "déca-," "hecta-," and "kilo-") violated the grammar of ancient languages, while the uneducated could make no sense of them at all. "These names," declared a delegate to the Convention, "novel and unintelligible to the large majority of our citizens, are not necessary for the maintenance of the Republic."[17] He did not know what he had escaped. Prieur de la Côte d'Or, who had taken an active part in the reform of weights and measures from the onset of the Revolution, had names of his own, derived, he said, from ancient languages and Low Breton. The irrationality of the Convention did not extend to adopting "kilicymbe," "myriadore," "ladedix," "pèzeprime," "centicadil," or "decidol," and Prieur, bending to the political wind in 1795, drew up what became the definitive metric names. The many centicadils of bitterness he then swallowed and decidols of crow he then ate were to damage the work of the metric reformers.[18] All concerned might have spared themselves the trouble. For decades the people stubbornly opposed decimalized units and their jabberwocky names.

---

15. HAS, 1789, 1–18 (text of 1792); Bigourdan, Système métrique (1901), 78–82.
16. France, Agence temp. poids mes., Aux citoyens (an iii), 10.
17. Bigourdan, Système métrique (1901), 82 (text of 11 Aug 1795).
18. Bouchard, Prieur (1946), 297–299; Delambre, Grandeur (1912), 212.

Republican time

Republican zeal is not easy to curb, and the decimilization of everything measured or metered figured among the excesses of the French Revolution. The Paris academicians demonstrated solidarity with the regime by dividing a right angle into 100 revolutionary degrees, and each such degree into 100 minutes; and they found much pleasant recreation in recomputing the trigonometrical functions in what Delambre later extolled as "the vastest [calculation] that had ever been done, or even conceived."[19] The cost of conversion to decimal units had the merit of expressing angles in the arithmetic system in which computations on them were performed. Academicians and surveyors appreciated the scheme, which still exists in France along with the sexigesimal, or Babylonian division of the circle.

A less enduring but farther reaching reform tried to rescue the calendar and timekeeping from the Babylonians and the Church. Even before the Revolution, anti-clerical almanack makers attributed days not to saints but to people they considered as benefactors of mankind. One such production, the *Almanac des honnêtes hommes* of 1787, which jumbled together Moses and Voltaire, Descartes and Aristotle, Newton and Cassini, Christ and Campanella, earned the attention of the public executioner. "This monstrous assembly," said the King's attorney, "advertises a project formed some time ago to destroy the Christian religion by ridicule....I blush to give an account to this court of the absurd and revolting consequences of this work of impiety, atheism, and madness."[20] This reasoned discourse will give a measure of the seriousness of tampering with the calendar.

In the autumn of 1793 the Committee of Public Instruction undertook to tamper. It was emboldened by several considerations. For one, almanack makers, newspaper editors, and politicians had for some time dated their statements from the beginning of the revolutionary era or, more recently, from the declaration of the Republic on September 22, 1792. Some worked from January 1,

---

19. *Exposition* (an x), 13–14; Delambre, *Rapport historique* (1810), 8.
20. Quoted in Villain, *La Rev. française*, 7 (1884), 538, 540–541; in general for the old almanacks and the new calendar, ibid., 457–459, 535–553, and 8 (1885), 623–631, and Andrews, *Amer. hist. rev.*, 36 (1931), 517–522.

250 / THE MEASURE OF ENLIGHTENMENT

others from July 14, 1789; some called 1793 the fourth year, others the second year, of the new epoch. Again, the heating up of revolutionary rhetoric condemned the old calendar not only for its saints' days, but also for its preservation of ancient superstition in the form of the seven-day week. And then there were arguments drawn from the fountain from which the metric reformers had drunk: the old calendar with its months with crazy lengths, its 24-hour days, and its bizarre rules for intercalation, were an affront to both reason and nature. On September 20, 1793 the Committee heard proposals for reform drawn up by Gilbert Romme with the assistance of academicians Alexandre-Guy Pingré (an old astronomer), Guyton de Morveau (chemist), and Lagrange and Monge (mathematicians).[21]

The report opens with a reference to the metric reform, "one of the most important operations for the progress of the arts and the human spirit." The Committee must do no less for the calendar. "Arts and history, for which time is a necessary element or instrument, also ask you for new measures of duration, which are equally free from the errors that credulity and superstitious custom have transmitted to us through centuries of ignorance." By great good fortune, "perhaps unique in the world's annals," the declaration of the republic and the end of monarchy had occurred on the very day of the autumnal equinox of 1792. "Thus the equality of days and nights was marked in the heavens at the very moment when the representatives of the French people proclaimed civil and moral equality as the sacred foundation of its new government....Thus the sun passed from one hemisphere to the other on the same day that the people, triumphing over the oppression of kings, passed from a monarchical to a republican regime." Nature and politics identified the day of the true equinox as the first day of the revolutionary year and, retrospectively, the equinox of 1792 as the beginning of the republican era.[22]

It remained to apply the decimal. "You have felt all the advantages of decimal numeration. You have adopted it for all weights and measures, and for the Republic's money; we propose to you to

21. Villain, *Rev. fran.*, *8* (1885), 635; Andrews, *Amer. hist. rev.*, *36* (1931), 524; cf. Zernhavel, *Rhythms* (1981), 84–86, and Baczko, in Nora, *Lieux* (1984), *1*, 39.
22. Guillaume, *Proc. verb.*, *2* (1894), 440, 442; Villain, *Rev. fran.*, *8* (1885), 636, 639.

introduce it into the division of the month." Each will have 30 days divided into three parts or *décades* of ten days each. Thus, in one blow, the irrational unequal ancient months will perish, and also the astrological seven-day week. And that is not all, or enough. "Perfection will be complete when time is submitted to the simple and general rule of dividing everything by tens." The day should be divided decimally. For the rest, the year should be given its true complement of days by adding five to the twelve months in ordinary years and six when required to keep the new year on the first day of the first month.[23] All this became law on October 5, 1793, which required that public acts and documents thenceforth be dated in the new style. Decimalization of time would begin on the first day of the third year of the Republic.[24]

As this last formulation indicates, the new calendar gave rise to awkward expressions for dates. It would be convenient to have names for the months and perhaps also for the days, if a rational nomenclature could be devised and, what is much more, deployed. The problem prompted an instructive argument in the Committee of Public Instruction. Some members, inspired by the revolutionary almanack makers, proposed to name the months after revolutionary heroes or virtues. Their opponents objected that the choice would be arbitrary and, necessarily, peculiar to the recent experience of the French. That would not recommend the scheme to the rest of Europe. But if the months and days were numbered rather than named? "Your calendar, which would have been that of the French nation, will become the calendar of all people. They will never abandon the numerical order, which is the natural one.... Your philosophical calendar can become the basis of the universal republic."[25] The Committee at first voted for moral names then reversed itself and stayed numerical. It also asked a sub-committee to consider the matter further.[26]

A report was submitted in the name of the sub-committee by P.F.N. Fabre d'Eglantine on the third day of the second month of

23. Guillaume, *Proc. verb.*, 2 (1894), 442–445; Villain, *Rev. fran.*, 8 (1885), 640–642.
24. Guillaume, *Proc. verb.*, 2 (1894), 448–450, 582–584.
25. Member Duhem, at discussion of 5 Oct 1793, in Guillaume, *Proc. verb.*, 2 (1894), 585–586 (quotes).
26. Ibid., 580–581, 586–587.

the second year of the Republic. It recommended month names derived from meteorological conditions and French euphonics, to wit: months 1–3, autumn, Vendémiaire, Brumaire, Frimaire; months 4–6, winter, Nivôse, Pluviôse, Ventôse; months 7–9, spring, Germinal, Floréal, Prairial; months 10–12, summer, Messidor, Thermidor, Fructidor. The days of the décade would answer to "Primidi," "Duodi," etc., up to "Décadi;" and each day of the year would be associated with a different agricultural implement or product, vegetable, fruit, or tree. Thus the calendar would become a vehicle of natural history, a reminder "not of the nation's cult, but of its culture" (figure 5.1.1), an occasion for disquisitions on mineralogy, physics and husbandry.[27] The scheme became law on 4 Frimaire an ii (24 November 1793), together with a specification of the revolutionary day: ten days composed of 100 decimal minutes each composed of 100 decimal seconds. The names of the months became current, but not the day names, associations, or decimal divisions.[28]

Toward the end of Frimaire, the Committee on Public Instruction published a *Calendrier de la République Française*. It combined rhetoric taken from Romme's report of the preceding September with several technical tables: conversion of Gregorian dates to Fabre's calendar for an ii; concordance of revolutionary and Babylonian time; a table of occurrence of the true autumnal equinox, from 1792 to 1804, in sexigesimal and in decimal hours; and a comparison of intercalations on the Gregorian and revolutionary styles. As an indicator of the advantages of the new scheme, the *Calendrier* gave the epact (the moon's age) on 1 Vendémiaire an ii and calculated the age on 23 Prairial an ii. Since Prairial is the ninth month, and since earch month is about a half day longer than a mean lunation, the moon is $17 + 23 + 8/2 = 44$ days old on the day assigned, or rather $44 - 29.5 = 14.5$ days old, "to sufficient [accuracy] for domestic purposes." There was also guidance for the perplexed horologist. A single clock face divided into 100 parts could mark the hour by the hundredth of a circuit, the day by a tenth, and the décade by the

---

27. Millin, *Annuaire* (an ii), viii–xii. The bulk of this volume consists of articles explaining each of the names—372 in all—given by Fabre d'Eglantine to the months and days.

28. Guillaume, *Proc. verb.*, 2 (1894), 697–713, quote on 703, 873–875; Villain, *Rev. fran.*, 8 (1885), 654–655, 740–749, 751.

THE MEASURE OF ENLIGHTENMENT / 253

FIG. 5.1.1 Allegory of the reform of the calendar. The legend on the building reads "Temple de l'année, dédié aux mois et jours;" the new names of the months, Floréal, Prairial, Messidor, etc., appear on the tops of the columns. In front of the temple, Liberty, holding Reason by the hand, looks down at popes, tyrants, and saints writhing in the mud, while a half-naked nature presents to Liberty and Reason a crowd of agricultural laborers and their implements, after which many days in the new calendar were named. Millin, *Annuaire* (an ii), frontispiece.

whole, all indicated by one hand; a second hand could mark decimal minutes and a third decimal seconds. Conversion between old and new styles required only a single hand moving across a clock face marked in sexigesimal and decimal time in concentric circles (figure 5.1.2).[29]

A few clocks beating decimal seconds were made, primarily for the use of academicians. Lavoisier had one. Lenoir made some for Méchain and Delambre, and another instrument maker, Ferdinand Berthoud, produced one for help in determining the length of a seconds pendulum. The Commission temporaire observed that the length of a decimal-second seconds pendulum must be shorter than that of the usual sort, "which will perhaps be more favorable to observations, and will certainly make astronomical clocks more convenient, since their pendulum will be shorter." The system had the merit of variety. Biot used both decimal and sexigesimal watches in his experiments on the speed of sound down pipes. Apparently bored with the figures obtained from the usual timekeeper, he employed both sorts, "to vary the numbers observed."[30]

Alas! The new ways of telling time did not recommend themselves to those it was designed to serve. The people declined to liberate themselves from the calendrical thrall of the Church. They might give up Saints' days, might replace the feast of John the Baptist with one in honor of the modern-day Salomé, the good doctor Guillotine; but they would not give up their Sundays. The government fulminated against a Sunday Sabbath, ordered public offices to remain open, insisted upon shutting down public business, markets, and theaters, on décadis, and sponsored decadal celebrations of virtue and heroism. The people would have none of it: they had their habits, and they preferred 52 days of rest per year to 36. The Convention and the Directorate found it more difficult to enforce the civil calendar on their citizens than to defeat the armies of Europe. Nor did the academicians and their fellow travelers fully adopt to revolutionary time. In 1793, the Académie offered a prize for a "pocket watch, capable of finding the longitude at sea, whose

29. Guillaume, *Proc. verb.*, 2 (1894), 875–892.
30. France, Comm. temp. poids mes., *Instruction* (an ii), 33–34, and *Avis* (an iv), 8; Truchot, *ACP*, 18 (1809), 319; Biot, *MSA*, 2 (1809), 411. The Geneva watch trade, fallen on bad times, tried to find a market for decimal clocks; Baczko, in Nora, *Lieux* (1984), *1*, 49.

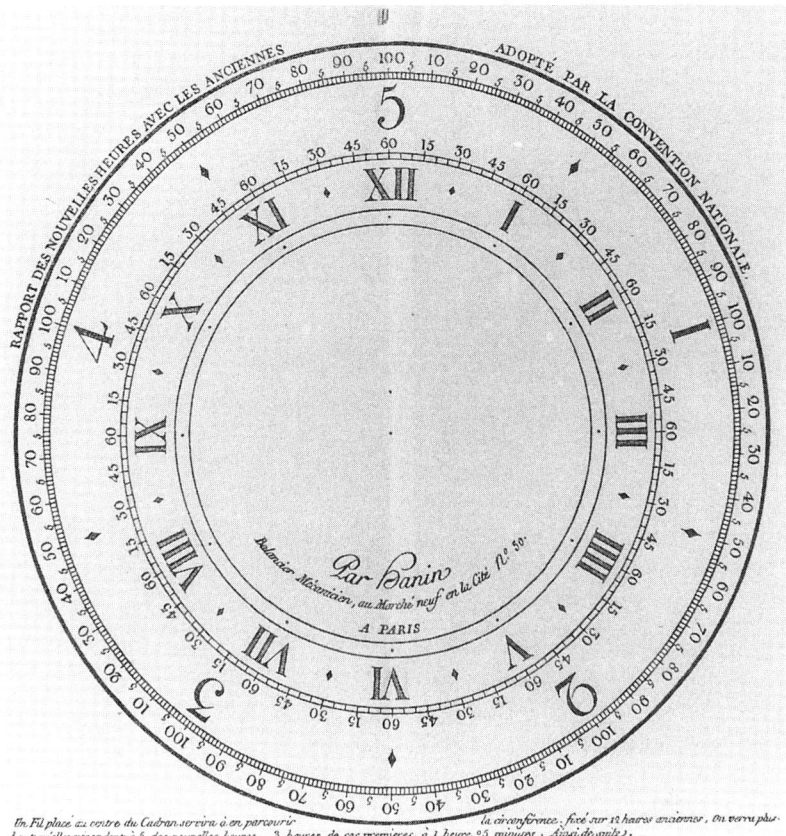

FIG. 5.1.2 Clock face, "adopted by the National Convention," showing conversion of old and new hours. The old hours I–XII, divided into 5-minute intervals, appear along the inner circles; the new hours 1–5 (for a half a day), divided into 100 minutes, occupy the outer circles. Guillaume, *Procès-verbaux* (1894), 2, 893.

divisions mark the decimal parts of the day." No one won. In 1798, the Academy reopened the competition, but dropped the requirement of decimilization of time, "to diminish as much as possible difficulties that might keep some competitors away." In 1802, as part of his accommodation with the Vatican, Napoleon dropped sanctions against religious observances on Sundays, and four years later, on 1 January 1806, the people had their way and the state repealed the calendar.[31] Revolutionary time had already been suspended, sine die, on 18 germinal an iii (April 7, 1795), as being of greater interest to science than to society.

The reasons given by the government for seeking repeal of the revolutionary calendar bear on the character of the metric reform. In systematizing weights and measures, they said, everything parochial had been discarded; nothing remained to identify the nation that produced it. But the calendar betrayed its origins, especially in the setting of the epoch by the destruction of the monarchy. Moreover, most knowledgeable savants opposed it because, despite the then recent progress of astronomy and the vast importance of instruments of angular measure, astronomers of the new era could not implement its calendar. As Delambre had argued in 1795, as soon as the political situation allowed him to criticize the work of the Republic, astronomers could not be depended upon to determine the sun's position to better than three seconds of arc, which corresponds to more than a minute of time. And, since it would take them several days to complete their calculations in close cases, their fellow citizens would not know for some time when the new year had begun. Also, taking the occurrence of the true equinox as the beginning of the year would introduce irregularities since the length of the year varies by as much as twenty minutes owing to perturbations of the earth's orbit by the constantly shifting planets.[32] Neither nature nor mathematics favored the revolutionary calendar. On 15 Fructidor an xiii (3 September 1805), Napoléon's Senate named its most illustrious savant, Laplace, who, as Minister of the Interior, had ordered his underlings to enforce the decadal festivals and the republican calendar, to report on the proposal to terminate the revolutionary era. Cleaving to his role as specialist, he located the chief fault of the new style in its method of intercalation. To be sure, there was another problem, indeed a serious inconvenience, of the revolutionary calendar: it put France out of step

31. Villain, *Rev. fran.*, 8 (1885), 754–757, 832, 835–838, 842–854; Guillaume, *Proc. verb.*, 2 (1894), 752, 804–806; *MIF*, 1 (an vi), i (quotes).
32. Froeschlé, in *Scientifiques et sociétés* (1990), 458, 461–464.

with all of Europe. The astronomer and the politician, the mathematician and the man of business, would all be best served if the Empire kept the days like everybody else. The only downside Laplace could see was the fear that the recall of the calendar would prompt the cancellation of the metric reform.[33] As will appear, the people had no more desire to trade the toise for the meter than to substitute *décadis* for Sundays.

2. ACTION OF THE ACADEMY

An anchor for revolutionary storms

In exchange for their salaries, status, and grants for special projects like geodetic surveys, the Paris academicians acted as technical advisers to the Crown. Their advice tended to be conservative and elitist, especially in respect of unsolicited proposals from inventors asking for state subventions or monopolies for their novelties. Experience confirmed what arrogance had suspected: most of the proposals were worthless and most of the proposers ignorant. The Academy suggested that artisans might be licensed, but only after passing an examination in geometry and other useful arts. During the Revolution, the Academy faced the hostility directed at all enclaves of privilege and the special anger of frustrated inventors, manufacturers, and would-be scientists who had suffered at its hands.[34]

Well into 1790 the Academy held its meetings, read papers, and planned projects as usual. Some academicians favored the new regime, but few wished it to go further; almost all lamented the disturbances that kept them from their work. Many recognized, however, that they would have to reorganize their company in keeping with the new style, especially after the establishment in 1791 of a patent law that took from them their chief public service.[35] As a partial offset, academicians became increasingly involved in

33. Villain, *Rev. fran.*, 8 (1885), 885–887; Crosland, *Society* (1967), 63.
34. Gillispie, *Science and polity* (1980), 97–99, 461–466; Hahn, *Anatomy* (1971), 118–121; Gillispie, in Clagett, *Critical problems* (1959), 257, 268–274.
35. Molard, *Description* (1811), 1, 7–27; Hahn, *Anatomy* (1971), 186–189; Guerlac, *Essays and papers* (1977), 467–468.

government technical projects. Of these, the most important, "repeatedly paraded as a prime example of science's potential value to the nation and a concrete instance of the Academy's proper function in society," was the reform of weights and measures.[36]

A month after the storming of the Bastille, academician Jean Baptiste Le Roy, physicist, mathematician, and one-time clockmaker, suggested that the Academy propose to the National Assembly a dissolution of the over-rich metrological heritage of the Republic. The proposal, probably drawn up by the Academy's secretary, the marquis de Condorcet, and certainly presented to the Assembly by Talleyrand, set aside all existing units in favor of the length of a pendulum that beat seconds at the 45th parallel of latitude, which passes just north of Bordeaux. Talleyrand proposed further that the British be invited to join in the determination and in promoting the result, "so that all nations might adopt it."[37]

On May 8, 1790, the Assembly considered the Talleyrand or Academy proposal together with several others to the same effect, notably one by Prieur, who expressly opposed using an arc of the meridian as a basis. "Besides the magnitude of the fundamental operation required, the difficulty of verifying it, and the impossibility of doing so daily, it is not easy to decide how exact the method might be." Here Prieur spoke as a military engineer familiar with surveying practice and with the lingering uncertainies over the earlier measurements of arcs. The same considerations caused Thomas Jefferson to give up his project of taking a decimal part of Cassini's degree as the basis of a new American unit: "the various trials to measure various portions of [meridians], have been of such various result, as to show there is no dependence on that operation for certainty."[38]

The seconds pendulum had been proposed as a standard setter for over a century. The pioneer geodecists preferred it: Picard, in 1671, proposed defining the pied as a third, and the toise as twice, the length of the seconds pendulum at a convenient place; Cassini observed that all domestic measures could usefully be referred to

36. Hahn, *Anatomy* (1971), 162–163.
37. Hahn, *Anatomy* (1971), 163–164; Bigourdan, *Système métrique* (1901), 14–15; Talleyrand, in Miller, *Speeches* (1790), 59.
38. Bouchard, *Prieur* (1946), 287–290; Pattison, *Beginnings* (1957), 48–49; Bigourdan, *Système métrique* (1901), 10.

such a unit, and even all European measures, since the length of a seconds pendulum is about the same throughout the Continent; La Condamine proposed international cooperation and a pendulum regulated at the equator, as determined by himself and "the hands of nature;" and Turgot, as controller of finance under Louis XVI, almost initiated a nationwide reform based on the seconds pendulum at 45°.[39] Just before the Revolution, the engineer Gaspar de Prony, who would play an important part in the metrication of France, declared the pendulum to be the ultimate arbiter of length; and after the promulgation of the meter, a compiler of dictionaries, no doubt lifting from his predecessors, held that "the length of the simple pendulum, an invariable quantity always easy to recover, seemed given by nature to serve as a measure in all countries."[40] The British had also considered the advantages of Cassini's suggestion, and of calibrating their yard by the pendulum, in order that "all future generations [may] obtain similar measures of length, capacity, and weight, and thereby render it altogether needless to cut them on stone, or to engrave them on brass, to perpetuate their existence."[41]

Miller was about to introduce proposals for the reform of English weights and measures based on the pendulum when a letter from Talleyrand inviting Britain to join France in finding the length of a seconds pendulum came to hand. The plan was simple, and a natural successor to the linking of the observatories of Paris and Greenwich; "all Europe would take [it] as a guarantor of rigorous exactitude."[42] While Parliament pondered the opportunity, the Revolution became too hot for Anglo-Saxon reformers, and the Paris Academy removed the rationale for collaboration by deciding not to take the pendulum as primary. Its decision seemed odd to many.[43] The rationale must be sought, not in the requirements of measurement, but in the circumstances into which the Revolution propelled the Academy.

39. Bigourdan, *Système métrique* (1901), 6–11; Cassini, *Grandeur* (1720), 250; La Condamine, *MAS*, 1747, 501–505, 511; Henry, *Correspondance* (1883), xxv, 234–235.
40. Prony, in Roy, *Description* (1787), xvi–xvii; Lunier, *Dictionnaire* (1806), 2, 615, s.v. "mesure."
41. Whitehurst, in *Works* (1792), iii (text of 1787); Miller, *Speeches* (1790), 44.
42. Talleyrand, in Miller, *Speeches* (1790), 71.
43. Miller, *Speeches* (1790), 43–44; Shuckburgh, *PT*, 88 (1798), 165–166.

The National Assembly accepted Talleyrand's proposal and sent it and a question about the most useful division of weights, measures, and monies to the Academy, which referred both matters to a committee composed of Borda, Lagrange, Lavoisier, Tillet, and Condorcet. The committee reported on October 27 that everything should be decimal.[44] It then handed what remained of its charge to a committee consisting of Borda, Lagrange, Laplace, Monge, and Condorcet. On March 19, 1791, these geometers reported that the pendulum was the poorer of the two universal and natural units they could imagine. They plumped instead for a piece of the Paris meridian and yet another measurement of it.

Their objections to the pendulum suggest a hidden agenda. In obtaining a length from pendulum beats, they wrote, a unit of time, which has nothing to do with distance, must be invoked; and this unit, the 86,400th part of a day, had the additional blemishes of being both arbitrary and nondecimal. "It is much more natural, in fact, to refer distances from one place to another [the academicians were thinking of a standard for cartography, not for commerce] to a quarter of a terrestrial circle rather than to the length of a pendulum." The committee did appreciate the greater convenience of the pendulum standard, for which they provided: the Academy would undertake to determine the length of a seconds pendulum at the Paris Observatory, to serve as a secondary reference, the primary to be one ten-millionth of the distance from pole to equator.[45]

Although the ambition of the academicians did not extend to measuring the full quarter meridian, it opened a task sufficiently large. Borda's committee proposed to redo the arc from Dunkirk to Perpignan and to extend it to Barcelona; to lay out new baselines to observe the pendulum; to determine the weight of an exactly measured volume of distilled water at the temperature of melting ice; and to compare all the old units in use in France with the new standards. They saw no advantage in British cooperation: "we have excluded from our advice every arbitrary determination, we have used only the common property of all nations.... In a word, if the

44. Borda et al., *MAS*, 1788, 1–6; Méchain and Delambre, *Base* (1806), *1*, 14; Bigourdan, *Système métrique* (1901), 16.

45. Borda et al., *MAS*, 1788, 1–6 (decimalization), 7–16 (arc over pendulum), texts of 1790. Cf. Bigourdan, *Système métrique* (1901), 17–18; Méchain and Delambre, *Base* (1806), *1*, 14–19.

memory of all our work disappeared and only the results remained, they would disclose nothing to show what nation conceived the idea and carried it through."[46] The Academy as a whole did not readily accept the recommendations of its committee of interested mathematicians: some objected that the arc had received enough attention; others, that the pendulum was easier. In the end, however, the Academy endorsed the recommendations and sent them to the National Assembly.[47]

The speciousness of the argument that the choice of Borda's committee was the most satisfactory and least arbitrary stands forth from their anticipation of the charge that enlightened people everywhere might not regard a section of the meridian through Paris and lying almost entirely within France as a unit dictated by nature. They argued: the section should extend equal distances on either side of the 45th parallel because there the seconds pendulum and the size of a degree have their mean values; it is a mere coincidence that the 45th parallel runs through France. But all meridians are bisected at 45°. Why take one through Paris? Because only there do meridians have arcs bisected at 45° that terminate at either end at sea level and that are short enough to measure. "There is nothing here that can give the slightest pretext for the reproach that we wished to assert any sort of dominance." Or, as Laplace put the point in a lecture in 1795, "had savants from all countries come together to fix the universal measure, they would not have made a different choice."[48] This flim-flam was perfectly clear to Jefferson: "The element of measure adopted by the National Assembly excludes, *ipso facto*, every nation on earth from a communion of measurement with them; for they acknowledge themselves, that a due portion for admeasurement for a meridian crossing the forty-fifth degree of latitude, and terminating at both ends at the same level, can be found in no other country on earth but theirs."[49]

46. Borda et al., *MAS*, 1788, 13–16, 19, text of 1790.
47. Bigourdan, *Système métrique* (1901), 19–21.
48. Borda et al., *MAS*, 1787, 15–16, text of 1790; Bigourdan, *Système métrique* (1901), 56; Laplace, *Oeuvres* (1878), *14*, 141, 145 (quote).
49. Jefferson to William Short, 28 July 1791, quoted by Hellmann, *Osiris, 16* (1931), 286. The academicians' grantsmanship has succeeded with some historians: Bassot, in France, Bureau des Longitudes, *Annaire*, 1901, D. 2, D. 16; Gillispie, in *DSB*, 15, 334–335, s.v. "Laplace;" Konvitz, *Cartographie* (1987), 47.

Biot wrote in 1803, when surveying the progress of science since the Revolution: "if the reasons that the Academy presented to the Constituent Assembly were not altogether the true ones, that is because the sciences also have their politics: sometimes to serve men one must resolve to deceive them." Biot, whose democratic ideology inclined him to openness, gave as the hidden agenda the Academy's wish to settle the shape of the earth once and for all.[50] According to Delambre, Borda convinced his committee to opt for the arc because his circle, as suggested by the Paris-Greenwich measurement of 1787, made possible a determination of the meridian far more accurate than Lacaille's. This consideration left a trace in the committee's report to the Academy in March 1791. Today's instruments, it said, are so good that future improvements would not sensibly change the length of the meter that they determine; "or at least the length of time separating us from an age when everyday transactions would require and could attain such a precision is so great in comparison with the life of a man as to amount to infinity itself."[51]

To these objectives—the old scientific imperative and the desire to vindicate and promote French instrumentation—should be added the social and strategic concerns of forging bonds with the new state. The metric project had high priority since the abolition of feudal rights had raised the gathering of rents and taxes in kind to a new level of confusion and litigation.[52] During the several years the project would last, the Academy could expect to enjoy strong government support and a useful flow of cash. On August 8, 1791, some 100,000 livres, more than the Academy's annual state subvention, was placed at its disposal; estimates of the entire cost of the project ran from 300,000 livres into the millions; and one can scarcely criticize the Academy if it saw in this commitment a pledge

50. Biot, *Essai* (1803), 355–356. Cf. Moreau, in *Scientifiques et sociétés* (1990), 120–121, 125–127, 143; Legendre, *MAS*, 1788, 753; and Rodriguez, *Phil. mag.*, 41 (1813), 22: "At the commencement of the French Revolution, men of science took advantage of the general impulse...and they proposed making a new measurement...."
51. Delambre, *Rapport* (1810), 5; Crosland, in Bugge, *Science* (1969), 20; Borda et al., *MAS*, 1788, 15, text of 1791.
52. Kula, *Measures* (1986), 236–237, 243.

on the part of the government to see it through troublous political times.[53]

Fall and rise

The National Assembly approved the Academy's revised proposal on March 30, 1791. The Academy immediately divided its work among five commissions: triangulations and latitude determinations (Cassini, Méchain, Legendre), baselines (Meusnier, Monge), pendulum of Paris (Borda, Coulomb), weight of water (Lavoisier, Haüy), and comparison of old and new measures (Tillet, Brisson, Vandermonde).[54] The whole enterprise was to be directed by Borda, Condorcet, Lagrange, and Lavoisier. Soon death and disinclination reduced this extraordinary mobilization of the brains of France. All geodetic work fell to Delambre and Méchain, Borda and Cassini took on the pendulum, and Lavoisier the water. The first year went by making instruments. Only one Borda circle, that of 1787, existed in 1791; two more, readable to 3 or 4 seconds of arc, were ready in the summer of 1792, when Méchain set out for Spain and Delambre started north. The instrument answered its advertisements: although difficult to use in a cramped church tower, its precision with enough repetitions was, in Delambre's words, "nearly incredible."[55]

Great accuracy was also achieved in the pendulum experiments, which took place during the summer of 1792 in a pit in the Paris Observatory. The general technique, by no means original with the metric project, was to observe coincidences between a swing of a simple pendulum of length L and that of a clock that accurately beat seconds. Let t_1 be the clock time when both pendulums move through the midpoint of their swing together, and let t_2 be the time of the next coincidence. In the interval $\Delta t = t_2 - t_1$, therefore, the number of swings of the two pendulums differs by one and the half period of the simple pendulum is $T = \Delta t / (\Delta t \pm 2)$. (A seconds pendulum has a period of 1/2, that is, half a swing lasts a second.) Thence the length of the seconds pendulum L_s can be deduced from

---

53. Hahn, *Anatomy* (1971), 163–164, 253; Favre, *Origines* (1931), 121–130; Bigourdan, *Système métrique* (1901), 32.
54. Ibid., 22–26.
55. Méchain and Delambre, *Base* (1806), 1, 20, 23, 43–44, 97–99; Delambre, *Rapport* (1810), 6.

the formula $T^2:1 = L:L_s$. If Δt is large, L_s can be obtained very accurately.

Borda designed the apparatus, which was made by Etienne Lenoir, who also supplied the repeating circles. The pendulum bob hung from a 12-foot steel cord suspended from a knife-edge mounted on a subterranean wall in the Paris Observatory. Borda and Cassini observed coincidences through a telescope pointed perpendicularly to the wall (figure 5.2.1). The largest departure from the mean of the twenty coincidences they recorded was one part in 100,000. They corrected their computed L_s for the dependence of the period on the amplitude of the swing, on temperature, on air pressure, on the flexure of the support and the steel cord, on the moment of inertia of the bob, and on much else. The result: L_s = 440.5593 lines of the toise used in the expedition to Peru.[56] "Thus," wrote the Commissioners who presided over the metric project in 1794, "the pendulum can be considered the depository of the unit of measure, or even a method of measuring the earth; and nothing is more suited to instil admiration of physics and geometry than seeing an undertaking that would appear to require travel from one end of the world to the other with great machines reduced to a very simple experiment done in one place with an instrument of modest dimensions."[57] That would have made an inexpensive as well as a reasonable standard. Lenoir's bill for the circles, the rules, the water apparatus, and the pendulum was just over 34,000 livres. The Academy then estimated—this is from a progress report of May 2, 1792—that the total project would cost 300,000 livres.[58]

During the following summer, Méchain and Delambre set off on their adventures. Other academicians started preliminary work on the units of volume and weight. The politicians were also at work, however, and the Academy's elitism grew ever more obnoxious to sensitive patriots. On 8 August 1793, the National Assembly decided to close the Academy along with other unrepublican corporations. The suppression came only a week after the Assembly had given cause for reassurance to those who expected that the

56. Borda and Cassini, in Méchain and Delambre, *Base* (1806), 3, 337–401, and in Wolf, *Mémoires sur la pendule* (1899), 1, 17–64. Cf. Biot, *Mélanges* (1858), 1, 74.
57. France, Comm. temp., *Instruction* (an ii), 29–30.
58. Bigourdan, *Système métrique* (1901), 22.

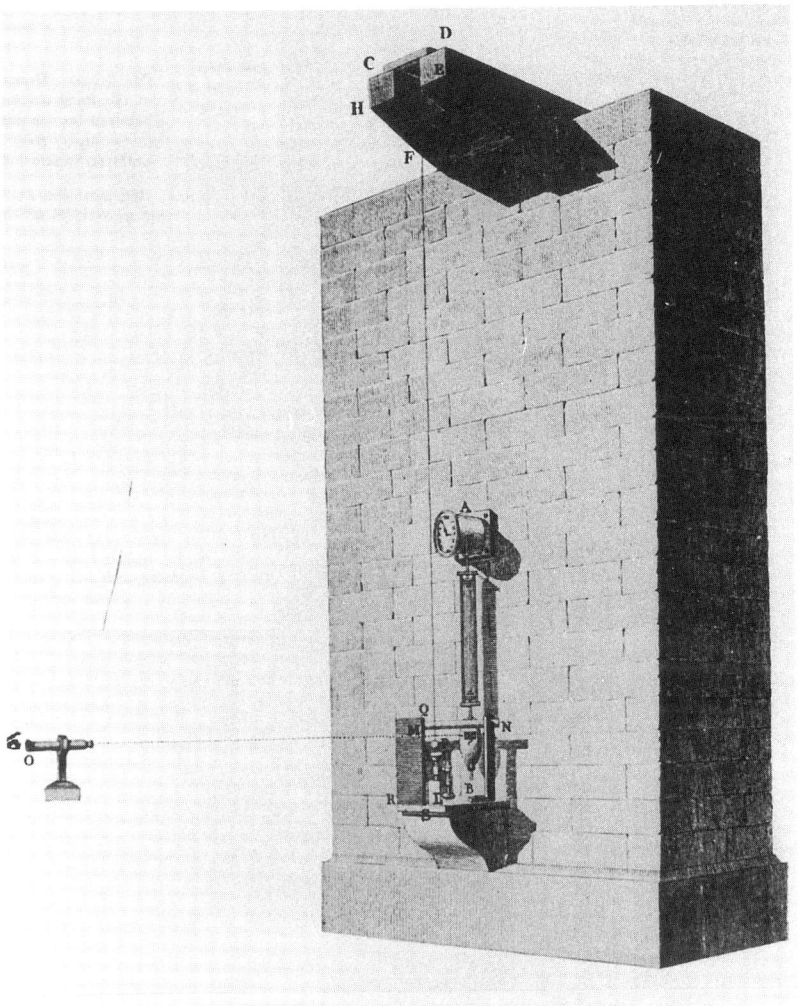

FIG. 5.2.1 Borda and Cassini's method for obtaining the length of a seconds pendulum. The observer counts the ticks between coincidences of the simple pendulum under study and that of a clock accurately beating seconds. Wolf, *Mémoires* (1889), *1*, pl. 2.

Academy's labors on measures in the national interest would protect it. That first week in August the Assembly had affirmed the decimal system and the meridianal definition of the meter, ordered the continuation of the work, and decreed that the Academy provide for the manufacture, distribution, and explanation of provisional meters for general use while it prosecuted its measurements. This provisional meter was defined as a ten-millionth of ninety times the average degree in France as determined by Lacaille, or 443.444 lines of the Peruvian toise. It differed from the definitive meter by about a quarter of a millimeter.

It is not advisable to assign an important job to a specialized government agency one week and to abolish the agency the next. The Assembly therefore immediately accepted the recommendation of the Committee of Public Safety that—because its work would erase the last vestiges of feudal divisions—the Academy's commission on weights and measures should be reestablished, with the same personnel, as an independent temporary commission (Commission temporaire des poids et mesures républicains). It consisted of Borda, Brisson, Cassini, Coulomb, Delambre, Haüy, Lagrange, Laplace, Lavoisier, Méchain, Monge, and Vandermonde.

The Academy's confidence in metrology as an anchor against political storms thus received partial justification. The job of overseeing the making and distribution of the provisional standards brought another large and pressing public responsibility to the academic rump and an important cash flow. The minimum cost of the standards, as estimated by Lenoir and others, would be some 200,000 livres. This minimum provided for copper standards for the geographical departments into which the Revolution had divided the old provinces and for iron standards for the prefectures. The Assembly preferred to avoid this whiff of discrimination and voted an additional 60,000 livres or so to make all the standards of copper except for the platinum exemplars.[59] The temporary commission expected the copper standards to be accurate to around one part in 100,000 for the meter and the grave (kilogram) and one part in 10,000 for the pinte (liter).[60]

---

59. Bigourdan, *Système métrique* (1901), 30–35; Champagne, *Role* (1979), 102–103; Bouchard, *Prieur* (1946), 293; Bugge, *Science* (1969), 206–207; Cappelli, *Rapports* (an x), 7; Hellman, *Osiris*, 13 (1958), 322–323.

60. Bigourdan, *Système métrique* (1901), 36–40.

The arrangements of August and September proved provisional in more ways than one. Lavoisier was arrested on November 28, 1793; the temporary commission requested his release as necessary to its work; the Committee of Public Safety of the Convention (into which the Assembly had transformed itself on declaring a republic) refused, and, for good measure, purged the temporary commission of Borda, Brisson, Coulomb, Delambre, and Laplace, who had not, in its opinion, shown a proper hatred for kings. The rejection of the appeal for Lavoisier and the order for the purge were drawn up by Prieur, who thus revenged himself on academicians who had opposed his political views and metrological proposals.[61]

What remained of the temporary commission did little more than print explanations of the new system and revolutionary rhetoric. "Soon [our] vision will no longer be affronted by those old weights and measures that still tell of the odious remains of times and things sullied by tyrants....Every child will know [the system]; and it will help diminish inequality among men."[62] There was no reason to put off enjoyment of the great benefit; the provisional meter would be perfectly satisfactory for commerce; the definitive would make the whole operation "more worthy of the powerful and enlightened nation that has undertaken it." Neither the rhetoric nor the practical business of furnishing provisional metersticks long protected the pitiful rump commission. In the spring of 1794, it requested 50,000 livres to pay its artisans; Prieur arranged to give 10,000 and to terminate its existence.[63]

We are familiar with the intervention of the military, in the person of General Calon, in reestablishing the metric survey. His provision of 110,000 livres for the work of Méchain and Delambre in May of 1795 raised the redoing of the Dunkirk-Perpignan line to the same financial importance as the mapping of captured territories, and fit well with proposals Prieur had made the previous March for the prompt completion of the metric project.[64] Prieur recommended setting up a temporary agency (Agence temporaire

61. Méchain and Delambre, *Base*, 1 (1806), 46–52.
62. Bigourdan, *Système métrique* (1901), 55–56 (address to the Convention, 19 Jan 1794).
63. France, Comm. temp., *Instruction* (an ii), xv–xvi; Bouchard, *Prieur* (1946), 295–296, 299–301, 461–464; Sarton et al., *RHS*, 3 (1950), 127.
64. Bigourdan, *Système métrique* (1901), 61.

des poids et mesures) to push through the manufacture of provisional standards and to oversee their deployment. He estimated that much could be done for 500,000 livres. He had to request another 500,000 livres the following September. Meanwhile the Army lent a hand by collecting old metal, chiefly church bells and discarded measures, to serve materially and symbolically as raw ingredients for metric measures.[65]

Prieur's program became law on April 7, 1795. It provided for using the "provisional" meter in everyday transactions even after the definitive platinum exemplar came into existence; confirmed the basic names, meter, liter, and gram; and ordered resumption of the measurement of the arc. It established a triumvirate under the Committee of Public Instruction to oversee the manufacture, distribution, and explanation of the provisional standards, using machines wherever possible, "so as to combine facility and swiftness with precision, and consequently to allow citizens to buy the new measures at a reasonable price;" and it specifically ordered the new agency to provide graphic representations of conversions requiring no calculations by users.[66] Later that month the scientific side of operations was entrusted to academic survivors of earlier commissions: Borda and Brisson were to bring in the exemplar of the copper provisional meter within a "décade" (ten days); Méchain and Delambre, to do their thing; Delambre, Laplace, and Prony, to fix a baseline near Paris; Borda, Haüy, and Prony, to determine the standard of weight; and Berthollet, Monge, and Vandermonde, to oversee the preparation of the platinum to be used in the definitive primary standards.[67]

3. RESPONSE OF THE PEOPLE

The Academy of Sciences resumed its existence late in 1795 as the principal part of the new Institut de France, a government

---

65. Kula, *Measures* (1986), 244.
66. France, Convention nationale, *Loi relative aux poids et mesures* (an iii), 4–7; Bigourdan, *Système métrique* (1901), 67–70. Cf. Hellman, *Osiris, 13* (1958), 324–327; Bouchard, *Prieur* (1946), 306–307. The agency consisted of François Gattey, A.M. Legendre, and C.E. Cocquebert.
67. Méchain and Delambre, *Base, 1* (1806), 58–64; Bigourdan, *Système métrique* (1901), 71–77.

monopoly of the most learned new republicans. On April 4, 1796, the Institut took over the metric project through a commission composed of the reborn academicians Berthollet, Borda, Brisson, Coulomb, Delambre, Haüy, Laplace, Legendre, Méchain, Monge, and Prony. They had developed skill in the political as well as the technical aspects of their charge. Laplace proposed the convening of an international body of savants to check and accept the prototypes of the new units. "You realize [Laplace wrote Delambre] that all this is only a formality, to enable them to consider the system as their own, to do away with national envy, and to make them adopt the measures." Borda opposed the meeting; Napoleon supported it; the Directory ordered it.[68] The irrepressible Talleyrand, now minister of foreign affairs, issued the invitations and had the satisfaction this time of omitting Britain altogether.

The international committee of experts convened appropriately in a military setting, in the Dépôt de la marine, on November 28, 1798, having waited in Paris for a few months for Delambre and Méchain to remeasure a baseline. These patient experts represented client states and friendly powers, to wit: the Bavarian, Cisalpine, Helvetian, Ligurian, and Roman Republics, and Denmark, Sardinia, Spain, and Tuscany. The meeting divided into three working committees, one to control the weight of the kilogram, a second to compare the scales used in the work with the old toise, and the third—composed of four members calculating independently—to deduce the length of the meter from the cornucopia of measurements made by Méchain and Delambre. Their calculations fixed the definitive meter over-precisely at 443.295936 lignes of the toise de Peru taken at 17.6°C, some 0.144 lignes shorter than the provisional meter by then distributed throughout France. (The difference was meaningless for commerce, since it amounted to one foot in a kilometer.) Independent observers had nothing but praise for Borda's circle and for the care with which the units of volume and weight were deduced from the meter and the density of water.[69]

68. Hellman, *Osiris*, 13 (1958), 333–335; Crosland, in Bugge, *Science* (1969), 197; Laplace to Delambre, 29 Jan 1798, in Laissus, *RHS*, 14 (1961), 287–288.
69. Crosland, in Bugge, *Science* (1969), 197–200; Bugge, ibid., 205–211; Méchain and Delambre, *Base* (1806), 1, 92–94. Praise for the circle: Bugge, *Science* (1969), 204; Delambre, *Rapport* (1810), 232–233, quoting Zach.

The Danish delegate retained enough independence to observe that although Méchain and Delambre had been as precise as was humanly possible, "it cannot be assumed that the determination of the meter is so absolutely and completely borrowed from nature that there is no doubt at all about its final accuracy." No echo of this discordant and irrelevant observation was heard in the speeches on June 22, 1799, when the joint committees formally presented the prototype standards to the people. In accepting them, the president of the upper legislative body, the Conseil des anciens, remarked that it was a work of genius to have found an inviolable base of measurement in nature; "that, citizens, is the immortal service that the National Institute has rendered to the French Republic, or rather, the benefit that it offers to the entire human race."[70] It remained only to convince the race to accept the benefit.

Resistance

The experts returned to their countries encouraged to propagate the metric system at home for the improvement of commerce and the tightening of ties to France. The sort of pressure they were under appears from a hint earlier given by the Convention's Committee of Public Instruction to a newly liberated territory. The Committee pointed to the great "advantage it would be to the union just established between the French and Batavian republics to spread the system of uniform and decimal measures beyond French territory."[71] The Cisalpine Republic, consisting of Lombardy and Emilia-Romagna, also felt the pressure early. Annexed territory, such as the Département du Mont-Tonnerre carved out of the Rhineland, felt a heavier hand. By a decree of 18 June 1801, the Département would be obliged to use the metric system exclusively; all old weights and measures found in shops and all meter sticks on which former units of length were indicated would be confiscated.[72]

In friendly but not subject territory the system was recommended for its combination of abstract science and practical utility. "We [in

---

70. Bugge, *Science* (1969), 204, and in Crosland, *Stud. Volt. eight. cent.*, 87 (1972), 305–308; P.C.L. Baudin, "Réponse," in Méchain and Delambre, *Base* (1806), 3, 652. The president of the lower legislative body, the Conseil des cinq cents, spoke similarly; ibid., 3, 649.

71. Quoted by Laissus, *RHS*, 14 (1961), 288.

72. *Tables de comparaison* (an x), xxxii.

the Kingdom of Sardinia] are obliged to the nation that has sustained with unexampled courage the most terrible political upheavals for having put forth, in the very midst of a frightful tumult, the most abstract speculations of mathematicians as a rule for every-day transactions." "Nature and not France brings [the new measures] to us [Spaniards]. Let us accept them following our natural ally with whom we have so many commercial relations....It behooves Spain [because it helped Méchain] to set other nations the example of adopting these units."[73]

None of this cajoling or commanding had much immediate effect. Even in France, where the new units were to oust the old in land transactions on 1 Vendémiaire an x (September 23, 1801), the people declined to benefit from the republican measures created ostensibly in their interest.[74] One of the orators at the presentation of the prototypes in July 1799 had imagined the satisfaction a peasant would feel in computing his holdings in square meters. "The field that supports my children is such and such a fraction of the globe. In that proportion I am a co-owner of the world."[75] In practice the peasant and his surveyor, builder, and draper, refused to give up their toise sticks. So brutish a rejection of the natural and the rational perplexed the savants. "There is no respect in which the reform of weights and measures has not been advantageous," declared the president of the Agricultural Society of Paris (Société d'agriculture de Paris) in 1809, "and, consequently, if reason always were listened to, the success of the reform would have been complete."[76]

Alas, the meter had only complicated ordinary life. Artisans brought home their measures in familiar units and sat up all night converting them into meters, and into a crowd of irrelevant decimals, in order to write estimates and contracts in the obligatory figures. Why did they not measure in meters to begin with and enjoy "the happy effects of the most useful of the gifts that scientists have been able to make to society"?[77] Because they were

73. Capelli, *Rapports métriques* (an x), 5; Ciscár, *Memoria* (1800), 33–34.
74. Gattey, *Eléments* (an x), 65. Cf. Bugge, *Science* (1969), 205; Crosland, *Isis*, *60* (1969), 229–231; and Favre, *Origines* (1931), 196–202.
75. Quoted by Crosland, *Stud. Volt. eight. cent.*, *87* (1972), 285.
76. C[hessiron], in *Nouveau cours*, *8* (1809), 299–300.
77. Ibid., 292.

creatures of habit, vain, traditional, pious, ignorant, and ungrateful.[78] And also canny and suspicious. According to a British metrologist, who had it from a French merchant, consumers opposed the new units on the theory that they would be an advantage to shopkeepers, and shopkeepers opposed them for fear that consumers would be able to figure out true costs. The government hoped that a play with words would overcome suspicion. On November 4, 1800, it allowed the use of old words for multiples of the new measures, for example, "toise" for "two meters," "pinte métrique" for "litre," and so on. The reformers thought the concession damaging and absurd: if the people could use Greek words like "chirurgien" and "apothécaire," why not "kilogramme"?[79]

Because the people did not like change and cared nothing for system. So stubborn and irrational were they that in 1812 the imperial government sacrificed the jewel of the reform, the key to democracy, that is, reckoning by tens. The sacrifice had the approval of Napoleon, who himself had difficulty with the metric system. The cartographers of the Dépôt de la guerre, who synthesized into a uniform "Carte de l'Empereur" the maps and plans of Germany scavenged by the ingénieurs géographes, had to put in scales in toises as well as meters; and Napoleon insisted that the maps made from Tranchot's data be drawn in the Cassini scale of 1:86400 rather than on the metric scale of 1:100000. "I don't care a fig for your decimal divisions," he said to the map makers, who knew who was in charge. "It is true," they said among themselves, "that this scale [of toises] does not agree perfectly with the decimal system; but it suits His Majesty."[80]

Two laws, of February 12 and March 28, 1812, indicate the state of affairs. The first ordered that the official system be taught in all schools throughout the empire, that it alone be used in public administration, markets, and commercial transactions. The second

78. Concocted from compliments in ibid.; Méchain and Delambre, *Base* (1806), 3, 651 (text of 1799); Ciscár, *Memoria* (1800), 38–39; and Garnier, *Traité* (1818), 281.

79. Kelly, *Metrology* (1816), 25–26; Gattey, *Eléments* (an x), 9–10; Chessiron, *Nouveau cours* (1809), 8, 296. The convenience of halves and doubles of the fundamental units was widely urged as an advantage of the reform, e.g., in *JP*, 47 (1799), 59, 60.

80. Quotes from Berthaut, *Ingénieurs* (1902), 2, 160, 148, resp.; ibid., 119, 163–171; cf. Fischer, *Napoleon* (1988), 90n, 279, 282. The Navy also only slowly adopted metric measure: Fleurieu, *Application* (an viii), 2–4, 9–21.

destroyed the official system by introducing "common measures" (*mesures usuelles*) divided by twos and threes: a toise of 2 meters divided into 6 pieds; an aune of 1.2 meters divisible into halves, quarters, eighths, and sixteenths, and also into thirds, sixths, and twelfths; and similarly for weights and coins. Our British metrologist enjoyed the spectacle: "Thus, after twenty years of trouble, mystery, and litigation, no advances are made, except that of having one common standard."[81] Otherwise confusion had only been compounded: the puzzled citizen had the provisional meter of 1795, the definitive of 1799, the compromised names of 1801, the common system of 1812, and, unofficially and omnipresent, the ancient units of the Kings of France. It is easy to change governments, not hard to reform currency, but almost impossible to alter the most common weights and measures. "Nothing has a greater tendency to grow worse, or more obstinately resists improvement." One took the long view. "This new metric system, all of whose parts fit together so simply, will no doubt triumph in the end over the obstacles that habit still opposes to its general adoption," so reads a text of 1820, "especially if it continues to be obligatory."[82]

Spread

A gauge of the rate of penetration of the new units is the pattern of their use in books requiring specification of units, like atlases, compendia of architectural drawings, and manuals of surveying, building, and engineering. Although such works employing old units exclusively were published into the 19th century,[83] most presented measurements and calculations in both meters and toises and their multiples. A good indication for the early years is the collection of architectural designs awarded prizes by the Institut de France and other government bodies between 1795 and 1803. Forty of these designs have units: three use toises only, two use meters only, and the balance use both. Before 1800 toises have priority when both scales are present; after 1800, meters.[84] A collection of

---

81. Kelly, *Metrology* (1816), xiii; ibid., 27–30, for text of the laws of 1812; Favre, *Origines* (1931), 203–205.

82. Respectively, Kelly, *Metrology* (1816), xvi, and Puissant, *Traité* (1820), 245.

83. See, for example, Viel, *Construction des édifices* (1803); Ducrest, *Vues nouvelles* (1803); Molard, *Description* (1811), 1; Krafft, *Recueil* (1812); Belidor, *Science des ingénieurs* (1813).

84. *Projets d'architecture* (1806).

plans for large structures published in 1823 shows a very considerable change: of seventy-six plates with scales, two have toises only, eleven have both, and sixty-three meters only; and the old measures occur exclusively in one section of the collection, on stone bridges.[85] The obvious extrapolation does not hold, however; a set of 105 house plans published in 1843 under the promising title *Paris moderne* suggests that builders of private dwellings then still preferred to express themselves in the measures of the old regime. About 30 percent of the plans use toises only; another 30 percent, meters only; and 40 percent, both. All the plans are dated, the earliest to 1815; the strongest showing of the toise against the meter occurs not at the beginning, but in the period 1835–1839. The meter takes over after 1840, the year in which the metric system at last became obligatory in France.[86]

Up-to-date technical manuals show the same equivocation as architectural drawings. An authoritative indicator is the huge treatise on the building arts (four volumes in five of text, three of plates) published between 1805 and 1816 by Jean-Baptiste Rondelet, chief architect of the church of Sainte-Geneviève and architectural consultant to the government. Rondelet's plates strongly favor the old measures; his text appears to be the work of a schizophrenic. The earliest volume (1805) uses pieds, pouces, and lignes and also metric units, sometimes converted but often not, in tables and in calculations; at one point it gives dimensions in meters and computes in toises.[87] In the volumes published in 1812, Rondelet usually proffers all dimensions in both the old and new style, but without the advantage of decimal notation: for example, "31 pieds 3/7 (10 mètres 209 millimètres)." His preference for the old appears from his way of obtaining specific gravity, which he understands as the weight in grams of a cubic centimeter of a substance. He measures the weight of a sample, in air and in water, in onces and gros, deduces the weight of a cubic pied, and converts to the metric system. The preference for the old units persists into the last volumes,

85. Bruyère, *Etudes* (1823).
86. Lemmonier and Normand, *Paris moderne* (1843); Bouchard, *Prieur* (1946), 313. The obligation was laid down in a law of 4 July 1837; it took effect on 1 January 1840.
87. Rondelet, *Traité* (1805), 3, 57, 66–67, 83 ff., 102, 164, 179–183, 269 ff., 388–396.

published in 1816. Then—after 1,839 pages of schizophrenic computations, the last of which provides the cost of old bricks in old money—Rondelet introduces some "notions about metric measures."[88] It is instructive that this instruction precedes a recomputation of costs in metric measures and decimalized currency. The relative ease of such computations was the great benefit that the reformers had advertised.

From about 1810 many technical manuals by professors and government officials used metric measures exclusively. For example, the professor of mathematics at the Ecole impériale militaire, Louis Puissant, and Pierre Pommiés, professor at the Lycée Napoléon, both influential teachers of surveying and geodesy, used meters exclusively and without explanation in many textbooks beginning in 1807.[89] Jean-Nicolas Hachette, professor at the Ecole polytechnique, brought the metric system into the preface of his textbook on machines, without comment, as the only satisfactory way "to express their effect in numbers." The chemist Jean-Antoine Chaptal, one-time minister of the interior and long-time educational reformer, naturally used nothing but modern measures in his survey of the state of French industry after the fall of Napoleon. Still there are hints of backsliding where least expected. The official surveyors of France might have worked exclusively in meters and understood the higher geodesy; but the local land measurer often enough made do with old units and little geometry into the reign of Louis Philippe. A manual for such measurers, "especially people who have not studied geometry," was published in 1833 by an inspector of the cadastral survey. It contains an explanation of the metric system, an injunction to use it, and tables for the easy conversion of old local measures into meters, "and vice versa."[90]

As it spread, the metric system did help to realize the reformers' ambition to improve the numeracy of Europeans. In the Ancien Régime, despite its baroque abundance of metrological units, most people had need only for a single system; during the Empire, because of the government's policy of driving the meter home,

88. Ibid., 1, 99, 101–102, 158–160; 4, 583–604, 609–644.
89. Puissant, *Traité* (1807), 139, 154; Pommiés, *Manuel* (1808), 166–169.
90. Chaptal, *L'industrie française* (1819); Lefevre, *Guide practique* (1833), i–vi, 15–21.

more and more citizens had to learn how to convert one set of measures into another. If the first advance of European "Rechenhaftigkeit"—"the inclination, habit, and ability to resolve the world into numbers and to bring these numbers together into an artificial system of income and outgo"—may be likened to literacy in one language, the domestication of the metric system made reckoners bilingual.[91]

The early treatises on metric calculations introduced not only decimal arithmetic, including the concept of place, but also the idea of significant figures and the operation of rounding off. An *Exposition abrégée du nouveau système*, published in the provinces in 1802 or 1803, may stand for the genre. It explains that the many decimals generated during conversion are artifacts of multiplication and division, and demands that they be dropped from the final answer.[92] Rounding off, says Rondelet, in his belated account of metric computations, is essential to the new system. Rounding off might in practice mean only rounding up, as readers of an *Arithmétique pratique* of 1800, which recommends changing 0.7411440 to 0.75, learned to do; a practice that would favor shopkeepers, who could round up each of their many small transactions.[93] Even arithmetic has its social bias. The utility of metric conversions in teaching numeracy appears further from popular textbooks of arithmetic that use such computations as exercises. This practice persisted long after the universal adoption of the metric system in France.[94]

According to the president of the Paris Agricultural Society, the great majority of people who could read and write in 1810 knew nothing more of arithmetic than addition and subtraction. He had a prescription for correcting this innumeracy and a vision of the happy consequences. "If decimal calculations could be introduced into primary schools along with the use of the new measures, not

91. Quote from Sombart, *Der Bourgeois* (1913), 164.
92. *Exposition abrégée* [1802], 23–24, 29–30; cf. the contemporary *Tables de comparaison* (an x), xiv–xxix; Capelli, *Rapports* (an x), 49–57.
93. Rondelet, *Traité* (1805), 4:2 (1816), 610–613; Poittier, *Arithmétique pratique* (an viii), 146–152. Gattey, *Eléments* (an x), 32–33, rounds up only if the part dropped is ≥ 0.5 of the last figure kept.
94. E.g., Lacroix, *Traité* (1805), 115–146; Garnier, *Traité* (1818), 1, 263–275, 281–293; and, for the later period, Thinon, *Leçons* (1860), 131.

only would the housewife be able to make all the computations she requires, but also the worker could measure without difficulty and, by adding the use of the rule and compass for tracing geometrical figures, he would be able to draw all his plans himself, and the farmer would have no problem with surveying."[95] This grand project has been realized in large measure. The male academic and quantitative spirit of the late Enlightenment found a fertile if capricious partner in *La belle Marianne,* the spirit of revolutionary France.

---

95. Chessiron, *Nouveau cours* (1809), *8,* 293.

BIBLIOGRAPHY

The following abbreviations are used:

AC	*Annales de chimie*
ACP	*Annales de chimie et physique*
AJP	*American journal of physics*
AS	*Annals of science*
BJHS	*British journal for the history of science*
HAS	Académie des sciences, Paris, *Histoire*
HSPS	*Historical studies in the physical and biological sciences*
JP	*Journal de physique*
MAS	Académie des sciences, Paris. *Mémoires*
MIF	Institut de France. Classe des sciences physiques et mathématiques. *Mémoires*
MSA	Societé d'Arcueil. *Mémoires*
PT	Royal Society of London. *Philosophical transactions*
RHS	*Revue d'histoire des sciences*
ZsV	*Zeitschrift für Vermessungswesen*

Abbri, Ferdinando. "The chemical revolution: A critical assessment." *Nuncius,* 4 (1989), 303–15.

Andrews, G.G. "Making the revolutionary calendar." *American historical review,* 36 (1931), 515–32.

Arbuthnot, John. *Miscellaneous works.* 2 vols. Glasgow: J. Carlisle, 1751.

Arnold, D.H. "The *Mecanique physique* of S.D. Poisson: The evolution and isolation in France of his approach to physical theory, 1800–1840." *Archive for the history of exact sciences,* 28 (1983), 243–367, and 29 (1984), 1–54, 287–307.

Arrighi, Gino, ed. *Carteggi di Giovanni Attilio Arnolfini. Quarantaquattro lettere inedite di Girolamo de la Lande, Ruggiero Guiseppe Boscovich e Leonardo Ximenes.* Lucca: Azienda Grafica, 1965.

Baczko, Bronislaw. "Le calendrier républicain" *Les lieux de mémoire.* Ed. Pierre Nora. Vol. 1. *La République.* Paris: Gallimard, 1984. Pp. 37–83.

Barnett, Martin K. "The development of thermometry and the temperature concept." *Osiris, 12* (1956), 269–341.

Barruel, Etienne. "Physique générale." Ecole polytechnique. *Journal*, 2e cahier, Floréal and Prairial [an iii] (published Nivôse, an iv), 128–44; 3e cahier, Messidor, Thermidor, and Fructidor, an iii (Prairial, an iv), 337–44; 4e cahier, Vendémiaire, Brumaire and Frimaire, an iv (Vendémiaire, an v), 623–45.

Barsanti, Danilo, and Leonardo Rombai. *Leonardo Ximenes: Uno scienziato nella Toscana lorenese del settecento.* Florence: Edizioni Medicea, 1987.

Bassot, Léon. "Notice historique sur la fondation du système métrique." France. Bureau des Longitudes. *Annuaire* (1901), D.1–D. 43.

Beccaria, G.B., and Domenico Canonica. *Gradus taurinensis.* Turin: Typographia regia, 1774.

Belidor, Bernard Forest de. *La science des ingénieurs, dans la conduite des travaux de fortification et d'architecture civile.* Ed. C.LM.H. Navier. Paris: Firmin Didot, 1813.

Bérard, J.E. "Mémoire sur les propriétés des différaetes espèces de rayons qu'on peut séparer au moyen du prisme de la lumière solaire." *MSA, 3* (1817), 5–47.

Bernleithner, Ernst. "Oesterreichs Kartographie zur Zeit des Graphen Ferraris." In *Cartographie* (1978), 129–47.

Berthaut, H.M.A. *La carte de France, 1750–1898.* 2 vols. Paris: Imprimerie du Service géographique, 1898–99.

Berthaut, H.M.A. *Les ingénieurs géographes militaires, 1624–1831.* 2 vols. Paris: Imprimerie du Service géographique, 1902.

Berthollet, C.L. *Essai de statique chimique.* Paris: F. Didot, 1803.

Berthollet, C.L., J.A. Chaptal, and J.B. Biot. "Sur un mémoire de M. Béraud, relatif aux propriétés physiques et chimiques des divers rayons qui composent la lumière solaire." *AC, 85* (1813), 309–25.

Bianca, Mariano, ed. *La scienza a Firenze. Itinerari scientifici a Firenze e provincia.* Florence: Alinea, 1989.

Bickerman, J.J. "Theories of capillary attraction." *Centaurus*, 19 (1975), 182–206.

Bigourdan, Georges. *Le système métrique des poids et mesures. Son établissement et sa propagation graduelle.* Paris: Gauthier-Villars, 1901.

Biot, J.B. "Sur la théorie du son." *JP*, 55 (1802), 173–82.

Biot, J.B. *Essai sur l'histoire générale des sciences pendant la révolution française.* Paris: Duprat and Fuchs, 1803.

Biot, J.B. "Extrait de la relation d'un voyage aérostatique." *JP*, 59 (1804), 314–20.

Biot, J.B. "Expériences sur la propagation du son à travers les corps solides et à travers l'air, dans les tuyaux très-élongés." *MSA*, 2 (1809), 405–23.

Biot, J.B. *Traité de physique expérimentale et mathématique.* 4 vols. Paris: Deterville, 1816.

Biot, J.B. "Sur une loi remarquable qui s'observe dans les oscillations des particules lumineuses, lorsqu'elles traversent obliquement les lames minces de chaux sulfatée ou de cristal de roche, taillées parallèlement à l'axe de crystallisation." *MSA*, 3 (1817), 132–47.

Biot, J.B. "Examen comparé de l'intensité d'action que la force répulsive extraordinaire du spath d'Islands exerce sur les molécules lumineuses de diverses couleurs." *MSA*, 3 (1817), 371–84.

Biot, J.B. *Mélanges scientifiques et littéraires.* 3 vols. Paris: Michel Lévy, 1858.

Biot, J.B., and D.F.J. Arago. "Sur les affinités des corps pour la lumière, et particulièrement sur les forces réfringentes des différents gaz." *MIF*, 1806, 301–87.

Borda, Charles. *Description et usage du cercle de réflexion, avec différentes méthodes pour calculer les observations nautiques.* 2nd edn. Paris: Firmin Didot, 1802.

Borda, Charles, J.L. Lagrange, A.L. Lavoisier, Mathieu Tillet, and M.J.A.N. Caritat de Condorcet. "Rapport." *HAS*, 1788, 1–6. Text of 27 Oct 1790.

Borda, Charles, J.L. Lagrange, P.S. de Laplace, Gaspar Monge, and M.J.A.N. Caritat de Condorcet. "Rapport...sur la choix d'une unité de mesures." *HAS*, 1788, 7–16. Text of 19 Mar 1791.

Boscovich, R.G., and Christopher Maire. *Voyage astronomique et géographique dans l'Etat de l'Eglise entrepris par l'ordre et sous les auspices du pape Benôit XIV.* Paris: Tilliard, 1770.

Boscovich, R.G. "De litteraria expeditione per pontificiam ditionem." Accademia della scienze, Bologna. *Commentarii*, 4 (1757), 353–96.

Bouchard, Georges. *Un organisateur de la victoire. Prieur de la Côte-d'Or, Membre du Comité de salut publique.* Paris: R. Clavreuil, 1946.

Boyer, Carl B. "Early principles in the calibration of thermometers." *AJP*, 10 (1942), 176–80.

Boyer, Jacques. "Le centenaire du système métrique." *Revue encyclopédique*, 9 (1859), 845–54.

Bradley, James. "A letter...giving an account of a new-discovered motion of the fix'd stars." *PT*, 35 (1729), 637–61.

Bradley, James. "A letter...concerning an apparent motion observed in some of the fixed stars." *PT*, 45 (1748), 1–43.

Bradley, James. *Miscellaneous works and correspondence.* Oxford: Oxford University Press, 1832.

Brisson, M.J. *Dictionnaire raisonné de la physique.* 7 vols. 2nd edn. Paris: Libraire économique, 1800.

Brown, Harcourt. *Science and the human comedy. Natural philosophy in French literature.* Toronto: University of Toronto Press, 1976.

Brown, Lloyd Arnold. *Jean Dominique Cassini and his world map of 1696.* Ann Arbor: University of Michigan Press, 1941.

Brown, Sanborn C. "Discover of the differential thermometer." *AJP,* 22 (1954), 13–17.

Brown, Sanborn C. *Benjamin Thompson, Count Rumford.* Cambridge, Mass.: MIT Press, 1979.

Bru, Bernard. "Laplace et la critique probabiliste des mesures géodésiques." In Lacombe and Costabel, *Figure* (1988), 223–44.

Brunet, Pierre. *Maupertuis.* Paris: A. Blanchard, 1929.

Brunet, Pierre. *La vie et l'oeuvre de Clairaut (1713-1765).* Paris: Presses universitaires de France, 1952.

Bruwier, Marinette. "Les 'Mémoires historiques, chronologiques et economiques' de Ferraris." In *Cartographie* (1978), 60–73.

Bruyère, L. *Etudes relatives à l'art des constructions.* Paris: Bance aîné, 1823.

Buchwald, J.Z. *The rise of the wave theory of light. Optical theory and experiment in the early nineteenth century.* Chicago: University of Chicago Press, 1989.

Buffon, G.L. Leclerc, Comte de. "Réflections sur la loi d'attraction." *MAS,* 1745, 493–500, 551–2.

Bugge, Thomas. *Science in France in the revolutionary era.* Ed. M.P. Crosland. Cambridge, MA: MIT Press, 1969.

Bunsen, Robert. "Caloric researches." *Philosophical magazine,* 41 (1871), 161–82.

Burr, Alexander C. "Notes on the history of the experimental determination of the thermal conductivities of gases," *Isis,* 21 (1934), 169–86.

Cajori, Florian. *The chequered career of Ferdinand Rudolf Hassler*. Boston: Christopher, 1929.

Capelli, Charles. *Rapports métriques*. Turin, Felix Buzan, an x.

Cardwell, D.S.L., ed. *John Dalton and the progress of science*. Manchester: Manchester University Press, 1968.

Carter, Alice C. "The Dutch barrier fortress in the 18th cnetury, as shown in the de Ferraris map." In *Cartographie* (1978), 259–71.

La Cartographie au xviiie siècle et l'oeuvre du comte Ferraris (1726–1814). Brussels: Credit communal de Belgique, 1978. (*Collection histoire pro civitate*, no. 54.)

Carvalho, Joacquim de. *Correspondência cientifica dirigida a Joāno Jacinto de Magalhānes, 1769–1789*. Coimbra: Universidad de Coimbra, 1952.

Cassini, G.D. *La meridiana del tempio di S. Petronio. Tirata, e preparata per le osservazioni astronomiche l'anno 1665. Rivista, e restaurata l'anno 1695*. Bologna: Erede di Vittorio Benacci, 1695.

Cassini, G.D. (Cassini I.) "De la figure de la terre." *MAS*, 1713, 187–98.

Cassini, G.D. "Les élémens de l'astronomie verifiez par Monsieur Cassini par le rapport de ses tables aux observations de M. Richer faites en l'isle de Cayenne." *MAS*, 1666–1699, 8 (1730), 53–117.

Cassini, Jacques. (Cassini II.) *De la grandeur et de la figure de la terre*. Paris: Imprimerie royale, 1720.

Cassini, J.D. (Cassini IV.) "De la jonction des observatoires de Paris et de Gréenwich, et précis des travaux géographiques exécutés en France, qui y ont donné lieu." *MAS*, 1788, 706–17.

Cassini de Thury, C.F. (Cassini III.) "Mémoire sur la prolongation de la perpendiculaire de Paris jusqu'à Vienne en Autriche." *MAS*, 1763, 299–325.

Cassini de Thury, C.F. "De la méridienne de Paris, prolongée vers le Nord, et les observations qui ont été faites pour décrire les frontières du Royaume." *MAS*, 1740, 276–92.

Cassini de Thury, C.F. *Relation des deux voyages faits en Allemagne par ordre du roi.* Paris: Nyon, 1765.

Cassini de Thury, C.F. *Relation d'un voyage en Allemagne, qui comprend les opérations relatives à la figure de la terre et à la géographie particulière.* Paris: Imprimerie royale, 1775.

Cassini, C.F. "Description des conquêtes de Louis XV, depuis 1745 jusqu'en 1748." In Cassini, *Relation* (1775), 123–94.

Cassini de Thury, C.F. *La description géometrique de la France.* Paris: J.C. Desaint, 1783.

Cavendish, Henry, et al. "The report of the committee appointed by the Royal Society to consider the best method of adjusting the fixed points of thermometers." *PT*, 67 (1777), 816–57.

Champagne, Ruth Inez. *The role of five eighteenth-century French mathematicians in the development of the metric system.* Ph.D. thesis, Columbia University, 1979.

Chapman, Allan. "The accuracy of angular measuring instruments used in astronomy between 1500 and 1850." *Journal for the history of astronomy*, 14 (1983), 133–37.

Chapman, Allan. "The design and accuracy of some observatory instruments of the seventeenth century." *AS*, 40 (1983), 457–71.

Chapman, Allan. *Dividing the circle: The development of critical angular measurement in astronomy 1500–1830.* Chichester, West Sussex: Ellis Horwood, 1990.

Chappert, André. *Etienne-Louis Malus (1775–1812) et la théorie corpusculaire de la lumière.* Paris: Vrin, 1977.

Chaptal, J.A.C. *De l'industrie françoise.* 2 vols. Paris: Renouard, 1819.

C[hessiron], L. "Mesures." In *Nouveau cours complet d'agriculture théorique et pratique*. Vol. 8. Paris: Deleville, 1809. Pp. 291–314.

Ciscár, Gabriel. *Memoria elemental sobre los nuevos pesos y medidas decimales fundadas en la naturaleza*. Madrid: Imprenta real, 1800.

Clairaut, A.C. *Théorie de la figure de la terre tirée des principes de l'hydrostatique*. Paris: Durand, 1743; 2nd edn, Paris: Courcier, 1808.

Clairaut, A.C. "Sur les explications cartésienne et newtonienne de la réfraction de la lumière." *MAS*, 1739, 259–75.

Clairaut, A.C. "Avertissement...sur le système du monde, dans les principes de l'attraction." *MAS*, 1745, 577–8.

Clairaut, A.C. "De l'orbite de la lune, en ne négligeant les quarrés des quantités de même ordre que le force perturbatrices." *MAS*, 1748, 421–40.

Clairaut, A.C. *Théorie de la lune*. St Petersburg: Académie impériale des sciences, 1752.

Close, Charles. *The early years of the Ordnance Survey*. Ed. J.B. Hanley. New York: A. Kelley, 1969.

Clotten, M. "Das Vermessungswesen im ehemaligen Königreich Hannover." *ZsV*, 10 (1881), 292–8, 387–98, 425–38, 445–59.

Cohen, P.C. *A calculating people. The spread of numeracy in early America*. Chicago: University of Chicago Press, 1982.

Colloque Gay-Lussac. Palaisseau: Ecole polytechnique, 1980.

Configliachi, Pietro. "Mémoire sur la force magnétisante du bord le plus reculé du rayon violet du spectre solaire." *JP*, 77 (1813), 212–35.

Cornell, E.S. "Early studies in radiant heat." *AS*, 1 (1936), 217–25.

Cornell, E.S. "The radiant heat spectrum from Herschel to Melloni, part I: The work of Herschel and his contemporaries." *AS*, 3 (1938), 119–37.

Costabel, Pierre. "Le 'calorique du vide' de Clément et Desormes." *Archives internationales d'histoire des science*, 21 (1968), 3–14.

Costabel, Pierre. "Science positive et forme de la terre au début du xviiie siècle." In Lacombe and Costabel, *Figure* (1988), 97–113.

Coulomb, C.A. "Construction et usage d'une balance électrique." *MAS*, 1785, 569–77. In Coulomb, *Mémoires* (1884), 107–15.

Coulomb, C.A. "Mémoire...où l'on détermine suivant quelles lois le fluide magnétique ainsi le fluide électrique agissent." *MAS*, 1785, 578–611. In Coulomb, *Mémoires* (1884), 116–46.

Coulomb, C.A. "Sur la manière dont le fluide électrique se partage entre deux corps conducteurs mis en contact." *MAS*, 1787, 421–67. In Coulomb, *Mémoires* (1884), 183–229.

Coulomb, C.A. *Mémoires.* Paris: Gauthiers-Villars, 1884. (Société française de physique. *Collection de mémoires relatifs à la physique*, 1.)

Crosland, M.P. "The origin of Gay-Lussac's law of combining volumes." *AS*, 17 (1961), 1–26.

Crosland, Maurice. *The Society of Arcueil.* Cambridge, MA: Harvard University Press, 1967.

Crosland, Maurice. "The first reception of Dalton's atomic theory in France." In Caldwell, ed., *John Dalton* (1968), 274–87.

Crosland, Maurice. "Introduction." In Bugge, *Science* (1969), 1–18.

Crosland, Maurice. "The congress on definitive metric standards, 1798–1799: The first international scientific congress?" *Isis*, 60 (1969), 226–31.

Crosland, Maurice. "'Nature' and measurement in eighteenth-century France." *Studies on Voltaire and the eighteenth century*, 87 (1972), 277–309.

Crosland, Maurice. *Gay–Lussac. Scientist and bourgeois.* Cambridge: Cambridge University Press, 1978.

Crosland, Maurice. "Chemistry and the chemical revolution." In Rousseau and Porter, *Ferment* (1980), 389–418.

Crosland, Maurice, and Crosbie Smith. "The transmission of physics from France to Britain, 1800–1840." *HSPS, 9* (1978), 1–61.

Crouch, T.D. *The eagle aloft. Two centuries of the balloon in America.* Washington, D.C.: Smithsonian, 1983.

Dalby, Isaac. "The longitudes of Dunkirk and Paris from Greenwich, deduced from the triangular measurement in 1787, 1788, supposing the earth to be an ellipsoid." *PT, 81* (1791), 236–45.

Dalton, John. "On the expansion of elastic fluids by heat." Manchester Literary and Philosophical Society. *Memoirs, 5:2* (1803), 595–602.

Danjon, André, and André Couder. *Lunettes et télescopes.* Paris: Editions de la Revue d'optique, 1935.

Daumas, Maurice. *Les instruments scientifiques aux xviie et xviiie siècles.* Paris: Presses universitaires de France, 1953.

Daumas, Maurice. "Precision of measurement and physical and chemical research in the 18th century." In A.C. Crombie, ed. *Scientific change.* New York: Basic Books, 1963. Pp. 418–30.

David, Jean-Claude. "Grimm, Lalande et le quart de cercle de l'Ecole royale militaire." *Dix-huitième siècle, 14* (1982), 277–87.

Débarbat, Suzanne. "La qualité des observations astronomiques de Picard." In Picolet, *Picard* (1987), 157–73.

Débarbat, Suzanne. "Coopération géodésique entre la France et l'Angleterre à la veille de la Révolution française: Echanges techniques, scientifiques et instrumentaux." *Echanges d'influences scientifiques et techniques entre pays Européens de 1780 à 1830.* Paris: CTHS, 1990. Pp. 47–76. (Congrès national des sociétés savantes, CXIV. Section histoire des sciences et des techniques. *Actes.*)

Dee, John. "Mathematicall preface." In H. Billingsley, tr. *The elements of geometry of the most auncient philosopher Euclide of Megara.* London: I. Daye, 1570.

Delambre, J.B.J. *Rapport historique sur les progrès des sciences mathématiques depuis 1789, et sur leur état actuel.* Paris: Imprimerie impériale, 1810.

Delambre, J.B.J. *Histoire de l'astronomie moderne.* 2 vols. Paris: Huzard-Coucier, 1821.

Delambre, J.B.J. *Histoire de l'astronomie au dix-huitième siècle.* Paris: Bachelier, 1827.

Delambre, J.B.J. *Grandeur et figure de la terre.* Ed. G. Bigourdan. Paris: Gauthier-Villars, 1912.

Delamétherie, J.C. "Discours préliminaire." *JP,* 46 (1798), 1–134.

Delaroche, François. "Observations sur le calorique rayonnant." *JP,* 75 (1812), 201–28.

Delaroche, François. "Observations sur le calorique rayonnant." Société philomathique. *Bulletin,* 3 (1812), 131.

Delaroche, François, and J.E. Bérard. "Mémoire sur la détermination de la chaleur spécifique des différens gaz." *AC,* 85 (1813), 72–110, 113–82.

Deluc, J.A. "An essay on pyrometry and aerometry, and on physical measures in general." *PT,* 68 (1778), 419–553.

Depuydt, L. Frans. "The large-scale mapping of Belgium, 1800–1850." *Imago mundi,* 27 (1975), 23–6.

Desaguliers, J.T. "A dissertation concerning the figure of the earth." *PT,* 33 (1725), 201–22, 239–55, 277–304, 344–5.

Desaguliers, J.T. *Course of experimental philosophy*. 3rd edn. 2 vols. London: A. Millar, 1763.

Desormes, C.B., and Nicolas Clément. "Détermination expérimentale du zéro absolu de la chaleur et du calorique spécifique des gaz." *JP, 89* (1819), 321–46, 428–55.

Devic, J.F.S. *Histoire de la vie et des travaux de J.D. Cassini IV*. Clermont: A. Daix, 1851.

Dhombres, Jean. "Mathématisation et communauté scientifique française (1775–1825)." *Archives internationales d'histoire des sciences, 36* (1986), 249–93.

Dhombres, Jean. "La théorie de la capillarité selon Laplace, mathématisation superficielle ou étendue?" *RHS, 42* (1989), 43–77.

Dhombres, Nicole, and Jean Dhombres. *Naissance d'un pouvoir: Sciences et savants en France, 1793–1824*. Paris: Payot, 1989.

Dhombres Firmas, L.A. "Extrait [des mémoires sur la formule barometrique, par L. Ramond]." *JP, 75* (1812), 253–75.

Donovan, Arthur. "Lavoisier and the origins of modern chemistry." *Osiris, 4* (1988), 214–31.

Dörflinger, Johannes. "Leben und kartographisches Werk der Ingenieur-Offiziere de Traux (unter besonderer Berücksichtigung von Maximilien de Traux)." In *Cartographie* (1978), 191–223.

Dörries, Matthias. "Prior history and after effects: Hysteresis and *Nachwirkung* in the 19th cnetury." *HSPS, 22:1* (1991), 25–55.

Ducrest, L. *Vues nouvelles sur les courans d'eau, la navigation intérieure et la marine*. Paris: Perronneau, 1803.

Dulong, P.L., and A.T. Petit. "Recherches sur les lois de dilation des solides, des liquides, et des fluides élastiques, et sur la mesure exacte des températures." *ACP, 2* (1816), 240–63.

Dulong, P.L., and A.T. Petit. "Des recherches sur la mesure des températures et sur les lois de la communication de la chaleur." *ACP*, 7 (1818), 113–54, 225–64, 337–67.

Dulong, P.L., and A.T. Petit. "Sur quelques points importants de la théorie de la chaleur." *ACP*, 10 (1819), 395–413.

Eckhardt, C.L.P.G. "Auszug aus Herrn Ministerialraths Eckhardt vorläufige Nachricht." *Astronomishe Nachrichten*, 12 (1835), 129–34.

Erxleben, J.C.P. *Anfansgründe der Naturlehre*. Ed. G.C. Lichtenberg. 3rd, 4th, 5th edns. Göttingen: J.C. Dietrich, 1784, 1787, 1791.

Evans, James, and Brian Popp. "Pictet's experiment: The apparent radiation and reflection of cold." *AJP*, 53 (1985), 737–53.

Evans, T.S. "Historical memoranda respecting experiments intended to ascertain the calorific powers of the different prismatic rays." *Philosophical magazine*, 45 (1815), 401–10.

Exposition abrégée du nouveau systême des poids et mesures, d'après le metre définitif. Avignon: Seguin, an x.

Farrar, K.R. "Dalton's scientific apparatus." In Cardwell, ed., *John Dalton* (1968), 159–86.

Fauque, Daniel. "Un instrument essentiel de l'expéditions pour la mesure de la terre: le quart du cercle mobile." In Lacombe and Costabel, *Figure* (1988), 209–21.

Favre, Adrien. *Les origines du système métrique*. Paris: Presses universitaires de France, 1931.

Feldman, T.S. "Applied mathematics and the quantification of experimental physics: The example of barometric hypsometry." *HSPS*, 15:2 (1985), 127–97.

Fink, K.J. "Actio in distans, repulsion, attraction. The origin of an eighteenth-century fiction." *Archiv für Begriffsgeschichte*, 25 (1982), 69–87.

Finn, B.S. "Laplace and the speed of sound." *Isis, 55* (1964), 7–19.

Fischer, E.G. *Physique mécanique*. Tr., with notes, by J.B. Biot. Paris: Bernard, 1806.

Fischer, Joachim. *Napoleon und die Naturwissenschaften*. Stuttgart: Steiner, 1988.

Fischer, J.K. *Physikalisches Wörterbuch, oder Erklärung der vornehmsten zur Physik gehörigen Begriffe und Kunstwörter so wohl nach atomistischer als auch nach dynamischer Lehrart betrachtet*. 7 vols. Göttingen: J.C. Dietrich, 1798–1806.

Fischer, J.K. *Geschichte der Physik seit der Wiederherstellung der Künste und Wissenschaften bis auf die neuesten Zeiten*. 8 vols. Göttingen: J.J. Röwer, 1801–08.

Flaugergues, Honoré. "Extrait d'un mémoire sur le rapport de la dilation de l'air avec la chaleur." *JP, 77* (1813), 273–92.

Fleurieu. C.P.C. de. *Application du système métrique décimal appliqué à l'hydrographie et aux calculs de la navigation*. 3rd edn. Paris: Imprimerie de la République, an viii.

Forbes, E.G. *The Euler-Mayer correspondence (1751–1755). A new perspective on eighteenth-century advances in the lunar theory*. New York: American Elsevier, 1971.

Forbes, E.G. "The geodetic link between the Greenwich and Paris observatories in 1787." *Vistas in astronomy, 28* (1985), 173–81.

Fox, Robert. "Dalton's caloric theory." In Cardwell, ed., *John Dalton* (1968), 187–201.

Fox, Robert. "The background to the discovery of Dulong and Petit's law." *BJHS, 4* (1968), 1 22.

Fox, Robert. *The caloric theory of gases from Lavoisier to Regnault*. Oxford: Oxford University Press, 1971.

Fox, Robert. "The rise and fall of Laplacian physics." *HSPS, 4* (1974), 89–136.

France. Comité d'instruction publique. *Tables de rapports entre les mesures républicaines et les mesures anciennes, le plus généralement employées en France.* Paris: Imprimerie de la République, 5e jour complétaire, an iii [1795].

France. Commission temporaire des poids et mesures républicaines. *Instruction sur les mesures déduites de la grandeur de la terre, uniformes pour toute la République, et sur les calculs relatifs à leur division décimale.* Paris: Imprimerie nationale exécutive du Louvre, an ii [1794].

France. Commission temporaire des poids et mesures républicaines. *L'agence temporaire des poids et mesures, aux citoyens rédacteurs de la feuille du cultivateur, en réponse à des objections contre la nomenclature des mesures nouvelles, inserées dans le no. 38 de ce journal.* Paris: Imprimerie de la République, Thermidor, an iii [1795].

France. Commission temporaire des poids et mesures républicaines. *Avis.* Paris: Imprimerie de la République, Frimaire, an iv [1795].

France. Convention nationale. *Loi relative aux poids et mesures.* Paris: Imprimerie de la République [1795].

Frängsmyr, Tore, J.L. Heilbron, and Robin E. Rider, eds. *The quantifying spirit in the eighteenth century.* Berkeley: University of California Press, 1990.

Frankel, Eugene. "The search for a corpuscular theory of double refraction: Malus, Laplace, and the prize competition of 1808." *Centaurus, 18* (1974), 223–45.

Frankel, Eugene. "J.B. Biot and the mathematization of experimental physics in Napoleonic France." *HSPS, 8* (1977), 33–72.

Freudenthal, Gad, ed. *Etudes sur Hélène Metzger.* Paris: Fayard, 1989.

Friedman, R.M. "The creation of a new science: Joseph Fourier's analytical theory of heat." *HSPS, 8* (1977), 73–99.

Friedrich, Klaus. "Die Bedeutung Franz Xavier von Zachs für die Entwicklung der astronomischen Geographie in Deutschland." *Petermanns geographische Mitteilungen*, 117:2 (1973), 147–53.

Froeschlé, Michel. "Calendrier républicain correspondait–il à une nécessité scientifique?" In *Scientifiques et sociétés* (1990), 453–65.

Galle, A. "Über die geodätischen Arbeiten von Gauss." In Gauss, *Werke* (1863), 11:2, Abh. 1.

Gallois, Lucien. "L'Académie des sciences et les origines de la carte de Cassini." *Annales de géographie*, 18 (1909), 193–204.

Gamond, M.A. Thomé de. *Atlas containing the plans and sections of the submarine tunnel between England and France*. London: Saville, 1870.

Garnier, Jean-Guillaume. *Traité d'arithmétique*. 4th edn. Gent: J.N. Houdin, 1818.

Gascoigne, John. *Cambridge in the age of the Enlightenment*. Cambridge: Cambridge University Press, 1989.

Gattey, F. *Eléments du nouveau systême métrique, suivis des Tables des rapports des anciennes mesures agraires avec les nouvelles*. Paris: Baily and Rondonneau, an x.

Gauss, C.F. *Bestimmung des Breitenunterschiedes zwischen den Sternwarten von Göttingen und Altona durch Beobachtungen am Ramsdenschen Zenithsector*. Göttingen: Vandenhoek and Ruprecht, 1828. In Gauss, *Werke*, 9, 1–58.

Gauss, C.F. *Werke*. 12 vols. Göttingen: Gesellschaft der Wissenschaften, 1868–1929.

Gauss, C.F. *Briefwechsel zwischen Gauss und Bessel*. Leipzig: Engelmann, 1880. (Gauss. *Werke*. Ergänzungsreihe, 1.)

Gay-Lussac, J.L. "Recherches sur la dilation des gaz et des vapeurs." *AC*, 43 (1802), 137–75.

Gay-Lussac, J.L. "Relation d'un voyage aérostatique." *JP*, 59 (1804), 454–62.

Gay-Lussac, J.L. "Premier essai pour déterminer les variations de température qu'éprouvent les gaz en changeant de densité, et considérations sur leur capacité pour le calorique." *MSA*, 1 (1807), 180–204.

Gay-Lussac, J.L. "Mémoire sur la combinaison des substances gazeuses, les unes avec les autres." *MSA*, 2 (1809), 207–34.

Gay-Lussac, J.L. "Extrait d'un mémoire sur la capacité des gaz pour le calorique." *AC*, 81 (1812), 98–108.

Gay-Lussac, J.L. "Sur le calorique du vide." *ACP*, 13 (1820), 304–8.

Gehler, J.S.T. *Physikalishces Wörterbuch*. 6 vols. Leipzig: Schwickert, 1795–1801.

Gehler, J.S.T. *Physikalisches Wörterbuch, oder Versuch einer Erklärung der vornehmsten Begriffe und Kunstwörter der Naturlehre*. 2nd edn. 11 vols. Leipzig: Schwickert, 1825–45.

Gerardy, Theo. "Der Briefwechsel zwischen Carl Friedrich Gauss und Carl Ludwig von Lecoq." Akademie der Wissenschaften, Göttingen. Math.-Phys. Klasse. *Nachrichten*, 1959:4, 37–63.

Gilbert, L.W. "Eine nöthige Verbesserung der Resultate Gay-Lussac's über die Ausdehnung der Gasarten und der Dämpfe durch Wärme." *Annalen der Physik*, 12 (1802), 396–8.

Gilles, Bernhard. *J.Ch.P. Erxlebens "Anfangsgründe der Naturlehre" als Spiegelbild der physikalischen Wissenschaft im letzten Viertel des 18. Jahrhunderts*. Thesis, University of Mainz, 1978.

Gillispie, C.C. "The *Encyclopédie* and the Jacobin philosphy of science: A study in ideas and consequences." In Marshall Clagett, ed. *Critical problems in the history of science*. Madison: University of Wisconsin Press, 1959. Pp. 255–89.

Gillispie, C.C. *Science and polity in France at the end of the ancien régime*. Princeton: Princeton University Press, 1980.

Gillispie, C.C. *The Montgolfier brothers and the invention of aviation.* Princeton: Princeton University Press, 1983.

Gillispie, C.C. "Scienze e istruzione nella Rivoluzione francese." *Intersezioni,* 9 (1989), 401–13.

Gillispie, C.C. "Scientific aspects of the French Egyptian expedition." American Philosophical Society. *Proceedings,* 133 (1989), 447–74.

Gliozzi, Mario. "Il calorico." *Cultura e scuola,* 14 (Oct-Dec 1975), 183–209.

Godin, Louis. "Méthode pratique de tracer sur terre un parallèle par un degré de latitude donné; et du rapport du même parallèle dans le sphéroide oblong, et dans le sphéroide applati." *MAS,* 1733, 223–32.

Goldfarb, S.J. "Rumford's theory of heat: A reassessment." *BJHS,* 10 (1977), 25–36.

Greenberg, John. "Geodesy in Paris in the 1730s and the Paduan connection." *HSPS,* 13 (1983), 239–60.

Greene, J.C. *American science in the age of Jefferson.* Ames: Iowa State University Press, 1984.

Gregory, Olinthus. "Brief memoir of Charles Hutton." *Imperial magazine,* 5 (1823), 207–27.

Gren, F.A.C. *Grundriss der Naturlehre.* 3rd edn. Halle: Hemmerde and Schwetsche, 1797.

Grigull, Ulrich. "Das Newtonische Abkühlungsgesetz." *Physis,* 20 (1978), 213–35.

Guerlac, Henry. *Lavoisier—The crucial year. The background to his first experiments on combustion in 1772.* Ithaca, NY: Cornell University Press, 1961.

Guerlac, Henry. "Chemistry as a branch of physics: Laplace's collaboration with Lavoisier." *HSPS,* 7 (1976), 193–276.

Guerlac, Henry. *Essays and papers in the history of modern science*. Baltimore: Johns Hopkins University Press, 1977.

Guicciardini, Niccolò. *The development of Newtonian calculus in Britain, 1700–1800*. Cambridge: Cambridge University Press, 1989.

Guillaume, J. *Procès-verbaux du Comité d'instruction publique de la Convention nationale*, vol. 2. Paris: Imprimerie nationale, 1894.

Guyton de Morveau, L.B. "Sur la dilatabilité de l'air et des gaz par la chaleur." *AC*, 1 (1790), 256–99.

Haasbroek, N.D. *Investigation of the accuracy of Krayenhoff's triangulation (1802–1811) in Belgium, The Netherlands, and a part of western Germany*. Delft: Rijkscommissie voor geodesie, 1972.

Hahn, Roger. *The anatomy of a scientific institution. The Paris Academy of Sciences, 1666–1803*. Berkeley: University of California Press, 1971.

Hahn, Roger. "Laplace and Boscovich." In M. Bossi and P. Tucci, eds. Bicentennial commemoration of R.G. Boscovich. *Proceedings*. Milan: Unicopli, 1988. Pp. 71–82.

Hall, A.R., and Laura Tilling, eds. *The correspondence of Isaac Newton*, vol. 5. Cambridge: Cambridge University Press, 1975.

Hammer, E. "Zur Geschichte der Basismessung." *ZsV*, 20 (1891), 446–8.

Hassenfratz, J.H. "Discours sur le cours de physique générale." Ecole polytechnique. *Journal*, 5e cahier (an vi), 236–42.

Hassenfratz, J.H. "Physique générale: De l'enseignement de cette science." Ecole polytechnique. *Journal*, 6e cahier (an vii), 372–408.

Haüy, R.J. *Traité élémentaire de physique*. 2nd edn. 2 vols. Paris: Courcier, 1806.

Heilbron, J.L. "Duhem and Donahue." In Robert S. Westman, ed. *The Copernican achievement*. Berkeley: University of California Press, 1975. Pp. 276–84.

Heilbron, J.L. "Introduction." In Wayne Shumaker, tr. *Propaedeumata aphoristica*. Berkeley: University of California Press, 1978. Pp. 1-99.

Heilbron, J.L. *Electricity in the 17th and 18th centuries*. Berkeley: University of California Press, 1979.

Heilbron, J.L. *Physics at the Royal Society during Newton's presidency*. Los Angeles: William Anders Clark Memorial Library, 1983.

Hellman, C. Doris. "Legendre and the French reform of weights and measures." *Osiris*, 1 (1936), 314-39.

Henry, Charles. *Correspondance inédite de Turgot et Condorcet*. Paris: Charavay frères, 1883.

Herivel, J.W. "Aspects of French theoretical physics in the 19th century." *BJHS*, 3 (1966), 109-32.

Herbst, Hans-Dieter. "Das Wechselverhältnis von Astronomie und Maschinenbau, dargestellt am Beispiel des Wirkens von Georg von Reichenbach (1771-1826)." *NTM. Schriftenreihe für Geschichte der Naturwissenschaften Technik und Medizin, 28:1* (1991), 61-72.

Herschel, William. "Investigation of the powers of the prismatic colours to heat and illuminate objects." *PT, 90* (1800), 255-83. In Herschel, *Scientific papers* (1912), 2, 53-69.

Herschel, William. "Experiments on the solar, and on the terrestrial rays that occasion heat." *PT, 90* (1800), 293-326, 437-538. In Herschel, *Scientific papers* (1912), 2, 77-98, 98-146.

Herschel, William. "Experiments on the refrangibility of the sensible rays of the sun." *PT, 90* (1800), 284-92. In Herschel, *Scientific papers* (1912), 2, 70-6.

Herschel, William. "Investigation of the powers of the prismatic colours to heat and illuminate objects." *Nicholson's journal*, 4 (1800-01), 320-6.

Herschel, William. *The scientific papers.* 2 vols. London: Royal Society and Royal Astronomical Society, 1912.

Home, R.W. "Introductory monograph." In *Aepinus's essay on the theory of electricity and magnetism.* Tr. P.J. Connor. Princeton: Princeton University Press, 1979. Pp. 1–224.

Home, R.W. "Physical principles and the possibility of a mathematical science of electricity and magnetism." In Métirier, *Poisson* (1981), 151–66.

Hornsby, Thomas. [A proposal to purchase astronomical instruments]. Oxford: privately printed, 1771.

Howse, Derek. *Nevil Maskelyne. The seaman's astronomer.* Cambridge: Cambridge University Press, 1989.

Hugenin, Marcel. "French cartography of Corsica." *Imago mundi,* 24 (1970), 123–37.

Hutton, Charles. *Philosophical and mathematical dictionary* [1795]. 2 vols. 2nd edn. London: the author, 1815.

Hylsop, Beatrice Fry. *French nationalism in 1789 according to the general cahiers.* Revised edn. New York: Octagon, 1968.

Jahn, G.A. *Geschichte der Astronomie vom Anfang des 19. Jahrhunderts bis zum Ende des Jahres 1842.* 2 vols. Leipzig: H. Hunger, 1844.

Jenisch, David. *Geist und Charakter des achtzehnten Jahrhunderts. Politisch, moralisch, aesthetisch und wissenschaftlich betrachtet.* 3 vols. Berlin: Akademie der Wissenschaften, 1800–01.

Jordan, W. "Oesterreichische Geodäsie." *ZsV,* 28 (1899), 52–60.

Kelly, P. *Metrology; or, An exposition of weights and measures, chiefly those of Great Britain and France.* London: Lockingston et al., 1816.

King, Henry C. *The history of the telescope.* Cambridge, MA: Sky publishing, 1955.

Kleinert, Andreas. "Physik zwischen Aufklärung und Romantik: Die 'Anfangsgründe der Naturlehre' von Erxleben und Lichtenberg." In Bernhard Fabian, Wilhelm Schmidt-Biggemann, and Rudolf Vierhaus, eds. *Deutschlands kulturelle Entfaltung. Die Neubestimmung des Menschen*. Munich: Kraus International, 1980. Pp. 99–113.

Koeman, Cornelis. "Instruments et méthodes pour les levers du terrain à l'époque du comte du Ferraris." In *Cartographie* (1978), 289–98.

Konvitz, J.V. "Redating and rethinking the Cassini geodetic surveys of France, 1730–1750." *Cartographica*, 19 (1982), 1–15.

Konvitz, J.V. *Cartography in France, 1660–1848. Science, engineering, and statecraft*. Chicago: University of Chicago Press, 1987.

Krafft, Fritz. "Der Weg von den Physiken zur Physik an den deutschen Universitäten." *Berichte zur Wissenschaftsgeschichte*, 1 (1978), 123–62.

Krafft, J. Ch. *Recueil d'architecture civile, contenant les plans, coupes et élévations des chateaux, maisons de compagne, et habitations rurales, jardins anglais, temples, chaumières, kiosques, ponts, etc.* Paris: Bance et al., 1812.

Kuhn, T.S. "The caloric theory of adiabatic compression." *Isis*, 49 (1958), 132–40.

Kula, Witold. *Les mesures et les hommes*. Tr. Joanna Ritt. Paris: Maison des sciences de l'homme, 1984.

Lacombe, Henri, and Pierre Costabel, eds. *La figure de la terre du xviiie siècle à l'ère spatiale*. Paris: Gauthier-Villars, 1988.

La Condamine, C.M. de. "Extrait des opérations trigonométriques, et des observations astronomiques, faites pour la mesure des degrés du méridien aux environs de l'équateur." *MAS*, 1746, 618–88.

La Condamine, C.M. de. "Nouveau projet d'une mesure invariable, propre à servir de mesure commune à toutes les nations." *MAS*, 1747, 489–514.

La Condamine, C.M. de. *Mesure des trois premiers degrés du méridien dans l'hémisphère austral*. Paris: Imprimerie royale, 1751.

Lacroix, S.F. *Traité élémentaire d'arithmétique, à l'usage de l'Ecole centrale des quatre nations*. Paris: Courcier, 1805.

Lafluente, Antonio. "L'aventure et la science dans l'expédition au Pérou (1735–1743)." In Lacome and Costabel, *Figure* (1988), 139–50.

Lafuente, Antonio, and Antonio Mazuecos. *Los caballeros del Punto Fijo. Ciencia, politica y aventura en la expedición geodésica hispanofrancesca al virreinato del Perú en el siglo xviii*. Barcelona: SERBAL/CSIC, 1987.

Lafuente, Antonio, and José L. Peset. "La question de la figure de la terre au xviiie siècle." *RHS*, 37 (1984), 235–54.

Lagarde, Lucie. "Contribution de l'abbé Picard à la cartographie." In Picolet, *Picard* (1987), 247–61.

Lagrange, J.L. "Nouvelles recherches sur la nature et la propagation du son." Accademia delle scienze, Turin. *Mélanges de physique et mathématique*, 2 (1760–61), 11–172.

Laissus, Yves. "Deux lettres de Laplace." *RHS*, 14 (1961), 285–96.

Lambton, William. "An abstract of the results deduced from the measurement of an arc on the meridian." *PT*, 108 (1818), 486–517.

Laplace, P.S. de. "Leçons de mathématiques données à l'Ecole Normale en 1795." In *Oeuvres* (1878), *14*, 10–177.

Laplace, P.S. de. *Exposition du système du monde*. Paris: Imprimerie du Cercle-Social, 1796.

Laplace, P.S. de. "Des réfractions astronomiques." In *Oeuvres* (1878), *4*, 349–417. Text of 1805.

Laplace, P.S. de. "Sur l'action capillaire." In *Oeuvres* (1878), *4*, 233–277. Text of 1806.

Laplace, P.S. de. "Sur la théorie des tubes capillaires." *JP, 62* (1806); *Oeuvres* (1878), *14*, 217–27.

Laplace, P.S. de. "Sur le mouvement de la lumière dans les milieux diaphanes." *MSA, 2* (1809), 111–42.

Laplace, P.S. de. "Considérations sur la théorie des phénomènes capillaires." *JP, 89* (1819), 259–64; *Oeuvres* (1878), 259–64.

Laplace, P.S. de. *Oeuvres complètes*. 14 vols. Paris: Gauthier-Villars, 1878–1912.

Lavoisier, A.L. "De la combinaison de la matière du feu avec les fluides évaporables, et de la formation des fluides élastiques aériformes," *MAS*, 1777, 420–32; *Oeuvres* (1878), 2, 212–24.

Lavoisier, A.L. "Mémoire sur la combustion en général." *MAS*, 1777, 592–600; *Oeuvres* (1862), 2, 225–33.

Lavoisier, A.L. *Elements of chemistry* [1789]. Tr. Robert Kerr. Edinburgh: William Creech, 1790.

Lavoisier, A.L. *Oeuvres*, vol. 2. *Mémoires de chimie et physique*. Paris: Imprimerie impériale, 1862.

Lavoisier, A.L., and P.S. Laplace. "Mémoire sur la chaleur." *MAS*, 1780, 355–408. In Lavoisier, *Oeuvres* (1862), 2, 283–333. Tr. as *Memoir on heat* by Henry Guerlac. New York: Neale Watson, 1982.

Lefevre, A. *Guide pratique et mémoratif de l'arpenteur, particulièrement destiné aux personnes qui n'ont point étudié la géométrie*. Paris: Bachelier, 1833.

Legendre, Adrien-Marie. "Suite du calcul des triangles qui servent à déterminer la différence de longitude entre l'Observatoire de Paris et celui de Gréenwich." *MAS*, 1788, 747–54.

Lemay, Pierre, and Ralph E. Oesper. "Pierre Louis Dulong, his life and work." *Chymia*, 1 (1948), 171–90.

Lemmonier, —, and L.M. Normand. *Paris moderne*. Paris: Normand, 1843.

Lemoine-Isabeau. "L'élaboration de la carte de Ferraris." In *Cartographie* (1978), 39–58.

Lervig, Philip. "Les expériences de Gay-Lussac sur l'expansion adiabatique des gaz." In J. Coyac et al., eds. *Gay-Lussac: La carrière et l'oeuvre d'un chimiste français durant la première moitié du xixe siécle*. Palaiseau: Ecole polytechnique, 1980. Pp. 203–11.

Leslie, John. "Observations and experiments on light and heat, with some remarks on the enquiries of Dr Herschel, respecting those objects." *Nicholson's journal*, 4 (1800–01), 344–50, 416–21.

Leslie, John. *An experimental inquiry into the nature and propagation of heat*. London: J. Mawman, 1804.

Leslie, John. *Tracts historical and physical...respecting the election of Mr Leslie to the professorship of mathematics*. 2 vols. Edinburgh: Creech et al., 1806.

Leslie, John. "Dissertation fifth: Exhibiting a general view of the progress of mathematical and physical science chiefly during the eighteenth century." *Encyclopedia britannica*. 7th edn. London: Edinburgh: A. and C. Black, 1842. Vol. 1. Pp. 691–793.

Lessing und die Zeit der Aufklärung. Göttingen: Vandenhoek and Ruprecht, 1968.

Levallois, Jean-Jacques. "L'Académie royale des sciences et la figure de la terre." *La vie des sciences*, 3 (1986), 261–301.

Levallois, Jean-Jacques. "Picard géodésien." In Picolet, *Picard* (1987), 227–46.

Levallois, Jean-Jacques. "L'Académie royale des sciences et la figure de la terre." In Lacombe and Costabel, *Figure* (1988), 41–75.

Levallois, Jean-Jacques. "La méridienne de Dunkerque à Barcilone et la détermination du metre (1792–1799)." In *Scientifiques et sociétés* (1990), 423–40.

Levere, Trevor H. "Lavoisier: Language, instruments, and the chemical revolution." In Levere and W.B. Shea, eds. *Nature, experiment, and the sciences*. Dordrecht: Kluwer, 1990. Pp. 207–23.

Lévy, Jacques. "Picard créateur de l'astronomie moderne." In Picolet, *Picard* (1987), 133–41.

Lévy, Maurice. "Allocution dans la séance publique." Académie des sciences, Paris. *Comptes rendus, 131* (1900), 1019–40.

Libes, Antoine. *Trattato elementare di fisica esposto in un ordine nuovo secondo le moderne scoperte.* 2 vols. Florence: G. Piatti, 1803.

Libes, Antoine. *Histoire philosophique des progrès de la physique.* Paris: the author, 1810.

Liesganig, J.X. *Dimensio graduum meridiani Viennensis et Hungarici, Augustoruum jussu et auspiciis suscepta.* Vienna: A. Bernardi, 1770.

Lilley, S. "Attitudes to the nature of heat about the beginning of the nineteenth century." *Archives internationales d'histoire des sciences, 1* (1948), 630–9.

Lodwig, T.H., and W.A. Smeaton. "The ice calorimeter of Lavoisier and Laplace and some of its critics." *AS, 31* (1974), 1–18.

Lorenz, Martina. "Rezeption und Bewertung des Newtonschen Gravitationgesetzes in Physikbüchern im Zeitalter der Aufklärung in Deutschland." *NTM. Schriftenreihe für Geschichte der Naturwissenschaft- Technik-Medezin, 27* (1990:1), 25–39.

Lorette, Jean. "La fortune de la carte de Ferraris. Son emploi par la France de 1792 à 1816." In *Cartographie* (1978), 77–117.

Lovell, D.J. "Herschel's dilemma in the interpretation of thermal radiation." *Isis*, 59 (1968), 46–60.

Lundgren, Anders. "The changing role of numbers in 18th-century chemistry." In Frängsmyr, *Spirit* (1990), 245–60.

Malus, Etienne. "Sur une propriété de la lumière réfléchie." *MSA*, 2 (1809), 143–58.

Malus, Etienne. "Sur une propriété des forces répulsives qui agissent sur la lumière." *MSA*, 2 (1809), 254–67.

Malus, Etienne. "Théorie de la double réfraction." Institut des sciences, lettres et arts, Paris. *Mémoires de mathématiques et de physiques présentés...par divers savans*, 2 (1811), 303–508.

Manley, Gordon. "Dalton's accomplishment in meteorology." In Cardwell, *John Dalton* (1968), 140–58.

Marguet, F.P. *Histoire de la longitude à la mer au xviiie siècle, en France.* Paris: Challamel, 1917.

Marquet, Louis. "Le pendule à secondes et les étalons de longueur utilisés par l'expédition à l'Equateur: la toise de Pérou." In Lacombe and Costabel, *Figure* (1988), 191–206.

Martin, J.P. *La figure de la terre: Récit de l'expédition française en Laponie suédoise (1736–1737).* Cherbourg: Isoète, 1987.

Maskelyne, Nevil. "Introduction to the following observations [of Mason and Dixon]." *PT*, 58 (1768), 270–4.

Maskelyne, Nevil. "Postscript." *PT*, 58 (1768), 323–8.

Makelyne, Nevil. "Concerning the latitude and longitude of the Royal Observatory at Greenwich, with remarks on a memorial of the late M. Cassini de Thury." *PT*, 77 (1787), 151–87.

Mauskopf, S.H. "Haüy's model of chemical equivalence: Daltonian doubts exhumed." *Ambix*, 17 (1970), 182–91.

May, W.E. "Early reflecting instruments." *Nautical magazine*, 145 (1945), 21–6.

May, W.F., and H.D. Howse. "How the chronometer went to sea." *Vistas in astronomy*, 20 (1976), 135–7.

Mayer, Tobias. "Nova methodus perficiendi instrumenta geometrica et novum instrumentum geometricum." Akademie der Wissenschaften, Göttingen. *Commentarii*, 2 (1752), 325–36.

Mayer, Tobias. *Tabulae motuum solis et lunae novae correctae; quibus accedit Methodus longitudinum promota*. London: Richardson, 1770.

McKeon, Robert M. "Les débuts de l'astronomie de précision. I. Histoire de la realisation du micromètre astronomique." *Physis*, 13 (1971), 225–88.

McKeon, Robert M. "Les débuts de l'astronomie de précision. II. Histoire de l'acquisition des instruments de l'astronomie et de géodésie munis d'appareil de visée optique." *Physis*, 14 (1972), 221–42.

McKie, Douglas, and N.H. de V. Heathcote. *The discovery of specific and latent heat*. London: Arnold, 1935.

Méchain, P.F.A., and J.B.J. Delambre. *Base du système métrique décimal, ou Mesure de l'arc du méridien compris entre les parellèles de Dunkerque et Barcelona*. 3 vols. Paris: Baudouin, 1806–1810.

Melhado, E.M. "Chemistry, physics, and the chemical revolution." *Isis*, 76 (1985), 195–211.

Melhado, E.M. "Metzger, Kuhn, and 18th-century disciplinary history." In Freudenthal, *Etudes* (1989), 111–34.

Métirier, Michel, Pierre Costabel, and Pierre Dugac, eds. *Siméon-Denis Poisson et la science de son temps*. Palaiseau: Ecole polytechnique, 1981.

Middleton, W.E. Knowles. *The history of the barometer*. Baltimore: Johns Hopkins University Press, 1964.

Middleton, W.E. Knowles. *Invention of the meteorological instruments.* Baltimore: Johns Hopkins University Press, 1969.

Miller, David P. "Essay review." *History of science,* 19 (1981), 284-92.

Miller, John Philip. "Between hostile camps." *BJHS,* 16 (1983), 1-47.

Miller, John Riggs. *Speeches in the House of Commons upon the equalization of the weights and measures of Great Britain; with notes, observations, etc., etc., also, a general standard proposed for the weights and measures of Europe...together with two letters from the Bishop of Autun to the author, upon the uniformity of weights and measures.* London: J. Debrett, 1790.

Miller, Samuel. *A brief retrosepct of the 18th century.* 2 vols. New York: T. and J. Swords, 1803.

Millin, Eleuthérophile. *Annuaire du républicain, ou Legende physico-économique.* Paris: Droubin, an ii.

Molard, C.P., et al., eds. *Description des machines et procédés spécifiés dans les brevets d'invention, de perfectionnement et d'importation, dont la durée est expirée.* 93 vols. Paris: Huzard, etc., 1811-74.

Moreau, Richard. "Jean-Baptiste Biot, voluntaire de la République." In *Scientifiques et sociétés* (1990), 117-44.

Mori, Attilo. "Studi, trattative e proposte per la costruzione di una carta geografica della toscana nella seconda metà del secolo xviii." *Archivio storico italiano,* 35 (1905), 369-424.

Morichini, Domenico. "Second mémoire sur la force magnétisante du bord extrême du rayon violet." *JP,* 77 (1813), 293-308.

Morris, R.J. "Lavoisier and the caloric theory." *BJHS,* 6 (1972), 1-38.

Mudge, William. "An account of the measurement of an arc of the meridian, extending from Dunlove...to Clifton." *PT,* 93 (1803), 383-508.

Muncke, G.W. "Wärme." In J.S.T. Gehler, *Physikalisches Wörterbuch,* *10* (Leipzig: Schmickert, 1841), 52–1178.

Newton, Isaac. *Mathematical principles of natural philosophy.* Ed. Florian Cajori. Berkeley: University of California Press, 1934.

Newton, Isaac. *Philosophiae naturalis principia mathematica.* 3rd edn. (1726). Ed. Alexandre Koyré and I.B. Cohen. 2 vols. Cambridge, MA: Harvard University Press, 1972.

Newton, Isaac. *Opticks.* Ed. I.B. Cohen. New York: Dover, 1952.

Nicholson, William. *An introduction to natural philosophy.* 3rd edn. 2 vols. London: J. Johnson, 1790.

Nicholson, William. *The first principles of chemistry.* 2nd edn. London: Robinson and Robinson, 1792.

Nikolic, Djordje. "Roger Boskovic et la géodésie moderne." *Archives internationales d'histoire des sciences,* 14 (1961), 315–35.

Nischer-Falkenhof, Ernst von. "The survey by the Austrian General Staff...during the years 1749–1854." *Imago mundi,* 2 (1964), 83–8.

Nordmann, Claude J. "L'expédition de Maupertuis et Celsius en Laponie." *Cahiers d'histoire mondiale,* 10 (1966), 74–97.

Oldenburg, Henry. *Correspondence.* Ed. and tr. A. Rupert Hall and Marie Boas Hall. 9 vols. Madison, WI: University of Wisconsin Press, 1965–73.

Olmsted, J.W. "The scientific expedition of Jean Richer to Cayenne (1672–1673)." *Isis,* 34 (1942), 117–28.

Olmsted, J.W. "The 'application' of telescopes to astronomical instruments, 1667–1669." *Isis,* 40 (1949), 213–25.

Olson, Richard. "A note on Leslie's cube in the study of radiant heat." *AS,* 25 (1969), 203–8.

Olson, Richard. "Count Rumford, Sir John Leslie, and the study of the nature and propagation of heat at the beginning of the 19th century." *AS, 26* (1970), 273–304.

Olson, Richard. *Scottish philosophy and British physics 1750–1880.* Princeton: Princeton University Press, 1975.

Oseen, C.W. *Johan Carl Wilcke, experimental-fysiker.* Uppsala: Almqvist and Wiksells, 1939.

Outram, Dorinda. "The language of matured power: The *éloges* of Georges Cuvier and the public language of 19th-century science." *History of science, 16* (1978), 153–78.

Paldus, Josef. *Die militärischen Aufnahmen im Bereiche der Habsburgischen Länder aus der Zeit Kaiser Josephs II.* Vienna: A. Hölder, 1919. (Akademie der Wissenschaften, Vienna. Phil.-Hist. Klasse. *Denkschriften, 63:2.*)

Paolucci, N., G. Tagliaferri, and P. Tucci. "Le vicende scientifiche ed extrascientifiche della realizzazione della prima carta geografica della lombardia con metodi astronomici." Congresso nazionale di storia della fisica, VIII. *Atti.* Milan: Gruppo nazionale di coordinamento per la storia della fisica, 1988. Pp. 383–409.

Pattison, W.D. *Beginnings of the American rectangular land survey system, 1784–1800.* Chicago: University of Chicago, Department of Geography, 1957.

Pepys, Samuel. *Diary.* Ed. Robert Latham and William Matthews. 11 vols. Berkeley: University of California Press, 1970–83.

Perrin, C.E. "Research traditions, Lavoisier, and the chemical revolution." *Osiris, 4* (1988), 53–81.

Perrin, C.E. "Chemistry as peer of physics: A response to Donovan and Melhado on Lavoisier," *Isis, 81* (1990), 259–70.

Petit, A.T. "Théorie mathématique de l'action capillaire." École polytechnique. *Journal, 9:16* (1813), 1–40.

Petit, A.T., and P.L. Dulong. "Recherches sur quelques points importans de la théorie de la chaleur." *ACP*, 10 (1819), 395–413.

Pfitzer, A. "Zur Geschichte des Rheinisch-Westfälischen Katasters. Johann Jacob Vorlaender—ein Vorkämpfer des preussischen Vermessungswesen." *ZsV*, 42 (1913), 1–7, 40–9.

Picard, Jean. *Mesure de la terre*. Paris: Imprimerie royale, 1671.

Picard, Jean. "Mesure de la terre [1671]." *MAS*, 1666–1699, 7 (1729), 133–90.

Picolet, Guy, ed. *Jean Picard et les débuts de l'astronomie de précision au xviie siècle*. Paris: CRNS, 1987.

Picon, Antoine. "Les ingénieurs et la mathématisation. L'exemple du génie civil et de la construction." *RHS*, 42 (1989), 155–72.

Pictet, Marc August. "Considerations on the convenience of measuring an arch of the meridian, and of the parallel of longitude, having the Observatory of Geneva for their common intersection." *PT*, 81 (1971), 106–27.

Playfair, John. *Outlines of natural philosophy*. 2 vols. Edinburgh: A. Constable, 1812–14.

Playfair, John. *The works*. 4 vols. Edinburgh: A. Constable, 1822.

Poisson, S.D. "Mémoire sur la théorie du son." Ecole polytechnique. *Journal*, 7:14 (1808), 319–92.

Poisson, S.D. "Extrait d'un mémoire sur la distribution de l'électricité à la surface des corps conducteurs." *JP*, 75 (1812), 229–37.

Poisson, S.D. "Sur la vitesse du son." *ACP*, 23 (1823), 5–16.

Pommiés, Michel. *Manuel de l'ingénieur du cadastre*. Paris: Imprimerie impériale, 1808.

Pottier, Adrien. *Arithmétique pratique et démontrée, pour réduire les anciennes mesures en nouvelles, par une méthode courte et propre à faciliter les personnes qui désirent prendre connaissance, par principes, de ce système*. Paris: Bernard Moutardier, an viii.

Prevost, Pierre. "Mémoire sur l'équilibre du feu." *JP, 38* (1791), 314–23.

Prevost, Pierre. *Du calorique rayonnant*. Paris: J.J. Paschoud, 1809.

Prieur, C.A. *Instruction sur le calcul décimal, appliqué principalement au nouveau système des poids et mesures*. Paris: Imprimerie de la République, Germinal an iii.

Pringle, John. *Six discourses, delivered...on occasion of six annual assignments of George Copley's medal. To which is prefaced the life of the author, by Andrew Kippis*. London: W. Straham and T. Cadell, 1783.

Projets d'architecture et autres productions de cet art qui ont merités les grands prix accordés par l'Académie, par l'Institut National de France, et par des Jury du choix des artistes ou du gouvernement. Paris: Detourelle, 1806.

Prony, Gaspar de. "Discours préliminaire." In William Roy. *Description des moyens employées pour mesurer la base de Hounslow-Heath*. Tr. G. Prony. Paris: Didot, 1787. Pp. i–xx.

Puissant, L. *Traité de topographie, d'arpentage et de nivellement*. Paris: Courcier, 1807.

Randall, Wyatt W. *The expansion of gases by heat. Memoirs by Dalton, Gay-Lussac, Regnault and Chappuis*. New York: American Book, 1902.

Ramond de Carbonnieres, L.F.E. *Mémoires sur la formule barométrique de la mécanique céleste, et les dispositions de l'atmosphère qui en modifient les propriétés*. Clermont-Ferrand: Landriot, 1811.

Ravetz, J.R. "The representation of physical quantities in 18th century mathematical physics," *Isis, 52* (1961), 7–20.

Rayleigh, J.W. Strutt, 3rd baron. *Scientific papers.* 6 vols. Cambridge: Cambridge University Press, 1899–1920.

Regnault, H.V. "Recherches sur la dilation des gaz." *ACP,* 4 (1842), 5–63, and *ACP,* 5 (1842), 57–83.

Reichmann, Eberhard. *Die Herrschaft der Zahl. Quantitatives Denken in der deutschen Aufklärung.* Stuttgart: B. Metzler, 1968.

Repsold, J.A. *Zur Geschichte der astronomischen Messwerkzeuge, 1450–1830.* 2 vols. Leipzig: Engelmann, 1908.

Richeson, A.W. *English land measuring to 1800: Instruments and practices.* Cambridge, MA: MIT Press, 1966.

Rider, R.E. "Measure of ideas, rule of language: Mathematics and language in the 18th century." In Frängsmyr, *Quantifying spirit* (1990), 113–40.

Rigaud, S.P. "Memoire." In Bradley, *Miscellaneous works* (1832), i–lxx.

Robison, John. *A system of mechanical philosophy.* 4 vols. Edinburgh: J. Murray, 1822.

Rodriguez, Joseph. "Observations on the measurement of three degrees of the meridian in England by Lieut. Col. William Mudge." *PT,* 102 (1812), 321–51.

Rodriguez, Joseph. "Observations on the measurement of three degrees of the meridian conducted in England by Lieut.-Colonel William Mudge." *Philosophical magazine,* 41 (1813), 20–31, 90–100.

Rondelet, J.B. *Traité théorique et pratique de l'art de bâtir.* 4 vols. in 5, + 3 vols. of plates. Paris: the author, 1802–1817.

Rosenberger, Ferdinand. *Die Geschichte der Physik.* 3 vols. Braunschweig: Vieweg, 1882–90.

Rosmorduc, Jean. "Des idées de Gay–Lussac sur la lumière." In *Colloque Gay–Lussac* (1980), 213–40.

Rousseau, G.S., and Roy Porter, eds. *The ferment of knowledge: Studies on the historiography of 18th-century science.* Cambridge: Cambridge University Press, 1980.

Rowlinson, J.S., ed. *J.D. van der Waals: On the continuity of the gaseous and liquid states.* Amsterdam: North-Holland, 1988.

Roy, William. "Experiments and observations made in Britain, in order to obtain a rule for measuring heights with the barometer." *PT, 67* (1777), 653–770.

Roy, William. "An account of the measurement of a base on Hounslow-Heath." *PT, 75* (1785), 385–480.

Roy, William. "An account of the mode proposed to be followed in determining the relative situation of the Royal observatories of Greenwich and Paris." *PT, 77* (1787), 188–226.

Roy, William. "An account of the trigonometric operation, whereby the distance between the meridians of the Royal observatories of Greenwich and Paris has been detertmined." *PT, 80* (1790), 111–270.

Rumford, Benjamin Thomson, Count. "An inquiry concerning the nature of heat and the mode of its communication." *PT, 94* (1804), 77–182; *Works* (1968), *1,* 323–433.

Rumford, Benjamin Thomson, Count. "Historical review of the various experiments of the author on the subject of heat." In *Works* (1968), *1,* 443–96. Text of 1804.

Rumford, Benjamin Thomson, Count. "Reflections on heat." In *Works* (1968), *1,* 301–22. Text of 1804.

Rumford, Benjamin Thomson, Count. "Research on heat, second memoir." *MIF,* 1806, 79–87; *Works* (1968), *1,* 434–41.

Rumford, Benjamin Thomson, Count. *The collected works.* Ed. S.C. Brown. 5 vols. Cambridge: MIT Press, 1968–70.

Sabra, A.I. *Theories of light, from Descartes to Newton.* London: Oldbourne, 1967.

Sadoun-Goupil, Michelle. "Gay-Lussac disciple de Berthollet?" In *Colloque Gay-Lussac* (1980), 59–75.

Sarton, George, et al. "Documents nouveaux concernant Lagrange." *RHS*, 3 (1950), 110–32.

Saussure, H.B. de. *Voyage dans les alpes.* 4 vols. Geneva: S. Fauche 1779–96.

Saveney, Edgar. "L'ancienne académie et les académiciens." *Revue de deux mondes*, 84 (1869:6), 199–226.

Schilling, C. *Wilhelm Olbers. Sein Leben und seine Werke, 2. Briefwechsel zwischen Olbers und Gauss.* 2 vols. Berlin: Springer, 1900–09 (Gauss. Werke. Ergänzungsreihe, 4.)

Schimank, Hans. "Stand und Entwicklung der Naturwissenschaften im Zeitalter der Aufklärung." In *Lessing und die Zeit der Aufklärung*. Göttingen: Vandenhoek and Ruprecht, 1968. Pp. 30–76.

Schmidt, Johann Jacob. *Biblischer Mathematicus, oder Erläuterung der heil. Schrift aus den mathematischen Wissenschaften.* Züllichau: G.B. Fromann, 1736.

Schmidt, Rudolf. *Die Kartenaufnahme der Rheinlande durch Tranchot und v. Muffling, 1801–1828. 1. Geschichte des Kartenwerkes und vermessungstechnische Arbeiten.* Cologne and Bonn: Hanstein, 1973–75. (Gesellschaft für Rheinische Geschichtskunde, *Publicationen*, 12.)

Schmidt, Rudolf. "Die Kartenaufnahme der Rheinlande unter Tranchot und v. Müffling 1801–1828." In *Cartographie* (1978), 275–86.

Scientifiques et sociétés pendant la Révolution et l'Empire. Paris: Comité des travaux historiques et scientifiques, 1990. (Congrès national des sociétés savants, CXIV, 1989. Section histoire des sciences et des techniques. *Actes.*)

Scott, E.L. "Richard Kirwan, J.H. de Magellan, and the early history of specific heat." *AS, 38* (1981), 141–53.

Sebastiani, Fabio. "La memoria voltiana interno al colore." *Physis, 23* (1981), 89–113.

Sebastiani, Fabio. "Le teorie caloristiche di Laplace, Poisson, Sadi Carnot, Clapeyron e le teoria dei fenomeni termici nel gas formulata da Clausius nel 1850." *Physis, 23* (1981) 397–438.

Sebastiani, Fabio. "Le teorie microscopico-caloricistiche dei gas di Laplace, Ampère, Poisson e Prevost." *Physis, 24* (1982), 197–236.

Seguin, Armand. "Observations générales sur le calorique." *AC, 3* (1789), 148–242.

Seguin, Armand. "Seconde mémoire sur le calorique." *AC, 5* (1790), 191–271.

Shapin, Steven, and Simon Schaffer. *Leviathan and the air pump. Hobbes, Boyle, and the experimental life.* Princeton: Princeton University Press, 1985.

Shuckburgh, G.E. "Account of the equatorial instrument." *PT, 83* (1793), 67–128.

Shuckburgh, G.E. "An account of some endeavors to ascertain a standard of weight and measure." *PT, 88* (1798), 133–82.

Sigrist, René. *Les origines de la Société de physique et d'histoire naturelle (1790–1822). La science genevoise face au modèle français.* Geneva: ATAR, 1990. (Société de physique et d'histoire naturelle de Genève. *Mémoires,* 45:1.)

Silliman, Robert H. "Fresnel and the emergence of physics as a discipline." *HSPS, 4* (1974), 137–62.

Skelton, R.A. "The origins of the Ordnance Survey of Great Britain." *Geographical journal, 128* (1962), 415–26.

Smeaton, John. "Observations on the graduations of astronomical instruments; with an explanation of the method invented by the late Mr. Henry Hindley, of York, clock-maker, to divide circles into any given number of parts." *PT*, 76 (1786), 1–47.

Smith, J.R. *From plane to spheroid: Determining the figure of the earth from 3000 B.C. to the 18th-century Lapland and Peruvian expeditions.* Rancho Cordova, CA: Landmark Enterprises, 1986.

Sombart, Werner. *Der Bourgeois. Zur Geistesgeschichte des modernen Wirtschaftsmenschen.* Munich and Leipzig: Duncker and Humblot, 1913.

Spurgin, C.B. "Gay-Lussac's gas-expansivity experiments and the traditional mis-teaching of 'Charles law'." *AS*, 44 (1987), 489–505.

Stafford, B.M. "Science as fine art: Another look at Boullée's cenotaph for Newton." *Studies in 18th century culture*, 11 (1982), 241–78.

Stimson, A. "The influence of the Royal Observatory at Greenwich upon the design of 17th and 18th century angle-measuring instruments at sea." *Vistas in astronomy*, 20 (1976), 123–30.

Tables de comparaison entre les mesures anciennes en usage dans le Département du Mont-Tonnerre et les nouvelles mesures républicaines. Mainz: Pfeipfer, an x.

Tagliaferri, Guido, and Pasquale Tucci. "La visita dei Paesi bassi nel diario di viaggio di Barnaba Oriani (1752–1832), astronomo milanese del settecento." *Incontri*, 4 (1989), 59–83.

Talleyrand-Péroigord, C.M. de. *Proposition faite à l'Assemblée nationale, sur les poids et mesures.* Paris: Imperimerie nationale, 1790.

Taton, René. "Picard et la *Mesure de la terre*." In Picolet, *Picard* (1987), 207–26.

Taton, René. "L'expédition géodésique de Laponie (avril 1736–août 1737)." In Lacombe and Costabel, *Figure* (1988), 115–37.

Ten, Antonio E. "Les expéditions de Méchain et Biot-Arago et le prolongement de la méridienne de Paris jusqu'aux iles baléares." In Lacombe and Costabel, *Figure* (1988), 245-64.

Ten, Antonio E. "Le problème du 45e parallel et les origines du système métrique décimal." In *Scientifiques et sociétés* (1990), 441-52.

Terrall, Mary. "Representing the earth's shape: The polemics surrounding Maupertuis' expedition to Lapland." *Isis, 83* (1992), 218-37.

Thinon, A. *Leçons sur le système métrique et sur les applications nouvelles de l'arithmétique; à la mesure des longueurs, des surfaces, des volumes, des angles, du temps, des poids spécifiques, au jaugeage des tonneaux, à la cubature des bois, etc.* 12th ed. Paris and Bar-le-duc, Dezobry and Magdelleine, 1860.

Thomson, Thomas. *A system of chemistry.* 4 vols. Edinburgh: Bell et al., 1802.

Tilling, Laura. "Early experimental graphs." *BJHS, 8* (1975), 193-213.

Todhunter, Isaac. *A history of the mathematical theories of attraction and the figure of the earth from the time of Newton to that of Laplace.* 2 vols. London: Macmillan, 1873.

Topping, Michael. "Measurement of a base line upon the sea beach, near Porto Novo, on the coast of Coromandel." *PT, 82* (1792), 99-114.

Tremery, Jean-Louis. "Observations sur les rapports qui lient la théorie du magnétisme à celle de l'électricité, et sur le condensateur de Volta." Société philomatique, *Bulletin, 3* (1813), 291-6. (Memoir of May 1813).

Tremery, Jean-Louis. "Observations sur les expériences à l'aide desquelles les physiciens démontrent la réflexion du calorique." Société philomathique, *Bulletin, 3* (1813), 323-8.

Truchot, P. "Les instruments de Lavoisier." *ACP*, *18* (1879), 289–319.

Tweedie, Charles. *James Stirling; a sketch of his life and works along with his scientific correspondence.* Oxford: The Clarendon Press, 1922.

Vannereau, Marie-Antoinette. "L'oeuvre cartographique des Cassini." In *Cartographie* (1978), 227–34.

Viel, C.F. *De la construction des édifices publics sans l'emploi du fer, quel en doit être l'usage dans les batimens particuliers.* Paris: H.L. Perronneau, 1803.

Villain, Georges. "Etude sur le calendrier républicain." *La Révolution française*, *7* (1884), 451–9, 535–53; *8* (1885), 623–56, 740–58, 830–54, 883–8.

Volpicelli, P. "Del calorimetro a ghiaccio e suoi usi." *Giornale arcadico di scienze, lettere ed arti*, *60* (1833), 50–71.

Volta, Alessandro. "Del modo di render sensibilissima la più debole elettricità sia naturale, sia artificale." *PT*, *72* (1782), 237–80.

Volta, Alessandro. *Le opere.* 7 vols. Milan: Hoepli, 1918–29.

Vuillemin, Jules. "Sur la généralisation de l'estimation de la force chez Laplace." *Thalès*, *9* (1958), 61–75.

Wallis, H.M. "Geographie is better than divinitie: Maps, globes, and geography in the days of Samuel Pepys." In Norman J.W. Thrower, ed. *The compleat plattmaker. Essays on chart, map, and globe making in England in the seventeeth and eighteenth centuries.* Berkeley: University of California Press, 1978. Pp. 1–43.

Wallis, Helen. "Cartography in Great Britain in the 18th century." In *Cartographie* (1978), 165–75.

Wargentin, Peter. "Concerning the difference in longitude of the Royal observatories at Paris and Greenwich, resulting from the eclipses of Jupiter's first satellite." *PT*, *67* (1777), 162–86.

Warmé, Vulfran. *Eloge historique de M. Delambre.* Amiens: Caron-Duquenne, 1824.

Weiss, Berghard. *Zwischen Physikotherapie und Positivismus. Pierre Prevost (1751-1839) und die korpuskular-kinetische Physik der Genfer Schule.* Frankfurt: P. Lang, 1988.

Westphal, A. "Die geodätischen und astronomischen Instrumente zur Zeit des Beginns exakter Gradmessungen." *Zeitschrift für Instrumentenkunde,* 4 (1884), 152–66, 189–202.

Westphal, A. "Basisapparate und Basismessungen." *Zeitschrift für Instrumentenkunde,* 5 (1885), 257–74, 333–45, 373–85, 420–32.

Whitehurst, John. "An attempt toward obtaining invariable measures of length, capacity, and weight, from the mensuration of time, independent of the mechanical operations requisite to ascertain the center of oscillation, or the true length of pendulums." In John Whitehurst, *The works.* London: W. Bent, 1792. Item 2, pp. i–xiii, 1–34.

Whitman, Walt. *The complete writings.* Ed. R.M. Bucke et al. New York: Putnams, 1902.

Widmalm, Sven. "Accuracy, rhetoric, and technology: The Paris-Greenwich triangulation, 1784–88." In Frängsmyr, *Quantifying spirit* (1990), 179–206.

Widmalm, Sven. *Mellam kartan och verkligheten. Geodesi och kartläggning, 1695–1860.* Uppsala: Uppsala University, Institutionen för idé- och lärdomshistorie, 1990.

Wilcke, J.C. "Rön om varmens spänstighet och fördeling, i anledning af ångors upstingande och kyla, uti förtunnad luft." Royal Academy of Sciences, Stockholm. *Handlingar,* 2 (1781), 143–63.

Williams, Edward, William Mudge, and Isaac Dalby. "An account of the trigonometrical survey carried out in the years 1791, 1792, 1793, and 1794, by order of his Grace the Duke of Richmond." *PT,* 85 (1795), 414–591.

Williams, Edward, William Mudge, and Isaac Dalby. "An account of the trigonometrical survey carried on in the years 1795 and 1796, by order of the Marquis Cornwallis." *PT, 87* (1797), 432–541.

Wolf, C.J.E. *Mémoires sur le pendule, précédés d'une bibliographie.* 2 vols. Paris: Gauthier-Villars, 1889–1891.

Wolf, C.J.E. *Histoire de l'Observatoire de Paris de sa foundation à 1793.* Paris: Gauthier-Villars, 1902.

Young, Thomas. *A course of lectures on natural philosophy and the mechanical arts.* London: J. Johnson, 1807.

Young, Thomas. "[Review of works by Malus et al.]" *Quarterly review, 11* (1814), 42–56.

Zach, Franz Xaver von. "Zusätze des Herausgeber über die Lemberger Sternwarte." *Monatliche Correspondenz zur Beförderung der Erd- und Himmels-Kunde, 4* (1801), 550–8.

Zach, Franz Xaver von. "Beweis, dass die Oesterreichische Gradmessung des Jesuiten Liesganig sehr fehlerhaft, und zur Bestimmung der Gestalt der Erde ganz untauglich sey." *Monatliche correspondez zur Beförderung der Erd- und Himmelskunde, 8* (1803), 507–27; *9*, 32–8, 120–30.

Zach, Franz Xaver von. "Mémoire...sur le degré du méridien mesuré en Piémont par le P. Beccaria." Turin: F. Galletti, 1811.

Zach, Franz Xaver von. *Briefe Franz Xaver freiherrn von Zach...an P. Martin Alois David.* Ed. Otto Seydl. Prague: Gesellschaft der Wissenshaften, 1938.

Zach, Franz Xaver von. *Briefe Franz Xaver von Zachs in sein Vaterland.* [Mostly to Ludwig Schedius in Pest.] Ed. Peter Brosche and Magda Vargha. Budapest: L. Eotvos University, 1984.

Zerubavel, Eviatar. *Hidden rhythms. Schedules and calendars in social life.* Chicago: University of Chicago Press, 1981.

INDEX

Aberration, stellar: 43, 225, 226
Accuracy: of angular measure, 37–44, 228; in geodetic measurement, 186, 187, 189–195, 200–201, 209, 222–224; of sextant, 49; of repeaters, 51, 57, 61; required for lunar method, 47, 49; of baselines, 190, 192, 194–195, 205, 207, 210–211, 223, 227, 237; specific heats, 101, 110, 112; of Coulomb's magnetic measurements, 69, 70; of his electric ones, 76, 79; in calibrating thermometers, 94–96; of finding dilation of mercury, 99; of capillary rise, 163; of length of seconds pendulum, 264; of metric standards, 266
Achard, Franz Karl: 26, 29–30
Adiabatic expansion, law of: 185
Aepinus, Franz Ulrich Theodor: 10, 22, 72
Affinity: caloric, 14; cohesion, 27; capillarity, 164
Air condenser: 10
Alembert, Jean Le Rond d'.: on gravity, 150, speed of sound, 166; earth's shape, 218, 228; flattening, 224; decimalized trigonometry, 251
Amontons, Guillaume: 173
Analogy: and physical models, 18; between electrical fluid and caloric, 10–11, 13, 14, 15, 16, 84–85, 116, 124n; radiant heat and light, 22, 127, 134–135; and electric fluid, 120, 121; temperature and tension, 85–87, 94–96; gravitation and electricity and magnetism, 15, 67, 80–81; electricity and vacuum, 116–117; caloric and gas, 117; ice calorimeter and torsion balance, 102–104
Animal electricity: 16
Annalen der physik: 25–26
Annales de chimie: 25
Arago, François: 240; refraction, 61–62; extends Paris meridian, 240
Arbuthnot, Sir John: 245
Architectural drawings: 273–274
Arcueil, Société d': establishment, 140; program, 112, 141–147; dead hand, 146–147

322 / INDEX

Arithmetic: 2
Astronomy, for travellers: 50
Attraction: as effect, not cause, 17; at Paris Academy, 155; of mountains, 228–231, 242. See also Force
Aubert, Alexandre: 87, 202
Aune: 243
Autumnal equinox: and Republican calendar, 249, 250, 252, 258
Auzout, Adrien: 36

Babbage, Charles: 148
Baden-Domlac, Margrave of: 193
Ballooning: 8
Banks, Joseph: radiant heat, 123, 134; Paris-Greenwich link, 204
Barometric coefficient: 31–32, 90. See also Hypsometry
Baseline measurements: 187, 235. See also Accuracy
Becarria, Giambattista: 228, 229
Benedict XIV, Pope: 226
Bennet, Abraham: 82, 83
Bentham, Jeremy: 17
Bérard, Jacques Etienne: specific heat of gases, 107–112; radiant heat, 135; γ, 176; and Arcueil, 140n
Bergeret, Jacques Onésine: 46
Bern, Ökonomische Gesellschaft: 205
Bernoulli, Johann I: 166, 222
Berthollet, Claude-Louis: 96, 112, 171, 268, 269; chemistry and physics, 27; light and heat, 22; as patron of science, 140–141, 142n, 146–147
Berthoud, Ferdinand: 254
Bertrand, Louis: 118
Berzelius, Jöns Jacob: 21
Bessel, Friedrich Wilhelm: 240n
Bible: 1
Biot, Jean Baptiste: 240; balloon ascent, 8–9; on proper physics, 18, 30; instrumentalism, 21–22, 142; as Laplacian physicist, 140n, 147, 164; on refractive power of gases, 9, 31, 61–62; radiant heat, 135, 137; speed of sound, 171–174; extension of Paris meridian, 236–238; decimalized time, 254; metric politics, 208, 262–263; Borda and Coulomb, 36; Saint Pierre, 148; Laplace, 164
Bird, John: 44, 46; graduation of arc, 42, 44; evaluated, 53, 54
Black, Joseph: 11, 72, 101
Blagden, Joseph: 46n, 202, 203
Bohnenberger, Johann Gottlieb von: 212
Bologna, San Petronio: 37
Borda, Charles: 47, 263, 266, 267; as model physicist, 30, 35–36, 61; repeating circle, 52–54; improved by Lenoir, 55–59; compared with Ramsden's theodolite, 59–60; used in physics experiments, 61–65; measures refraction, 61; promotes metric project, 260–261, 262, 268, 270;

determines seconds pendulum, 264–265; Borda's rules, 235. *See also* Repeater
Boscovich, Roger: 201; geodesy of Papal states, 226–228; attraction of mountains, 228–229; promotes geodetic measurements, 229–230
Bouguer, Pierre: on gravity, 150; and Peru, 223; Paris meridian, 226; attraction of mountains, 228
Boyle's law: 95, 169, 170, 172, 173, 174, 179, 183
Bradley, James: 222; uses Bird and Graham instruments, 42–43; discovers aberration and mutation, 42–43; tests octant, 49; on Bird, 53, 54; Paris-Greenwich link, 202, 205; attraction of mountains, 230n
Brisson, Mathurin-Jacques: 263, 266, 267, 268, 269; instrumentalism, 20
Buffon, Georges-Louis Leclerc, comte de: 150
Bugge, Thomas: surveys Jutland, 206n; on accuracy of metric survey, 269n, 270n
Bunsen, Robert: 104

Cahiers: 244
Cain: 243
Calendar. *See under* Gregorian and Republican
Calibration of thermometers: 87–90, 95–96
Calon, E.N. de: centralizes revolutionary cartography, 197, 234–235
Caloric: defined, 5 19; as principle of separation, 10, 14; in Lavoisier's system, 12–13; in Standard Model, 10, 25, 86; analogy to electric fluid, 10–11, 13–15; to magnetism, light, and gas, 117–119; instrumentalist use, 19, 20, 120; law of force of, 85–86; and speed of sound, 170–171; and Laplace's gas model, 177–178. *See also* Heat
Calorimetry, ice: 101–104
Canals, in hydrostatic analysis: 157–158, 161, 163–164, 216, 219
Capacity, electrical: 11, 66
Capillarity: 27, 103; Clairaut's theory of, 154–157; Laplace's, 159–165
Carbonnières, Louis Ramond de: 32–33
Carte de l'Empereur: 272
Cartography: of France, 185–193, 196–198; Flanders, 191, 196; Austria-Hungary, 198–199; Belgium, 199; Bavaria, 200, 212; Tuscany, 201; Scotland, 202; Switzerland, 205–206; Britain, 207; India, 207; Rhenish provinces, 207–28; Netherlands, 208–210; Thuringia, 210–211; Hanover, 208, 211, 213; Denmark, 206n, 213; Papal States, 226–228; and the military, 191, 193n, 194n, 195–198, 230

324 / INDEX

Cassini, Gian-Domenico (Cassini I): *meridiana*, 37; collaboration with Picard, 187; quadrant, 225n

Cassini, Jacques (Cassini II): 211, 221; and Newton, 217; Paris meridian, 189; Paris perpendicular, 190, 220; seconds pendulum, 259; accuracy, 224–225

Cassini de Thury, César-François (Cassini III): on traveller's astronomy, 50; maps France, 190–191, 225; expedition to Vienna, 192–195; gets longitude by light flashes, 194, 228; promotes Paris-Greenwich link, 201–202, 211; advises Ferraris, 199; Clairaut and Maupertuis, 222; evaluated, 195, 202

Cassini, Jean Dominique (Cassini IV): 263, 266; on Paris-Greenwich link, 60, 202; finishes map of France, 191–192; promotes survey of Tuscany, 201; and English instrument makers, 44–47, 203; seconds pendulum, 264–265

Catherine the Great, Empress of Russia: 72

Cavendish, Henry: electric force, 15, 81, 202; thermometry, 87–88

Celsius, Anders: 222

Chaptal, Jean-Antoine: 9, 135, 275

Charles, Jacques-Alexandre: indentified, 95, 100

Charles' law: 95

Chateaubriand, François René de: 149

Chemical Revolution: 25

Chemistry: as science of insensible motions, 28; and physics, 24–27

Clairaut, Alexis-Claude: on gravity, 150–151; refractive force, 153–154; capillary force, 156–159; Lapland expedition, 222; geoid, 241

Clément, Nicolas: 183; on caloric, 10; specific heat of gases, 106, 112–117

Cocquebert, C.E.: 268n

Colbert, Jean Baptiste: 185

Comte, Auguste: 142

Condorcet, Marie-Jean-Antoine-Nicolas Caritat, marquis de: 16–17, 245, 263; urges decimalization, 247; and metric reform, 258, 260

Cotes, Roger; 167n

Coulomb, Charles Augustin: 263, 266, 267, 269; dualist in electricity and magnetism, 21–22; as model physicist, 30, 35; measures magnetic force, 67–72, 81; electric, 15, 75–81; difficulty of his measurements, 15, 79–80; proof plane, 81–82

Coulomb's balance: 35, 79–80, 86

Coulomb's law: as exemplar of physics, 15; and Standard Model, 80; as grantor of

Laplace's approach, 141
Curvature, radius of: 218

Dalby, Isaac: 239
Dalrymple, Alexander: 202
Dalton, John: 92, 146, 147; gas laws, 91, 106; on latent heat, 107; criticized, 96
Daun, Leopold Joseph von: 198, 199
David, King of Israel: 1
David, Martin Alois: 49
Davy, Humphry: 121
Décade: 251, 257
Decimalization: as rational and liberating, 246-248; of the calendar, 250-253; circular measure, 249-250; the day, 251-256; watches, 254-255; spread by metric system, 276; relaxed in metric system, 273. *See also* Republican calendar
Dee, John: 2
Degree, length of: Picard's, 187; Cassini II, 189-190; Lapland, 224; Peru, 224, Lacaille's 225-226; Boscovich's, 228; Maryland, Hungary, Piedmont, 229-230; Mudge's, 240-241
Delambre, Jean-Baptiste-Joseph: 208, 209, 213, 254, 264, 267, 268, 269, 270; physics and chemistry, 27; calculations for invasion of England, 206; measures Paris meridian, 233, 231-236; calculates flattening, 235-236, 241, 243; on metric politics, 262; on Borda's circle, 263
Delaroche, François: specific heats, 107-114; radiant heat, 135-136; cooling, 137; γ, 176-177
Deluc, Jean André: on Laplace, 20; Coulomb, 80; quantification in physics, 29; thermometry, 87; gas expansion, 90
Desaguliers, John Theodore: 24, 217-219
Desormes, Charles Bernard: 10, 183; collaboration with Clément, 107, 112-117
Diachronic terms: 8, 25-26
Dirac delta function: 146
Distance forces: in Enlightenment, 3; of imponderables, 5; and instrumentalism, 15. *See also* Attraction; Capillarity; Electricity; Force
Dividing engines: 43-44, 49, 215
Dixon, Jeremiah; 229, 230
Dollond, Peter: 43, 44, 50
Dulong, Pierre Louis: 144; calibration of mercury and air thermometers, 96-101; prejudice for simplicity, 106; specific heat of metals, 106; cooling, 138; and Arcueil, 140n, 147

Earth, shape of: according to Newton, 192, 216-219; Huygens, 192; Cassini II, 189; Wallis, 190n; relevant geometry, 219; Poleni's pro-

326 / INDEX

posal, 220; disagreement over flattening, 226–227; not an ellipsoid, 239–241; and metric politics, 240–241
Eckhardt, C.L.P.: 211, 212, 240
Electrical fluids: number of, 21–22; quantity, 81–82; analogy to caloric, 10–11, 13–15, 124n; to vacuum, 117–118
Electricity: and force physics, 10, 13–15; law of force: 72–81; capacity, charge, and tension, 82–85; analogy to gravity, 15, 80–81, 85; as branch of chemistry, 28; and instrumentalism, 21–22
Electroscopes: 82–84
Enlightenment: buzzwords, 245
Epact: 252
Epailly, ingénieur géographe: 209, 211, 212, 213
Erxleben, Johann Christian Polykarp: 23
Euler, Leonhard: 150, 193
Evaporation: 13
Exchanges, law of: 120–123

Fabre d'Eglantine, P.F.N.: 252–253
Fallon, Ludwig August von: 230
Ferraris, Johann Joseph von: 199
Ferraris map: 199–200
Feudal measures: 243–245
Fischer, Ernst Gottfried: *Physique mécanique*, 18–19, 28; on Gren's *Grundriss*, 18, 31; physics and chemistry, 26; and Coulomb, 79–80
Fischer, Johann Carl: instrumentalism, 17, 22; on light and heat, 22; physics and chemistry, 26; physics and math, 30
Flamsteed, John: 37, 39
Flattening of earth: defined, 218; Newton's value, 219; later estimates, 226, 227, 237–238, 239, 241
Flaugergues, Honoré: 96
Force: magnetic, 66–72; electric, 72–81; refraction, 150–155, 157–160; short-distance, 145–146, 157–165; analogy between electric and gravitational, 80–81. *See also* Attraction; Capillarity; Distance forces
Fortin, Nicolas: 61–62
Fourier, Joseph: 149n
France, Agence temporaire des cartes: 234
—, Agence temporaire des poids et mesures: 248n, 267–268
—, Assemblée nationale: 247, 258; orders Delambre's release, 233; weights and measures, 258, 261, 263–266; closes Académie, 264
—, Bureau des longitudes: 32, 184, 237–241
—, Comité d'instruction publique: Republican calendar, 249–254; metric project, 248, 268, 270
—, Comité du salut publique: 247; supports Calon, 234; sets

up Commission temporaire, 266; condemns Lavoisier, 268
—, Commission du mouvement des troupes: 234
—, Commission temporaire des poids et mesures: 254, 264, 266; terminated, 267
—, Conseil des anciens; 270
—, Conseil des cinq cents: 270n
—, Convention nationale: 248, 255, 267
—, Dépôt de la guerre: 192, 234; under Calon, 235
—, Dépôt de la marine: 269
—, Directoire: 254, 269
—, École impériale militaire: 275
—, Institut de France: 61, 139; prize of 1812, 106; member of First Class, defined, 139–140. *See also* Paris, Académie
Franklin, Benjamin: 10
Frederick the Great: 198
French Revolution, and geodesy: 231–235
Fresnel, Augustin: 147

Gases: discovery of, as watershed, 8–9; and definition of physics, 24, 25; as elementary principle, 9, 13; model of, 13, 91; expansion of, 9–10, 86, 91–100; refractive power, 31, 61–62
Gasometer: 107
Gattey, François: 268n
Gay-Lussac, Louis Joseph: 96, 107; chemistry and mathematics, 6n; heat and gases, 9; thermal expansion of gases, 92–95, 116, 173–175; air thermometer, 101; specific heats of gases, 105–106, 178; capillarity, 163; instrumentalism, 21, 142–143; tendency to generalize, 105, 112; balloon ascent, 8–9; and Arcueil, 143n, 146–147
Gay-Lussac's law: 95, 181, 185
Gauss, Carl Friedrich: on repeater, 60n; Laplace's physics, 165; Zach, 195n, 211–212; least squares, 211; geodetic measurements, 211–213; invents heliotrope, 213; gives value of flattening, 241; defines geoid, 241
Great Britain, Commissioners of Longitude: 49, 52
—, Ordnance Survey: 206–207, 237
Gehler, Johann Samuel Traugott: 21, 29, 30
Geodesy: instruments, 57–59; and military, 206, 210; and peasants, 195, 209–210, 223, 227, 231–233, 237. *See also* Military; Repeater; Sextant; Theodolite
Geoffrey, Etienne: 149n
Geoid: 241
George III: 234
Gilbert, Ludwig Wilhelm: 92n
Glass, thermal expansion of: 97
Godin, Louis: 219, 220, 221, 223
Göttingen, University: 42, 47
Gotha, Observatory: 50

Graduation of arcs: 39, 40–44
Graham, George: 39–42, 221
Graphical displays: 128, 130, 136
Gravity, theory of: as cynosure of physics, 3–5, 14; questioned, 150–151
Gregorian Calendar: 254
Gregory, Olinthus: 240
Gren, Friedrich Albrecht Carl: instrumentalism, 18, 21, 28; on light and heat, 22; *Grundriss*, 24, 28; and mathematics, 31
Greenwich, Royal Observatory: 37, 39, 42
Greenwich-Paris link: 201–207. *See also* under Paris
Grimaldi, Francesco Maria: 187
Guillotine, Joseph: 254
Guyton de Morveau, Louis-Bernard: 90–91, 250

Hachette, Jean Nicolas: 275
Hadley, John: 47, 49, 50
Halley, Edmund: 49, 190n
Hassenfratz, Jean-Henri: on development of physics, 8, 24, 30; ice calorimeter, 104
Hassler, Ferdinand: 206
Haüy, René Just: 80n, 263, 266, 268, 269; instrumentalism, 17, 19; electricity, 22; text on physics, 26, 28, 30; capillarity, 164
Heat: and gases, 9–10; analogy to electricity, 14, 124n; to light, 22; motion theory of, 19, 21; measurement of, 95; nature of, 101; of fusion, 102; radiant, 136–137; convection of, 137; adiabatic, 171. *See also* Caloric; Radiant Heat; Specific heat
Heberden, William: 87
Heliotrope: 213
Henley, William: 82
Herschel, William: 22; discovers infra-red, 128–131; criticized, 134–135; positivism, 132; double stars, 240
Hobbes, Thomas: 2
Hooke, Robert: 37
Hornsby, Thomas: 40
Horsley, Samuel: 87
Hounslow Heath. *See* Roy
Humboldt, Alexander von: 31, 104–105
Hume, David: 134
Huygens, Christiaan: 143, 145; geodesy, 190
Hypsometry: 31–32, 96–98

Ice: 119–120
Ice calorimeter, 12; analogy to torsion balance, 15n, 103; failings, 103–104
Iceland spar: 63–65
Imponderable fluids: defined, 5, 16; criticized, 6–7; Napoleon's version, 16; Biot's, 142; and instrumentalism, 17, 19; difficulty in measuring, 65–66. *See also* Caloric; Electricity; Heat; Light; Magnetism
Ingénieurs géographes militaires: 195–197, 272
Instrumentalism: 3, 16–23; and

imponderables, 17, 19; caloric, 19–21; electricity, 21–22; gravity, 150–151; of British school, 127–128; in Laplacian physics, 141–146
Instrument makers: French, 44, 60; British, 46–47; German, 212–213. *See also* Bird; Graham; Lenoir; Ramsden; Reichenbach; Sisson; Troughton
Ivory, James: 241

Jefferson, Thomas: 261
Jenisch, David: on gases, 9; causes, 18; Lavoisier, 24; physics and mathematics, 29, 30
Jesuits, and geodesy: 187, 192, 201, 226–228, 230
Job: 226, 228
Josephinische Landesaufrahme: 199
Josephus, Flavius: 243
Joshua: 226n
Journal de physique: 26
Juan, Jorge: 223
Jurin, James: 154, 246n

Kant, Immanuel: instrumentalism, 16–17; physics and critical philosophy, 24
Kater, Henry: 207
Kelly, P. (British metrologist): 273n, 274n
Kepler, Johannes: 37
Knowles, Charles: 72
Krayenhoff, Cornelius Rudolphus Theodorus: 213; and repeater, 61; maps the Netherlands, 208–209

Lacaille, Nicolas-Louis de: 228, 234, 264; on sextant, 49n; longitude by light flash, 197n; redoes Picard's measurements, 225; and provisional meter, 266
La Condamine, Charles-Marie de: 221, 223; on earth's shape, 226; seconds pendulum, 259; weights and measures, 244
Lacy, Franz Moritz: 198
Lagrange, Joseph-Louis: 140, 149n, 250, 266; speed of sound, 166, 168–170; metric project, 260
Lalande, Joseph-Jérome: 46
Lamartine, Alphonse Marie Louis de Prat de: 149
Lambton, William: 241
Lamétherie, Jean Claude de: 17
Landriani, Marsilio: 26
Laplace, Pierre Simon de: 6, 9, 112, 134, 149n, 266, 267; on evaporation, 12, 13; chemical combination, 14; electricity, 21–22, 81; hypsometry, 31–32; logarithms, 32; gas expansion, 91; calorimetry, 101–104; refraction, 157–158; capillarity, 160–165; speed of sound, 165, 171–184; geodesy, 212, 228n, 238, 240, 241; metric projects, 247, 256, 260, 261; instrumentalism, 19, 30, 141–146; his physics criticized, 145–149, 165; as

330 / INDEX

patron of science, 4, 140–141, 141n; collaboration with Lavoisier, 12, 13, 101–104; with Volta 12–13
Lapland expedition: 221–222, 229
Latent heat: introduced by Wilcke, 11; explained away by Dalton and Leslie, 107; defended by Delaroche and Bérard, 109–110; analogy to electricity, 13. See also Heat
Latitude: definition for elliptical earth, 218; accuracy in determining, 219
Lavoisier, Antoine Laurent: 8, 10, 134, 254, 263, 266; on specific and latent heat, 12; caloric, 12–13, 19, 20; electrification, 13; chemical combination, 14; light and heat, 22, 101; gas expansion, 91; ice calorimeter, 101–104; and metric reform, 247, 260; instrumentalism, 19, 20; on relations between chemistry and physics, 24–26; collaboration with Laplace, 12, 13, 101–104; with Volta, 12, 13; execution, 267
Lavoisier, Mme: 120
laws: simple, prejudice for, 105–106; Laplacians on, 141–144; of exchanges, 118–121; of adiabatic expansion, 183
least squares: 240–241
Lecoq, Carl Ludwig von: 212
Legendre, Adrien-Maris: 203, 263, 268n
Le Monnier, Pierre-Charles: 41, 222
Lenoir, Etienne: 47; identified, 55; makes Borda circles, 208; decimalized clocks, 254; and seconds pendulum, 264; metric project, 266. See also Repeater
Lenses: added to instruments, 36; achromatic, 43
Le Roy, Jean Baptiste: 258
Leslie, John: instrumentalism, 17, 125, 133; on latent heat, 107; differential thermometer, 123–124; radiant heat, 124–127, 137–138; criticizes Herschel's experiments, 132–134, and Laplace's physics, 148; on cooling and convection, 137–138; criticized, 96, 138n
Levy, Maurice: 7
Lichtenberg, Georg Wilhelm: instrumentalism, 17, 21; definition of physics, 23–24, 30
Libes, Antoine: 28n
Liesganig, Joseph Xaver, S.J.: 192, 198
Light: analogy to magnets, 16; to heat, 22; refraction of, 151–154
Logarithms: 32–33, 96, 138
London, Royal Institution: 121
—, Royal Society: 234; and thermometers, 86–87; map of Britain, 206; Paris-Greenwich link, 201–202; commissions

Mason and Dixon, 229; measurement at Schehallien, 231
Longitude: lunar method, 47, 49, 52–53; by Jupiter's moons, 187, 221; by light flashes, 194, 195, 212, 228; between Paris and Greenwich, 201–207
Louis XVI: 234

Macaulay, Thomas: 149
Magnetism: analogy to light, 16; to gravity, 66, 67; artificial, 67
Maire, Christopher: 201, 227
Malus, Etienne: double refraction, 63–65; radiant heat, 135; light rays, 143–144; his instrumentalism, 143–145; as Laplacian physicist, 140n, 158
Manliness and mathematics: 246
Mapping. *See* Cartography
Maraldi, Giacomo: 189
Maraldi, Giovanni Domenico: 190
Maskelyne, Neville: thermometry, 87; on Paris-Greenwich link, 202–205; on Lapland expedition, 223–224; on Mason and Dixon, 230; Schehallien measurements, 231
Mason, Charles: degree of meridian, 229–230; Schehallien, 231
Mathematics: as language of Enlightenment, 6; and physics, 29–31, 36; French emphasis on, criticized, 147–149; as liberating and manly, 245–247
Magnetic fluids: number of, 22; quantity, 69, 81
Maupertuis, Pierre Louis Moreau de: 191, 220, 224; and Lapland expedition, 223–224
Mayer, Tobias: on Bird telescope, 42; lunar method, 47, 49; repeater, 50–51
Méchain, Pierre-François-André: 203, 254, 264, 265, 266, 267, 268, 269, 270, 271; measures Paris meridian, 231–233; recalled by Calon, 235–236; proposes extension, 234
Meniscus: 155
Mercury, dilation of: 95–97, 98
Meter: defined, 260; provisional, 266, 267, 268; final value, 269
Metric project: and Borda circles, 263, 269; cost, 266, 267, 268; and numeracy, 246, 276; promoted by military, 267; and by force of arms, 270–271; legal enforcement, 270, 271, 272, 274; resisted by French, 270–273; undecimilized, 272–273; acceptance and spread, 269–270, 273–277; extension to Balearics and Shetlands, 236–238
Meusnier de La Place, Jean-Baptiste-Marie-Charles: 263
Michel [ingénieur géographe]: 200

Micrometer eyepiece: 36–37
Milan, Brera Observatory: 46
Military: and ballooning, 8; mapping, 191, 192, 193n, 194n, 195–201, 208, 210, 234–235; secrecy in cartography, 196, 197, 199; and metric project, 267–268
Miller, John Riggs: 244, 245; and seconds pendulum, 259
Moisture and gas expansion: 87–91
Monge, Gaspar: 140, 250, 260, 263, 266, 268, 269
Montesquieu, Charles-Louis de Secondat, baron de: 245
Montjouy: 232, 233
Morichini, Domenico Lino: 16
Mudge, William: 237; geodetic measurements, 239–240
Müffling, Friedrich Carl Ferdinan von: 210, 211, 212; on repeating circle, 61, 213, on heliotrope, 213; on flattening, 240; training, 210
Muncke, Georg Wilhelm: 21, 241
Munich: 212–213
Musschenbrock, Petrus van: 24

Names, metric: 248, 268, 272
Napoleon I: 239, 249; on imponderables, 16; astronomy, 36; mathematics, 148–149; as patron of science, 140; plans invasion of Britain, 206; favors cartography, 207–208; and Tranchot's survey, 209–210; Concordat, 256; metric system, 269, 272
Newton, Isaac: 2, 5, 6, 134, 154, 188, 190, 222; and Flamsteed, 39; spectrum, 127, 132; cooling, 137–138; refraction, 151–153; speed of sound, 166–170; earth's shape, 214–217; and manly science, 245–247
Nicholson, William: 19, 28–29
Nollet, Jean-Antoine: 87, 89
Numeracy: 2–3; growth, 246–247; and metric system, 275–276
Nutation, of the earth: 223, 224, 241; discovered, 43

Octant: 47–48. *See also* Sextant
Olbers, Heinrich Wilhelm Matthias: on repeater, 60n; on Zach, 195n
Oriani, Barnaba: 200n
Orry, Philibert: 190

Parallax, stellar: 37
Paris, Académie des sciences: 61; mapping of France, 185–186; prize competitions, 23, 254–256; speed of sound, 173, 174, 177; and Newton's theories, 220; expeditions to Lapland and Peru, 221–235; reform of weight and measures, 246, 257–260; promotes decimal system, 246, 249; opposes pendulum, 259–260; politics of metric project, 260–262; suppressed and

resurrected, 139, 234, 248, 268–269
—, Ecole polytechnique: 7–8, 24, 96, 142, 275; and physics instruction, 20, 27
—, Observatory: 36, 44, 61, 62; and the Cassinis, 187, 191–192; cartography, 188; seconds pendulum, 259, 263–265
—, Panthéon: 233
—, Société d'agriculture: 271, 276–277
Paris-Greenwich link: 201–207, 259
Paris meridian: measured by Cassinis, 188–190; by Lacaille, 225; by Méchain and Delambre, 231–237; extended to Formentera and the Shetlands, 237–238; politics of the metric measurement, 261–263
Parrot, Georg Friedrich: 80
Passau, Prince Bishop of: 193–194
Peace of Versailles: 205
Peace of Vienna: 200, 210
Peasants and geodesy: 195, 209, 210, 223, 227, 233, 235
Pendulum: *See* Seconds pendulum
Pepys, Samuel: 1–2
Peru expedition: 221, 223–224, 229
Petit, Alexis-Thérèse: identified, 96; calibration of thermometers, 96–101; as Laplacian physicist, 144–147; on specific heat of metals, 106; on cooling, 138
Phlogiston: 8, 12
Photometer: 133–134
Physica generalis: 27
Physica particularis: 28, 29
Physics: meanings of, 23–24; and chemistry, 24–27; and mathematics, 29–31; as science of sensible motions, 28–29
Picard, Jean: 222, 224; puts lenses in surveying instruments, 36–37, 188; his instruments, 38–39; mapping Paris region, 184–189; and aberration, 224; Newton, 217; seconds pendulum, 186, 258
Pictet, Marc-Auguste: on relation between chemistry and physics, 26; calibration of thermometers, 86b; on Ramsden's theodolite, 44, 59–60n; radiant heat and cold, 118–120
Pingré, Alexandre-Guy: 250
Pinte: 243
Playfair, John: 165
Poleni, Giovanni: 220
Pommiés, Pierre: 275
Poisson, Siméon Denis: and electricity, 21–22, 148; as Laplacian physicist, 140n, 147; on speed of sound, 172, 174–176, 183–184
Prevost, Pierre: radiant heat, 119–121; on Laplace's physics, 148
Priestely, Joseph: 82; electrical force, 81; Copley medal, 87

Prieur de la Côte d'Or, Claude Antoine: on expansion of gases, 91; favors decimal system, 247; and metric project, 248, 258, 267–268; and Lavoisier, 267
Prieur du Vernois. *See* Prieur de la Côte d'Or
Pringle, John: 86
Prony, Gaspar de: 259, 268, 269
Proof plane: 82
Proust, Joseph Louis: 146, 147
Puissant, Louis: 273n, 275; calculates flattening, 239

Quantification: in the Bible, 1; in physical science, 2–3; its quickening, 5–6
Quantity: of magnetic fluid, 69, 81; of electric, 81–82

Radiant heat: 16, 22–23; analogies to electricity and light, 119, 127, 135; and radiating surface, 121, 123, 125–126; Rumford's theory of, 121, 123–124; Leslie's, 125–126, 137–138; Newton's, 137–138; Herschel's discoveries, 127–132; attacked by Leslie, 132–134; confirmed by French, 135–137; and thermoscopes, 120–124; photometers, 132–134; polarization, 135; Laplace's gas model, 178–180. *See also* Heat
Ramsden, Jesse: 205, 206; dividing engine, 43, 49, 215; theodolite, 44; sextant, 200; dilatoriness, 46, 203. *See also* Sextant, Theodolite
Rayleigh, John Williams Strutt, 3rd baron: 165
Reason: as Enlightenment ideal, 245; and metric reform, 270–271
Reichenbach, Georg von: 213, 229–230
Refraction: by gases, 31, 61–62; double, 63–65; force of, 157–158
Regnault, Victor: 94–95
Repeater (repeating circle): Mayer's, 50–52; Borda's, 52–53; 269; in physics, 59–60; adopted by French military, 197–198; by Krayenhoff, 208; by German geodecists, 210–213; by Svandberg, 238; by the metric project, 229, 262–263; and theodolite, 239
Republican calendar: anticlericalism, 249–257; establishment, 251; month and day names, 252; repeal, 256–257. *See also* Decimalization
Republican time. *See under* Decimalization
Repulsion: 17. *See also* Attraction
Riccioli, Giambatista: 187
Richer, Jean: 188, 214
Richmond, Duke of: 206–207
Robison, John: 72–75
*Rochon, abbé: 118, 127
Rodriguez, Joseph: 240, 262n
Rome, Gregorian College: 227

Rømer, Olaus: 39
Romme, Gilbert: 250, 252
Roy, William: temperature correction to barometers, 90; on sextant, 50; map of Britain, 203–205, 208; on earth's shape, 239; measurements criticized, 205
Rufo, Father, instrument maker: 227
Rumford, Benjamin Thomson, Count: attacks Prevost's theory, 120–122; canon-boring experiment, 121; thermoscape, 122–123; surface effects on radiation, 121, 123–124; theory of radiant heat, 121, 123–124; positivism, 134

Saint Pierre, Bernardin de: 148–149
Saussure, Horace Bénédict de: 32, 118
Saxe-Gotha, Duke of: 46
Scheele, Carl Wilhelm: 117–118
Schehallien, Mount: 231
Scheiermacher: 212, 240
Schumacher, Heinrich Christian: 213
Seconds pendulum: as standard of length, 258–260; determined by Borda and Cassini, 10, 263–265; and decimalized time, 254
Seguin, Armand: 20
Seven Years War: and cartography, 191, 196, 198, 200, 203; and numeracy, 248

Sextant: 49–50, 200, 201n. See also Ramsden; Theodolite
Sharp, Abraham, 39
Short, James, 223
Shuckburgh-Evelyn, George August William: 40, 43–44
Significant figures: 31, 63, 276. See also Numeracy
Simon, Paul Ludwig: 80
Sisson, Jeremiah: 40
Smeaton, John: 39
Snel, Willebrord: 186, 187
Snel's law: 152–153
Société d'Arcueil. See under Arcueil
Sudner, Johann Georg von: 212
Sombart, Werner: 276n
Sound, speed of: 166; Newton's theory, 166–170; Lagrange's, 170–172; Laplace's, 171, 175–184; Biot's, 171–174; Poisson's, 174–175; and adiabatic heating, 171–172; and γ, 175–184
Specific heat: analogy to electricity, 11; in general, 101–104; of gases, by Gay-Lussac, 105–106; by Delaroche and Bérard, 107–114; by Clément and Desormes, 112–117; of vacuum, 107, 114–118; of air and water compared, 110, 113; at constant pressure, 110, 175–176; at constant volume, 114, 175–176; γ, 175–177, 180–183
Standard Model: defined, 5; and imponderables, 6–7, 16; caloric, 10, 86; instrumentalism, 17, 134–135;

quantification, 29; Coulomb's law, 80
Struve, Friedrich Georg Wilhelm: 243
Surveying and the military. *See under* Military
Svandberg, Jöns: 222
Swinden, Jan Hendrik van: 208, 209
Symmer, Robert: 10
Synchronic terms: 25–26; defined, 8

Talleyrand de Perigord, Charles Maurice de: 271; on metric reform, 245, 258–259
Telescope mountings: 40–42
Temperature: and barometer, 90; absolute, 116. *See also* Thermometer
Tension, electrical. *See* Electrical tension
Thenard, Louis-Jacques: 164
Theodolite: 213n; Ramsden's, 44–45, 203–207, 240; compared with repeater, 59–60, 239
Thermal expansion of gases: 87–91
Thermometer: calibration of, 87–90; air and mercury compared, 96–100; differential, 123–127
Thermoscope: Rumford's, 120–122; Leslie's, 123–126. *See also* Photometer
Thomson, Thomas: on "chemistry," 28; ice calorimeter, 104

Thompion, Thomas: 39
Tillet, Mathieu: 260, 263
Time, decimal. *See under* Decimalization; Revolutionary calendar
Toise de Pérou: 247, 266; and meter, 271
Torsion balance: 15, 102
Tralles, Johann Georg: 205–206
Tranchot, Jean Joseph: 240, 272; on repeater, 61; theodelite, 206; maps Corsica, 196; Rhineland, 209–211
Troughton, Edward: dividing machine, 44; sextant, 49; repeater, 60–61
Troughton, John: 40
Turgot, Etienne-François: 259

Ulugh Beg: 211

Vandermonde, Alexandre-Théophile: 263, 266, 268
Van der Waals, J.D.: 165
Vegobre, Louis de Manoël de: 80n
Venus, transit of: 193
Vienna: engineering school, 198
Villeneuve, Perny de: 208
Volta, Alessandro: battery, 16n; single-fluid theory, 21; electrical capacity, 83–86; collaboration with Laplace and Lavoisier, 13; on Lavoisier, 24; Coulomb, 80
Voltaire: 221, 224

Wallis, John: 190n
War of the Austrian Succession: 191, 195

Weights and measures: in ancien régime, 243–245; revolutionary, 248; and seconds pendulum, 258–260
Whitehurst, John: 259n
Whitman, Walt: 7
Wilcke, Johan Carl: 10; heat theory, 11; calorimetry, 101, 102
Williams, Edward: 239
Wittenberg, Duke of: 193
Wollaston, William Hyde: 22

Ximenes, Leonardo: 201

Yard, English: 259–260
Year, length of: 256
Young, Thomas: on aether, 121–123; against Laplacian physics, 147–148, 165

Zach, Franz Xaver von: on Humboldt, 31; Troughton, 60–61; Cassini III, 195; Lecoq, 200–201n; Reichenbach, 213; Beccaria, 229; Liesganig, 230; evaluates repeaters, 60–61, 271n; sextant, 50; Peruvian results, 224n; recruits Bohnenberger, 200; Hassler, 206; Müffling, 211; Gauss, 211–212; on cartography and national honor, 208; criticized, 195
Zach, Anton: 241
Zenith sector: 37–38, 42